Jenny Helwick 24.95

Clinical Genetics Handbook

Clinical Genetics Handbook

NATIONAL GENETICS FOUNDATION INC.
Ruth Y. Berini, Executive Editor
Eva Kahn, Editor

under a grant from the Henry J. Kaiser Family Foundation

Medical Economics Books
Oradell, New Jersey 07649

Library of Congress Cataloging-in-Publication Data

Clinical genetics handbook.

"Under a grant from the Henry J. Kaiser Family Foundation."
Bibliography: p.
Includes index.
1. Medical genetics—Handbooks, manuals, etc.
I. Berini, Ruth Y. II. Kahn, Eva. III. National
Genetics Foundation (U.S.) IV. Henry J. Kaiser
Family Foundation. [DNLM: 1. Genetics, Medical—
handbooks. 2. Hereditary Diseases—handbooks.
QZ 39 C641]
RB155.C5725 1986 616'.042 86-17932
ISBN 0-87489-410-7

Cover design: Penina M. Wissner

ISBN 0-87489-410-7

Medical Economics Company Inc.
Oradell, New Jersey 07649

Printed in the United States of America

Copyright © 1987 by Medical Economics Company Inc., Oradell, N.J. 07649. All rights reserved. None of the content of this publication may be reproduced, stored in a retrieval system, or transmitted in any form or by any means (electronic, mechanical, photocopying, recording, or otherwise) without the prior written permission of the publisher.

Contents

Contributors

Arthur D. Bloom, MD†
Professor of Pediatrics and of Human Genetics and Development
Director, Division of Human Genetics
Columbia University
College of Physicians and Surgeons
New York, NY

Eliot A. Brinton, MD†
Assistant Professor
Biochemical Genetics and Metabolism
The Rockefeller University
New York, NY

Jessica G. Davis, MD*
Director, Child Development Center
Chief, Division of Cytogenetics
North Shore University Hospital–Cornell Medical Center
Manhasset, NY

Deborah deLeon, MS*
Genetic Counselor
National Genetics Foundation
New York, NY

Mony deLeon, EdD†
Assistant Professor of Psychiatry
New York University Medical Center
New York, NY

Lois Forehand, MS*
Genetic Counselor
National Genetics Foundation
New York, NY

DeWitt S. Goodman, MD†
Tilden-Weger-Bieler Professor
Department of Medicine
Columbia University
College of Physicians and Surgeons
New York, NY

James W. Hanson, MD*†
Associate Professor of Pediatrics
Chairman, Division of Medical Genetics
University of Iowa
Iowa City, IA

Richard J. Havlik, MD†
Chief, Clinical and Genetic Epidemiology Section
Epidemiology and Biometry Program
Division of Heart and Vascular Diseases
National Heart, Lung and Blood Institute
Department of Health, Education and Welfare, NIH
Washington, DC

William G. Johnson, MD†
Assistant Professor of Clinical Neurology
Columbia University
College of Physicians and Surgeons
New York, NY

Eva Kahn, MS*
Genetic Counselor
National Genetics Foundation
New York, NY

Irene H. Maumenee, MD†
Associate Professor of Ophthalmology/Medicine
The Johns Hopkins University School of Medicine
Baltimore, MD

James L. Mills, MD*
Epidemiology and Biometry Research Program
National Institutes of Child Health and Human Development, NIH
Washington, DC

John J. Mulvihill, MD†
Clinical Genetics Section
Clinical Epidemiology Branch
National Cancer Institute, NIH
Washington, DC

Gilbert S. Omenn, MD, PhD*†
Dean, Schools of Public Health and Community Medicine
University of Washington
Seattle, WA

John Rainer, MD†
Chief, Department of Medical Genetics
New York State Psychiatric Institute
New York, NY

Arthur Robinson, MD†
Director, Cytogenetics Laboratory
National Jewish Hospital
Denver, CO

Gladys Rosenthal, MS*
Genetic Counselor
National Genetics Foundation
New York, NY

Jerome I. Rotter, MD*
Assistant Professor of Medicine and Pediatrics
Division of Medical Genetics
UCLA School of Medicine–Harbor UCLA Medical Center
Torrance, CA

Robert J. Ruben, MD†
Chairman, Department of Otorhinolaryngology
Albert Einstein College of Medicine of Yeshiva University
New York, NY

I. Herbert Scheinberg, MD†
Professor of Medicine
Albert Einstein College of Medicine of Yeshiva University
New York, NY

Sheila Traut, MS*
Genetic Counselor
National Genetics Foundation
New York, NY

Dorothy Warburton, PhD*
Director, Cytogenetics Laboratory
Associate Professor of Clinical Pediatrics, (in Human Genetics and Development)
Columbia University
College of Physicians and Surgeons
New York, NY

Judith Weinblatt, MS*
Genetic Counselor
National Genetics Foundation
New York, NY

Carol N. Williams, PhD†
Assistant Director
Center for Alcohol Studies
Brown University
Providence, RI

Hans Zellweger, MD*
Professor Emeritus
Department of Pediatrics
Division of Medical Genetics
University of Iowa
Iowa City, IA

*Author †Reviewer

Preface

In recent years we have enormously increased our understanding of the fundamental processes of heredity in general and in man in particular. These advances, which continue to develop at an almost explosive rate, are already being applied to the practice of medicine, especially for the prevention and early diagnosis of serious disease. Concurrently, medicine and the public have become aware of the overwhelming influence the genetic constitution has on fetal death, congenital malformations, cancer, and other chronic diseases contributing to human tragedy. Further illumination has come from the recognition of the interaction between the environment and the genes in the expression of the genotype. Modification of the environment, the major purpose of all medical therapy, is also applicable to genetic disease, especially before the onset of irreversible symptoms.

The primary-care physician has a unique role in genetic medicine. Because of his intimate and long-term care of individuals and families, he can be alert to familial "susceptibilities" as well as to the hopes and aspirations, including reproductive, of his patients. This handbook is aimed at helping him to be aware of the large variety of genetic problems as well as to distinguish those that are not primarily genetic.

The handbook is a manual, not a textbook. Detailed discussions on genetic research, as well as materials relating to clinical symptomology and medical management have been deliberately kept to a minimum, since they can be found in standard texts. Clinical relevance has been the criterion for inclusion of any item. Options relating to aspects of care are those, which, in the genetic context, should be brought to the physician's attention for consideration.

Medical genetics needs to become part of family medicine, with the medical geneticist being utilized more as a consultant than as a primary physician. I am convinced that medicine is on the verge of revolutionary changes and that primary-care physicians will find this book helpful in keeping abreast of clinically significant advances, so as to render the most effective care to their patients.

ARTHUR ROBINSON, MD

Director, Cytogenetics Laboratory
National Jewish Hospital
Denver, Colorado

Acknowledgments

The editors thank Arthur Robinson, MD, Director of the Cytogenetics Laboratory at National Jewish Hospital in Denver, for serving as medical advisor and for writing the preface. We also acknowledge with gratitude the contributions of the many primary care physicians in private and group practices whose reviews of our format and chapters make this book relevant to everyday clinical practice.

How To Use This Book

This Clinical Genetics Handbook has been written to enable the primary-care physician to have ready access to genetics information when practical questions arise in a clinical setting. Its purpose is to supplement standard medical texts with a focus on the genetic aspects of medicine. Both the content and the format are designed to facilitate the application of genetic services in the course of primary practice and to supply explicit background information for consideration of referral. The book covers the genetic disorders most likely to be of interest or concern to the primary care physician and his patients, rather than trying to catalogue the entire spectrum of genetic disease.

The content is arranged by organ systems to reflect possible clinical presentation. It has both categoric chapters, providing an overview of multiple disorders that may present as malfunction in a particular system, and specific chapters on individual disorders or issues. Specific disorders are discussed separately for one or more of the following reasons: the disorder is common in the general population or in a subgroup; it represents a common concern, frequently raised by patients; it is preventable or subject to effective treatment through available technology; it illustrates a classic example of inheritance; the disorder/topic is eminently suited to counseling in primary practice; it requires ongoing support from the family physician, due to the devasting nature of the problem (irrespective of tertiary referral).

The format of the book is designed for easy reference to relevant material.

Chapters are divided into sections that are clearly labeled with appropriate headings to enable the physician to find his topic quickly; subtopics are set off for rapid identification. The format is flexible, depending on the nature of the chapter, but generally includes the following sections:

Introduction: General description, overview, important features, overall incidence or prevalence

Etiology and pathogenesis: Mechanisms for cause and expression

Genetic characteristics and mode of inheritance: Mode of inheritance with risk figures, tabulated risk figures, and/or empiric risk tables—items relevant from a genetic point of view

Clinical notes: Short salient comments of interest related to genetic aspects, including diagnosis, onset, course, prognosis, and management

Procedures for diagnostic confirmation: A checklist with emphasis on procedures developed for clinical genetics

Considerations in management: Checklists of services appropriate for the disorder in genetic, psychosocial, and medical management

Suggested reading: Background and additional literature

Emphasis is placed on genetic characteristics of clinical relevance and on clinical information germane to genetic considerations in diagnosis, management, and counseling

Presentation of information is brief and in tabular form, when appropriate, including tables on common genetic disorders that may suggest consideration of specific disorders associated with various presenting features

Cross references and the index can be used to extract the maximum information from the book

The chapters on the process and the principles and methods of genetic counseling are not intended to be a course in basic genetics, but rather, a compilation of pertinent facts and useful guidelines to facilitate the explanation of genetic concepts to patients.

Section One

Applied Genetics

Chapter 1

Genetic Counseling as a Function of the Primary-Care Physician

Like all medical consultation, genetic counseling is a communication process, involving diagnosis, explanation, and options. It is essentially the same communication process as accompanies other serious diagnoses, except that the information and options may have a bearing on the health of family members other than the individual being counseled. Its purpose is to impart to patients and families the most diagnostic information available about a particular genetic disorder, and then to help them interpret and cope with the information.

Although referral to a genetics center may be indicated for complex cases, genetic health care, including counseling for most common problems, may well be best provided in primary health care, where rapport with the patient and family has been established. The primary-care physician may wish to consult with specialists in other disciplines, including genetics, for diagnosis and consultation to obtain the most specific current information available, but the actual counseling may be most effective coming from the family's own doctor.

INDICATIONS FOR GENETIC COUNSELING

Genetic counseling is appropriate for any condition in which a genetic etiology is known or suspected. Ruling out genetic causes should not preclude counseling. For example, it is as important for parents to know that their child does not have a genetic disorder, as it is for them to know the recurrence risks when the problem is unequivocally genetic. Table 1-1 lists common indications for genetic counseling.

DEFINITION OF GENETIC COUNSELING

Originally, diagnosis and presentation of recurrence risks for use in reproductive planning were considered the sole aims of genetic counseling. However, it has become clear that, aside from welcome news of little or no risk, few patients can integrate direct scientific information sufficiently well to cope not only with the problems, but with possible critical decision-making. Consequently, genetic counseling has been expanded to include attention to elements that may impinge on the proper assimilation and utilization of information.

As in other matters of strong emotional impact relating to health care and lifestyle, it is important to take into account a number of potentially confounding factors: the patient's ability to absorb complicated information under often devastating emotional conditions; the need to adapt information to the patient's level of understanding; the importance of considering potentially conflicting social, religious, and psychologic factors. By encouraging the patient to express feelings and conflicts, and by providing emotional, as well as medical support, the genetic counselor can promote effective use of the information provided.

With these principles in mind, genetic counseling has been defined as a communication process, intended to help affected individuals and/or their families to:

- Comprehend the medical facts (diagnosis, prognosis, and management)
- Understand the contribution of heredity to the disorder and to its recurrence risks in specific relatives
- Understand the options available for dealing with this risk (such as prenatal diagnosis, reproductive alternatives)
- Choose that course of action most compatible with their family goals, values, and religious beliefs
- Make the best possible adjustment to the condition and its implications

THE PROCESS OF GENETIC COUNSELING

The concept of a process is fundamental to achieving one's objectives. Four major components comprise genetic counseling services: diagnosis, informative counseling, supportive counseling, and follow-up.

Diagnosis. An accurate diagnosis is the first objective. A whole spectrum of diagnostic strategies may be utilized. Taking a detailed history is characteristic. Although genetic disease often presents as the first case in a family, a positive family history, either previously known or newly elicited by recognition of related symptoms, can be most informative. Medical records and autopsy reports of affected relatives can confirm a suspected diagnosis. All this, together with clinical evaluation and test results for the proband and/or other family members, will help to establish a definitive diagnosis. It can also determine whether a disease is truly familial and, if so, how it is transmitted within the family. Only then can precise information be provided.

Unfortunately, a satisfactory diagnosis cannot always be made based on our present level of understanding. However, even then, careful investigation and exclusion of high-risk possibilities may acceptably resolve concerns for many family members. Alternatively, an inheritance pattern provided by pedigree analysis may suggest recurrence risks in a given family, even in the absence of a specific diagnosis.

Informative Counseling. The communication of the information under difficult circumstances is the essence of genetic counseling. The nature of the information provided depends on the initial indication for the genetic workup, the diagnosis, and the counselee's chief concerns. However, especially if the counseling is indicated by the birth or new diagnosis of an affected child, the family is often so overwhelmed by anxiety that the genetic particulars of a disorder are farthest from their minds. Thus, it is important to start by explaining the *diagnosis* (including any uncertainties), describing the *natural history of the disease*, discussing the *prognosis*, and outlining options for *treatment and management*.

Only when this information is somewhat accepted, and coping has begun, can the family be confronted with the reality that it may happen again. For this reason, and for support, it is useful to visit the mother and child frequently when they are still in the hospital, to answer questions as they arise, and to encourage the expression of feelings as a normal part of psychologic healing.

Nonverbal conveyance of information may be more powerful than words. Some authorities, for example, recommend encouraging parents to see and hold their malformed infant, even if they are reluctant, because actual malformations are rarely as horrible as the fantasies of the parents.

Advising of *recurrence risks* is particularly crucial for those couples wishing to reproduce again; they require a realistic assessment of clinical risks for future offspring. It is critical, as well, to establish the reproductive risks for couples who came to counseling through other indications, and for those who may be at risk for a late-onset disorder.

Aside from explaining a poor prognosis and preparing a family for a lethal outcome, the citing of high recurrence risks may be the most difficult task for the counselor. Fortunately, the counseling often provides reassurance: dispelling patients' dire misconceptions, explaining cases where no increased risk is found,

TABLE 1-1
Indications for Genetic Counseling

- Family history of a known genetic disorder or recurrent pathologic condition
- Birth defects—single anomalies, multiple defect patterns, metabolic disorders
- Mental retardation or developmental delay
- Chronic neurologic or neuromuscular childhood disorders
- Short stature and other growth disorders
- Dysmorphic features
- Ambiguous genitalia or abnormal sexual development
- Carrier status for a genetic disease with increased incidence in specific population groups—sickle cell, Tay-Sachs, thalassemia
- Infertility, sterility, or fetal wastage
- Exposure to potentially mutagenic or teratogenic agents
- Pregnancy at age 35 or older
- Genetic risks in consanguinity
- Adult-onset disability of genetic origin
- Behavioral disorders of genetic origin
- Cancer, heart disease, and other common conditions with a genetic component

reporting risks for preventable problems, or stating risks for disorders with a low burden (less severe clinical effects).

Specific risk figures and mode of inheritance are discussed in the chapters on disorders and in Chapter 2. For any given counseling situation, such risks and mode of inheritance are based on the diagnosis, the distribution of affected individuals in the family, the results of carrier testing in family members, and a review of the medical/genetic literature.

Mendelian risks can range as high as 50% for a fully penetrant autosomal dominant disorder, like Huntington's disease, or even higher in rare situations. They may not be increased at all, as for a couple concerned about the recurrence of an X-linked recessive disorder like hemophilia, when the affected individual is a relative of the unaffected potential father. For disorders with a multifactorial inheritance, like neural-tube defects or diabetes, empiric risk figures must be relied upon. Current research activities raise the hope that new findings will allow more accurate estimates for certain disorders.

It may be impossible to inform families of a precise pattern of inheritance, when many genetic forms of the disease may exist. Hydrocephalus, for example, may occur sporadically, or it may be inherited as an autosomal recessive trait or as an X-linked trait. Such multiple genetic-risk possibilities should be conveyed to the family so they will know the range in the risk figures. If some genetic pattern of the disease is more frequently seen than others, a weighted overall-risk figure should be used.

When a diagnosis cannot be established, qualified risk counseling may be the only alternative. When a nongenetic etiology is identified, one can assure counselees that there is no evidence of increased genetic risk in the family.

Sporadic cases may be the result of nongenetic causes or new mutations. If it is a new mutation, there will be no increased risk for subsequent children of those parents. Less well known is the fact that the affected individual will still be at risk to pass on the gene to his/her offspring.

How the risk information is viewed depends not only on the risk, but on the *burden* conferred by the disorder. Of course, a 5% risk of a severely disabling disease may be considerably more important than a 50% risk for a trivial clinical problem, such as isolated polydactyly. Therefore, the degree of risk must be viewed in relation to the severity of the disease.

Counselees' perceptions of the *risk/burden relationship*, in turn, depend on their own philosophic and religious beliefs, educational background, life-style, prior experience and previous notions, as well as on the counseling content. How they react to the risk information depends on these perceptions. For example, some deaf couples have been greatly relieved upon learning that the risk for having a deaf child is less than 100%, namely that there is a chance to have a hearing child, while other families may abandon reproductive plans when they find that their risk for having a deaf child is significantly greater than the low risk for the general population.

Once the nature of the disorder has been clarified and the risk for recurrence described, the family can consider available *options* for dealing with their situation. The responsibility of the counselor is to advise families of their choices so that they can make an informed decision, which is most appropriate to their own circumstances.

Depending on the disorder, there may be a number of ways to prevent further manifestations or the recurrence in other family members. While some of the choices offered may involve carrier testing, diagnostic workups, monitoring, or preventive management for individuals at risk, the options presented in genetic counseling are primarily reproductive in nature. They range from decisions on whether to have more children at all, to the use of artificial donor insemination or adoption to reduce the risks. Birth control methods, including vasectomy and tubal ligation, may need to be discussed.

For many disorders advances in clinical genetics have made further reproduction at decreased risk a viable option. Some of the most useful reproductive information that can be provided concerns prenatal diagnosis. Ultrasound, fetoscopy, and, especially, amniocentesis have allowed for the intrauterine diagnosis of well over 200 fetal conditions. Chorionic villus sampling, still available only at selected centers, promises the potential for many prenatal diagnostic tests in the first trimester. In addition, monitoring for consideration of potential prenatal therapy can be offered in special cases.

Supportive Counseling is intended to help individuals and families cope with the psychologic burdens imposed by the discovery of a genetic disorder or birth defect. The chronic handicapping nature of many of these makes the need for supportive care particularly acute for families who have to deal not only with the medical problems, but with their personal anguish, as well as possible guilt feelings about the genetic nature of the problem and the potential stigma imposed by relatives and the community. Few decisions have more emotional complexity than those related to genetic risks and reproduction. Whatever reproductive decisions couples may make, support and reinforcement are crucial to their acceptance of and ability to live with their choices.

Follow-up. A follow-up letter summarizing the counseling session is useful to patients, enabling them to digest the information under less pressure. In addition, families should be encouraged to return if there is evidence that reinforcement and working through of emotional and psychologic obstacles are required for retention and useful interpretation of the counseling.

Often it is helpful for families to discuss their concerns, feelings, and options at the outset with a psychiatrist, social worker, or clergyman. Referral to support groups for ongoing care may be indicated. Among these are the various psychologic services, community resources, clergy, parent groups, and categoric disease foundations (see Section 4, Resources for clinical genetics).

Finally, long-term follow-up must be considered an integral part of genetic counseling. Genetic disorders and birth defects are lifelong, and risks recognized at the initial workup do not disappear. This type of follow-up counseling helps insure understanding and retention of information for future considerations. In addition, medical information may change over time, leading to revisions in diagnosis, refinements in treatment, or other changes in our understanding of the problem. Also, during an extended time period, new family members may be added through birth or marriage, and individuals may grow up and begin to plan new families. Thus, new needs may arise for re-examining the genetic implications of the family history.

SUGGESTED READING

Fraser FC: Genetics as a health care service. *N Engl J Med* 1976;295:486.

Harper PS: *Practical Genetic Counselling,* Baltimore, University Park Press, 1981.

Kelly PT: *Dealing with Dilemma*, New York, Springer-Verlag, 1977.

Kessler S (ed): *Genetic Counseling: Psychological Dimensions,* New York, Academic Press, 1979.

Chapter 2

Genetic Counseling—Principles and Methods

In the United States genetic counseling has been available to some extent since the 1950s. On the basis of genetic concepts and experience over the years, principles and methods have been developed as practical tools for effective delivery of genetic counseling as a service. Whereas Chapter 1 concentrates on the process of genetic counseling, this chapter provides a brief outline of these underlying concepts and procedures.

PEDIGREE CONSTRUCTION

When a genetic workup is indicated, the construction of a pedigree provides a practical graphic record of the family health history and the progress of the investigation. It can be viewed at a glance and is easily annotated and updated. Relationships of family members are represented schematically, with patterns of transmission of familial disorders emerging clearly as the symbols for those affected are drawn in. Figures 2-1 and 2-2 show the common symbols that are used in pedigree construction and a sample working pedigree with typical notations, respectively.

The known family history, thus recorded, may confirm the inheritance pattern suggested by a specific diagnosis in affected family members, facilitating the explanation of risks for various relatives. When the diagnosis is unclear, or needs confirmation, the pedigree will identify those relatives whose medical history may be relevant. Diagnostic medical records, family photographs, or autopsy reports can then be requested. Some relatives may be available for laboratory tests or clinical evaluation. If this information does not provide a precise diagnosis or is unavailable, the reported constellation of affected and unaffected family members may still suggest a specific mode of inheritance or rule one out, even though counseling may have to be qualified.

PATTERNS OF INHERITANCE AND RISK ASSESSMENT

Genetic disorders are those that are caused in whole or in part by an alteration in one or more genes or in the amount of chromosomal material. Although some genetic conditions do not tend to run in families and hence are not hereditary, such as monosomy X or Turner's syndrome, many do; and they carry a predictable recurrence risk for other family members. The risk for a specific genetic disorder to occur or recur in a family depends on the mode of inheritance combined with other factors—including population incidence, ethnicity, relationship to an affected individual, degree of penetrance, environmental influence on the disorder, and reproductive fitness of affected persons. It may be modified by laboratory results, such as carrier detection and prenatal diagnosis.

Genetic conditions are classified according to the type of genetic lesion into the following categories: *chromsosome disorders, single-gene or mendelian disorders, and polygenic or multifactorial disorders*. On the basis of their mode of transmission, single-gene defects are further subdivided into *autosomal dominant, autosomal recessive,* and *X-linked conditions*. All have characteristic inheritance patterns and recurrence risks.

The chapters in this volume review genetic characteristics relevant to particu-

FIGURE 2-1 Symbols Commonly Used In Pedigree Construction

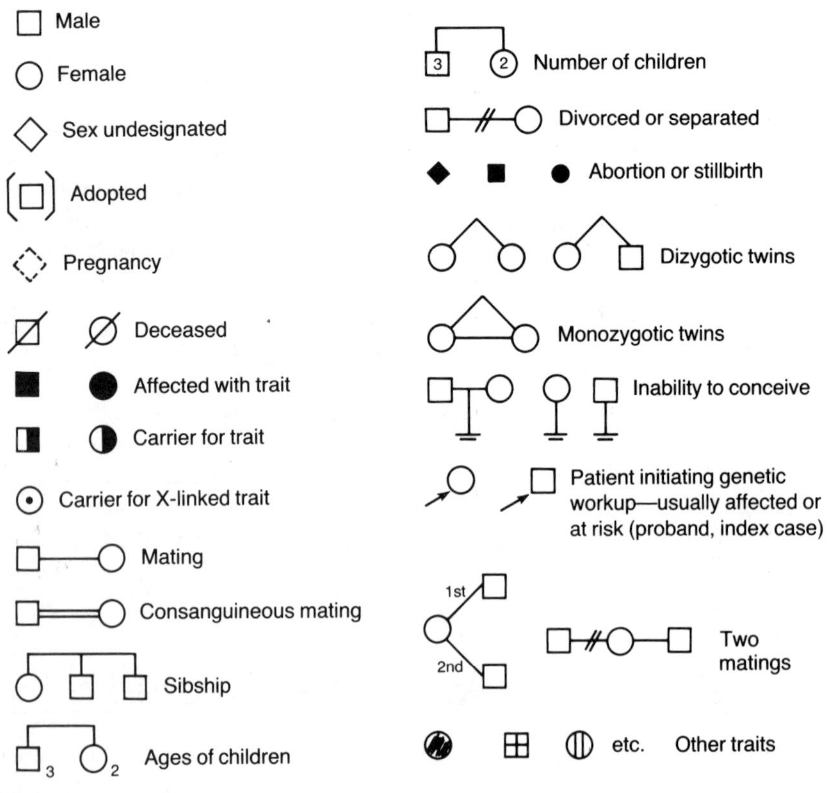

FIGURE 2-2 Sample Working Pedigree

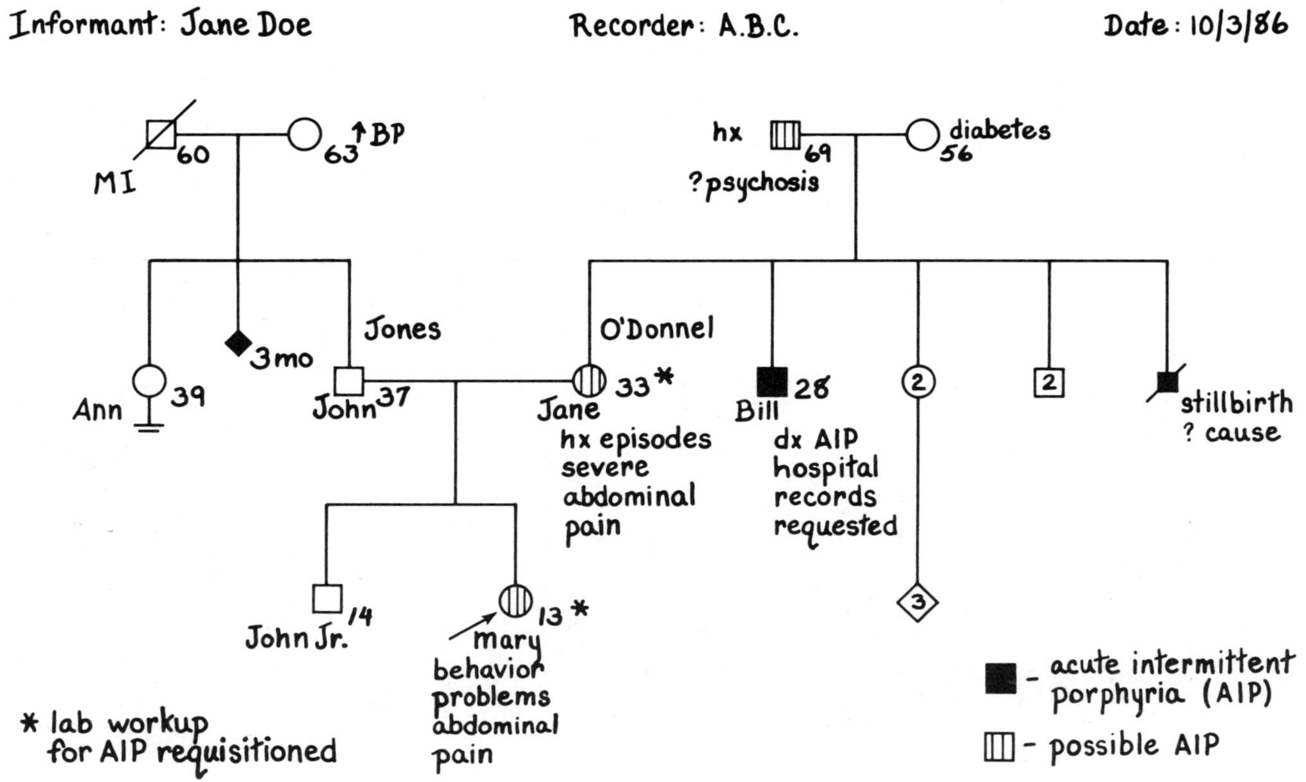

FIGURE 2-3 Autosomal Dominant Inheritance

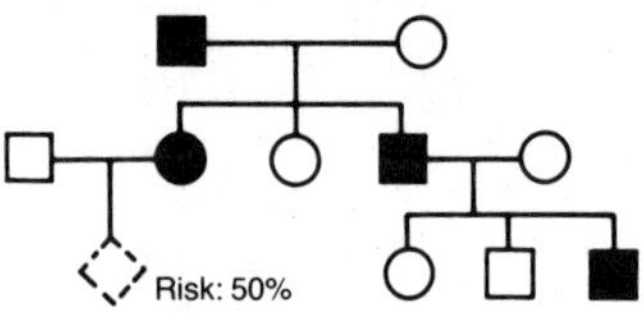

Typical Autosomal Dominant Pedigree

Genetic characteristics:

An autosomal dominant disorder is one in which the heterozygote is symptomatic, i.e., an abnormality is evident when a mutation is present in only one of a gene pair on an autosomal (all except X and Y) chromosome.

The disorder is transmitted directly from one generation to the next by both sexes, and both sexes are equally at risk.

The chance for an affected individual to pass on the mutant gene is 50% for each conception.

An unaffected individual cannot pass on the gene.

Wide variability in clinical expression is common in many autosomal dominant disorders.

Offspring of minimally affected individuals may be severely affected.

Some disorders exhibit reduced penetrance, or "skip" generations, i.e., the disorder is not expressed at all in some individuals who carry the gene (usually identified by virtue of having passed on the gene to affected offspring).

The risk for offspring of a gene carrier to be clinically affected is 50% times the penetrance of the disorder (percent of gene carriers that develop clinical symptoms).

Nonpenetrance can be viewed as an extreme form of variable expression, in that the particular mutant gene is not detectable by current methods in clinically asymptomatic individuals.

A careful examination may identify minimal expression, revealing the presence of the gene.

Late onset in some dominant disorders complicates counseling, since inheritance of the gene often cannot be determined prior to onset of symptoms (Huntington's disease . . .). Since age of onset may vary or range over a long period, it may be difficult to say when the age of risk has been passed.

lar disorders. Chromosomal anomalies are covered in a separate chapter. Figures 2-3 to 2-6 show typical pedigrees and pertinent facts about the other modes of transmission, as well as simple diagrams to help explain the types of mendelian inheritance to patients (especially when the drawing is made in their presence). Tables 2-1 to 2-3 summarize recurrence risks in the three principal types of mendelian inheritance.

Sporadic cases of disorders with unknown etiology pose special problems for genetic counseling. Sometimes, further investigation of the medical and pregnancy histories or of seemingly unrelated conditions in other family members may reveal a clue to an etiologic agent or to a family pattern. A literature search may provide information on similar cases. Although efforts at syndrome identification are yielding a growing body of literature, too often even the best-informed spe-

Homozygosity of dominant disorders (both genes of a pair are mutant) is extremely rare, unless the disorder is mild and common (e.g., familial hypercholesterolemia) or two affected individuals are likely to marry (e.g., achondroplasia). Expression of the disorder in the homozygous state is generally lethal or very severe.

Isolated cases of dominant disorders may be the result of a new mutation. Careful examination of first-degree relatives may reveal minimal expression and establish familial incidence, ruling out a new mutation.

If the presence of a new mutation can be established, the risk to siblings becomes negligible, as opposed to 50% in familial cases. However, the risk to offspring of the affected individual is 50% , as in any clear-cut dominant condition.

New mutations are more likely to happen with increased paternal age, but none are considered to have highly increased incidence.

Transmission of Mutant Gene

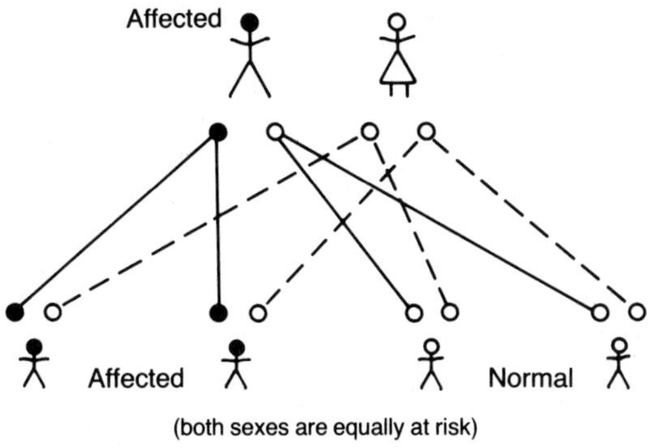

Each person has two genes for any given trait and passes on one of them randomly to each child, who then has one from each parent (● = mutant gene; O = normal gene). A child with one mutant gene is affected. On average, 50% will be affected, 50% normal.

cialists cannot give the parents reliable recurrence-risk information. However, it is less dangerous to give qualified counseling than to offer more specific figures associated with a mistaken or uncertain diagnosis.

FURTHER CONSIDERATIONS IN RISK ASSESSMENT

In all types of inheritance, precise diagnosis is the cornerstone of accurate counseling. Of the genetic concepts discussed below, some complicate and others facilitate the achievement of this precision.

Pleiotropy refers to a single mutant gene or a pair of mutant genes having diverse effects in the expression of a disorder. Conversely, *genetic heterogeneity*

FIGURE 2-4 Autosomal Recessive Inheritance

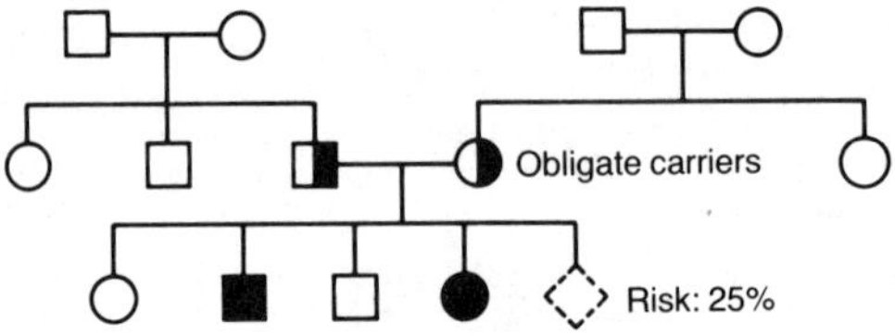

Typical Autosomal Recessive Pedigree

Genetic characteristics:

An autosomal recessive disorder is one which is fully expressed only in the homozygous state, i.e., a mutation is present on both genes of a pair on an autosomal chromosome. The mutant gene has been inherited from both parents, who are carriers.

Carriers, with only one mutant gene (heterozygotes) are usually not clinically symptomatic and are often identified only through the birth of an affected child (obligate carriers).

The disorder is generally found only in sibships, with males and females equally at risk, and no other relatives affected (except in inbred families).

When both parents are carriers, the risk for occurrence is 25% for each conception, which also has a 50% risk for carrier status and a 25% chance for two normal genes.

Unaffected siblings of affected individuals have a two in three risk of being carriers.

Carrier testing, if available, definitively identifies carriers, and thus, whether a mating is at risk for affected offspring.

Prenatal diagnosis, if available, establishes whether a fetus is affected.

When carrier testing is unavailable, risk estimates for other family members can be given only when the diagnosis is definite. They depend on relationship to the affected individual and population heterozygote frequency, except in consanguineous marriages (see Chapter 53) or with disorders where affected individuals are likely to marry each other.

In most disorders, the risk for offspring of affected individuals is not very high, except in consanguineous or assortative matings, because of the low population frequency of carriers (notable exceptions are disorders found with increased frequency in certain population groups, for example, sickle cell anemia in blacks). However, all offspring are carriers.

Unless the disorder is a common one, the risk for offspring of healthy siblings is only minimally

refers to different mutant genes or different mutations in a particular gene producing similar clinical effects. As specific a diagnosis as possible is important, because recurrence risks, clinical course, and prognosis may differ greatly. Increasingly, investigators are discovering that conditions once lumped into a few categories are actually collections of distinctly separate disorders (note the growing lists of subtypes among the inborn errors of metabolism or the connective-tissue disorders).

As more information becomes available on the biochemical and biophysical parameters of genetic disease, more precise diagnostic tests follow. In addition, increased (except in consanguineous matings). For other relatives the risk becomes negligible.

Consanguinity is increased among the parents of individuals with a recessive disorder; more so, the rarer the condition. Conversely, consanguinity noted in the parents of a patient with an unidentified disorder, suggests possible autosomal recessive inheritance.

Autosomal recessive disorders are subject to less variability than dominant ones, but genetic heterogeneity (different mutations) accounts for many disorders that resemble each other but may have very different prognoses.

When two affected individuals have children, all will be affected if both parents have the same gene mutation. If the parents have mutations on different genes, none of the children will be affected. Such assortative mating is not uncommon for some conditions—albinism, hearing loss. . . .

Many sporadic cases of recessive disorders are noted, since modern families are small. Risk counseling can be given only when the diagnosis is definite. With the 25% risk for an affected child, many carrier couples will have no affected children at all and may not be identified.

Transmission of Mutant Gene

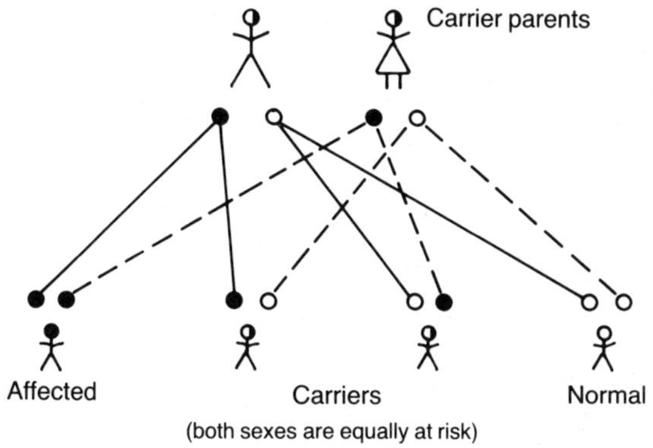

Each person has two genes for any given trait and passes one of them on randomly to each child (● = mutant gene; O = normal gene). A child with two mutant genes is affected; a child with one mutant and one normal gene is a carrier. On average, 25% will be affected, 50% carriers, and 25% normal.

other specifically genetic diagnostic strategies are used to identify and define genetic disease. Among them are *marker studies* based on the *linkage* of two genes located so close to each other on the chromosome that they are rarely separated in reproduction. One can be easily characterized through a polymorphism and the other is potentially mutant. In families with sufficient polymorphisms, studies can track the transmission of the mutant gene for carrier testing and prenatal diagnosis.

Recombinant DNA techniques are used to discover restriction-enzyme cleavage sites that are linked to potentially mutant genes and that exhibit sufficient

FIGURE 2-5 X-Linked Inheritance

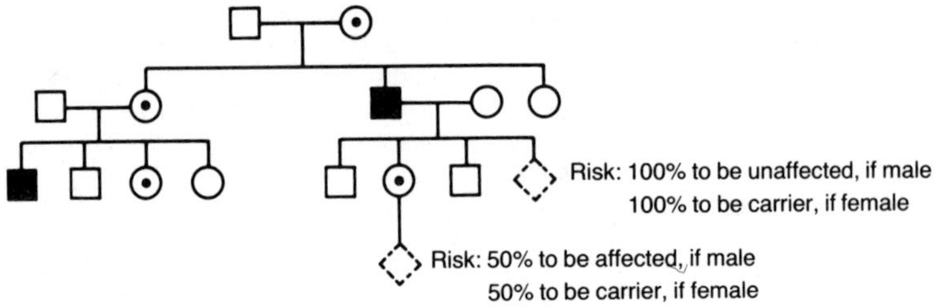

Typical X-linked Recessive Pedigree

Genetic characteristics:

An X-linked disorder is one in which the mutant gene is located on the X chromosome. Since females have two X chromosomes and males only one (hemizygous), inheritance patterns differ from the autosomal disorders.

Most X-linked disorders are recessive, i.e., the disorder is generally expressed only in the absence of a normal X chromosome. Therefore, typically, males are affected and females are carriers (all but the last of the characteristics listed below refer to X-linked recessive inheritance).

Female carriers may be mildly symptomatic in some X-linked recessive disorders due to unequal inactivation of their X chromosomes; this can be useful for carrier detection.

The hallmark of X-linked inheritance is that there is no male to male transmission, since males transmit an X chromosome only to daughters. Thus, affected males have only unaffected sons and carrier daughters.

Carrier females have a 50% risk with each conception of transmitting the mutant gene to offspring. Sons inheriting the gene will be affected; daughters will be carriers.

Unaffected males cannot transmit the disorder (notable exception—fragile X syndrome).

Females can express the full-blown disorder when the father is affected and the mother is a carrier. This is very unlikely, except in common conditions (e.g., G6PD deficiency in populations of Mediterranean or African ancestry). Even rarer is the affected female who happens to have only one X chromosome (Turner's syndrome).

Women are considered obligate carriers if they have more than one affected son, an affected male relative and a proven carrier daughter, or an affected son and an affected brother or maternal uncle.

Women are at risk for being carriers if they have one affected son, one affected brother, one affected maternal uncle, or a sister with an affected son.

polymorphism for studies in appropriate families. As more such probes are developed, additional families will be subject to this type of analysis. More specific diagnosis is possible without family linkage studies, when a restriction enzyme is identified that cleaves a gene at the exact site of the mutation (now available for sickle-cell disease).

The concept of *genetic association* is easily confused with linkage. This term

Where carrier testing is not available, the risk estimate for carrier status for female relatives in families with diagnosed cases of X-linked recessive disease can be modified on the basis of other family history or laboratory findings by use of Bayesian analysis. For example, a woman whose maternal grandmother is a known carrier has a 25% unmodified risk to be a carrier, but only a 10% modified risk in a family where her mother had two unaffected and no affected sons.

An isolated case of a diagnosed X-linked disorder may be the result of a new mutation in the affected individual or in the mother, or may be the result of carrier status transmitted to the mother by her mother. Clarification of this point can make the difference between 50% (mother is a carrier) and negligible (new mutation) in the recurrence risk for the next male pregnancy and in the risk for daughters or sisters to be carriers. Unfortunately, carrier status of the mother may be very difficult to determine, unless reliable carrier testing is available.

When specific prenatal diagnosis is not available for an X-linked recessive disorder, prenatal sex determination may be offered, with the option to carry only female pregnancies to term.

X-linked dominant disorders are very rare. Since they may be expressed by both males and females, the inheritance pattern may resemble autosomal dominant inheritance, but with no male to male transmission. The disorder may be lethal in males (e.g., incontinentia pigmenti).

Transmission of Mutant Gene

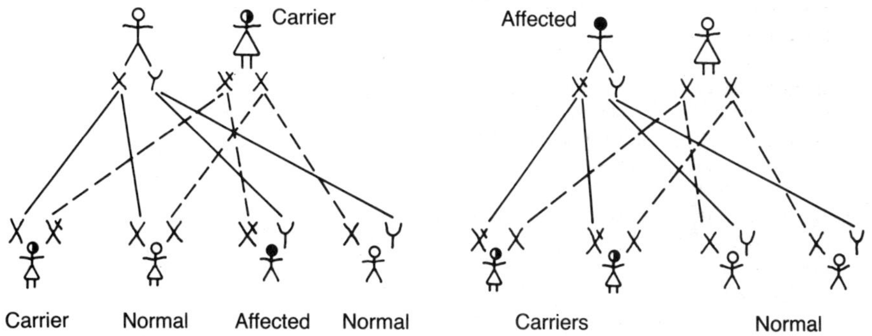

Women have two X chromosomes and men have one X and one Y. Each parent passes on one sex chromosome to each child. The sex of the child depends on whether the X or the Y was passed on by the father. An individual with one X chromosome containing a mutant gene (X̸) and no normal X is affected (men); one with a normal X in addition to the one with the mutant gene, is a carrier (women).

refers to some particular genotypes (e.g., HLA haplotypes) being found more often in individuals with a genetic, usually multifactorial, disorder, than in the general population. A specific polymorphism, thought to contribute to the predisposition of the disorder in such cases, has nothing to do with proximity on a chromosome. Such association has some predictive value, but is not yet considered useful for routine diagnosis.

When a genetic disorder is relatively common in the general population or in a subpopulation—such as Tay-Sachs' disease in Ashkenazi Jewish populations, sickle-cell disease among blacks, thalassemia in populations of Mediterranean ancestry—*carrier screening* makes sense, especially if prenatal diagnosis can be offered. Most of the time the screening provides reassurance that the marriage is not at risk for affected children. In the few marriages where both spouses are found to be carriers, the testing provides an alert for follow-up and for potential prenatal diagnostic evaluation. Unfortunately, no reliable carrier test has yet been developed for cystic fibrosis, which is the most common recessive disorder among Caucasians.

FIGURE 2-6 Polygenic/Multifactorial Inheritance

Common Conditions With Multifactorial Inheritance

Cleft lip/palate
Congenital heart defects
Essential hypertension
Neural tube defects
Pyloric stenosis

Genetic characteristics:

Many disorders that occur in familial clusters are known to have a genetic component, but do not follow clear mendelian inheritance patterns or give evidence of a chromosomal etiology.

The terms polygenic and multifactorial reflect the assumption that many genes, as well as environmental factors, contribute to the development of the disorder, but generally little is known about specific causes.

Given categories of disorders may represent a group of clinically similar, but etiologically heterogeneous conditions. As long as these cannot be distinguished from each other, the disorder must be viewed as a group. With improved classification, subgroups with clear-cut causes may be identified, and risk estimates for the remaining category may change.

Recurrence risks are based on empiric data from population studies of individual disorders. They differ, depending on the heritability of the specific disorder and the population in which it was studied. They tend to be higher with higher population incidence. Recurrence-risk tables are given in chapters on multifactorial disorders. They share a number of characteristics:

- The risks to sibs or offspring are lower than mendelian risks, often falling somewhere near 3% when there is one affected individual in the family.
- Recurrence risks increase when more family members are affected.
- Recurrence risks decrease as the relationship to the affected individual becomes more distant, usually becoming negligible for third-degree relatives.
- Recurrence risks tend to be higher when the severity in the index case is high.
- Parental consanguinity increases the recurrence risk.

Some multifactorial disorders are more common in one sex (e.g., pyloric stenosis is more likely in males). Recurrence risks are greater when the affected family member belongs to the sex with the lower incidence.

Pedigrees in small families may resemble those of mendelian disorders, even when such an inheritance is unlikely by virtue of a certain diagnosis. Occasionally, extensive pedigrees do provide evidence of mendelian inheritance (usually dominant) in a specific family.

TABLE 2-1
Summary of A Priori Risks for Autosomal Dominant Transmission

AFFECTED FAMILY MEMBER(S)	RISK IN EACH PREGNANCY FOR OFFSPRING TO INHERIT MUTANT GENE*	
	For all autosomal dominant disorders (%)	
One parent	50	
One parent and offspring	50	
Two offspring	50	
Both parents	50 for heterozygote like parent 25 for homozygote (usually lethal or very severe)	
Offspring and other relative(s) of parent, who are obligate carriers of the mutant gene	50	
	Risk differs with characteristics of disorder	
	In disorders with complete penetrance before reproductive age or with definitive carrier test; parents not carriers of mutant gene	In disorders with incomplete penetrance and/or variable expressivity or late onset, when presence of mutant gene in parent cannot be ruled out (%)
One offspring	Negligible	50
Sibling of parent	Negligible	25
Aunt/uncle of parent	Negligible	12.5
Mother/father of parent (grandparent of offspring)	Negligible	25
Niece/nephew of parent	Negligible	12.5

*Whether, when, and how severely affected a gene carrier may be depends on modifying factors, such as degree of penetrance, variability of expression, and age of onset. For some disorders Bayesian analysis, based on modifying factors, can be used to estimate reduced probabilities for carrier state and/or risk to be affected.

Neonatal screening, mandated in most states in the United States, identifies infants actually affected with one or more of a number of disorders. The screening is particularly crucial for conditions requiring prompt institution of therapy to prevent severe morbidity, such as the mental retardation in phenylketonuria or thyroid deficiency.

In some cases pedigree analysis may be hampered by inconsistencies with clinical or laboratory findings. Possible *nonpaternity* may provide an explanation. In other cases, the question of paternity (usually for legal purposes) sometimes constitutes the sole reason for a medical contact. *Paternity testing,* based on genetic polymorphisms, including blood-group antigens, red cell enzymes, serum proteins, and particularly HLA antigens, can be used to prove exclusion of paternity in a high proportion of cases. Paternity may be established as highly probable, but the result can never be considered totally conclusive.

REPRODUCTIVE OPTIONS

Diagnostic information, prognosis, and treatment or management of genetic disease are medical services that depend on the specific disorder; they often do not differ from medical care in general. The discussion of risks with the patient, offspring, to relatives sets genetic counseling apart from most other medical consultations. The perceived gravity of a risk depends not only on the actual figures,

TABLE 2-2
Summary of Risks for Autosomal Recessive Transmission

Risks for Disorder

CARRIER STATUS OF PARENTS	RISK IN EACH PREGNANCY FOR AFFECTED OFFSPRING (%)
Both parents carriers	25
One parent carrier, other parent not affected	0
One parent affected, other parent carrier	50
One parent affected, other parent not carrier	0
Both parents affected, all offspring affected	100

Risks for Carrier Status*

AFFECTED RELATIVE(S) OF UNAFFECTED INDIVIDUAL	RISK FOR UNAFFECTED INDIVIDUAL TO BE CARRIER (%)
Son/daughter	Obligate carrier 100
Sister/brother	2 in 3 (66.6)
Mother/father	Obligate carrier 100
Niece/nephew	50
Grandparent	50
Aunt/uncle	33.3
First cousin	25
Affected relative(s) in more extended family	Increased over population carrier frequency; more specific risks can be derived from pedigree analysis
No affected relatives	Generally equal to population carrier frequency, which tends to be low, except in certain disorders in high-risk populations (e.g., cystic fibrosis in Caucasians, sickle-cell disease in blacks)
	Increased over population carrier frequency when spouse is known carrier or at increased risk to be one, and spouses are related by blood; more specific risks can be derived from pedigree analysis

Formula for calculating risk for an affected child:

(one parent's risk to be carrier) × (other parent's risk to be carrier) × (1/4)

*Carrier testing, when available, can remove the element of risk, by confirming or excluding carrier state.

TABLE 2-3
Summary of A Priori Risks for X-Linked Recessive Transmission

AFFECTED RELATIVE OF FEMALE PARENT*	RISK FOR FEMALE TO BE A CARRIER (%)	RISK FOR EACH FUTURE SON TO BE AFFECTED (%)
One son		
new mutation } ∞	Negligible } †	Negligible } †
no new mutation } ∞	100 } †	50 } †
Two sons	100 (obligate carrier)	50
One son and other affected maternal relative(s)	100 (obligate carrier)	50
Father	100 (obligate carrier)	50
Maternal grandfather	50	25
One brother or sister's son		
new mutation } ∞	Negligible } †	Negligible } †
no new mutation } ∞	50 } †	25 } †
Maternal uncle		
new mutation } ∞	Negligible } †	Negligible } †
no new mutation } ∞	25 } †	12.5 } †
Paternal uncle, paternal grandfather, brother's son	Not increased over general population risk	

*Unaffected males, regardless of affected relatives, cannot pass on the gene for an X-linked recessive condition (exception: fragile X syndrome). Affected males pass on the gene to all their daughters (obligate carriers) and none of their sons.

∞Mother of affected individual may have inherited the mutant gene, or the new mutation may have occurred in her.

†In the absence of accurate carrier testing it is not possible to distinguish between these two possibilities. With serious disorders it may be prudent to assume carrier status for the mother of one affected son. Bayesian analysis of the pedigree and inconclusive carrier information can be used to estimate an intermediate risk. Prenatal diagnosis, if available, can obviate the need for such calculations.

but to a large extent on such intangibles as the counselees' interpretation of the potential burden, their ethical and moral convictions, their prior experiences, and their psychologic needs. In view of this, genetic counseling generally involves a concerted effort to promote a clear understanding of the facts, encouragement for counselees to explore their own feelings, and help in arriving at decisions most appropriate for their specific situation (see Chapter 1).

Most of these decisions are of a reproductive nature, involving a choice by potential parents among the options that may be appropriate for their case. Of course, the outcome of a genetic workup may indicate either no increased risk or an acceptable increased risk, obviating any change in planning. When risks and/or burdens are not acceptable to counselees, the options offered may include one or more of the following.

Prenatal diagnosis, if feasible for the disorder in question, reduces the risk for affected offspring to a direct answer about the presence or absence of the condition. If termination of an affected fetus is acceptable to the parents, a family of unaffected children can be planned (see Chapter 3).

Some counselees will elect to plan *no further reproduction*. Some will *delay* any planning until they can evaluate the impact of the existing problem or clarify their own reactions. Contraceptive counseling may include the consideration of permanent, as well as temporary, measures. Sterilization may be considered, especially of the affected partner with an autosomal dominant disorder or a chromosome abnormality, or of the female carrier of an X-linked condition. When both parents are carriers of an autosomal recessive disorder, sterilization for one of the partners may be elected as a simpler solution than other methods of contraception, but one must remember that possible future offspring with another partner might not be at risk for the rare disease.

When transmission of a genetic liability from the male constitutes the risk factor (autosomal dominant or chromosome abnormality) or contributes to it (e.g., autosomal recessive), *artificial insemination by donor* may be elected to avoid increased risks. Since acceptance of this option can be a serious psychologic threat to the male partner and thus to the marriage, it is generally recommended only when both partners have given it serious consideration and are in wholehearted agreement.

If donor insemination is elected, a careful medical and family history of the donor, and, if possible a chromosome analysis, are important to ensure that no known additional genetic risk can be identified in the pedigree. The information also provides a family health history for future use by the child. With rare autosomal recessive conditions, the likelihood that the donor carries the same mutant gene as the couple is very small. Nevertheless, if a carrier test is available, he should be tested. For the more common diseases, such as Tay-Sachs' or sickle cell, carrier testing is essential. When carrier testing is not feasible for a common autosomal recessive disorder like cystic fibrosis (carrier rate among Caucasians is 1/20), donor insemination is not generally considered an acceptable option.

The female counterparts of donor insemination are *surrogate motherhood* and *embryo transfer*. Both are subject to the same caveats as donor insemination. They could be used, additionally, for women who are carriers of X-linked disorders; and are available, although on a limited basis.

Adoption, once a widely used option for childless couples, has the benefit of avoiding the transmission of mutant genes by either parent without placing responsibility on one. However, with the shortage of available infants, especially from reputable agencies, the chances of raising a family in this way are slim. In addition, when a potential parent actually has a genetic disorder or is at risk for developing one, the likelihood of being offered an adoptive child decreases. If adoption is a viable option, consideration of the genetic background of the adoptive child is important. A family health history of the biologic parents may reveal genetic risks and/or serve as a biologic background for future considerations.

If adoption is not available or potential adoptive parents are apprehensive about making a permanent commitment, *foster parenthood* may provide an interim solution.

MANAGEMENT AND TREATMENT IN GENETIC DISEASE

All too often, genetic disease is still equated with hopelessly incurable disease. Although genetic lesions cannot be undone, and there is no cure in that sense, recent advances have led to effective treatment and improved management for numerous conditions. Current research, particularly in the areas of gene therapy and enzyme replacement, holds promise for more. Individual chapters discuss relevant treatment modalities for specific disorders.

On the whole, unfortunately, many genetic disorders are still subject only to symptomatic treatment and supportive care, but advances in diagnosis, prevention, and management are so rapid, that the hope for substantially reducing the toll of genetic disease has become a realistic goal.

SUGGESTED READING

Desnick RJ, Grabowski GA: Advances in the treatment of inherited metabolic diseases, in Harris H, Hirschorn K, (eds): *Advances in Human Genetics,* vol 11. New York, Plenum Press, 1981.

Friedman TF: *Gene Therapy: Fact and Fiction in Biology's New Approaches to Disease.* New York, Cold Spring Harbor, 1983.

Harper PS: *Practical Genetic Counseling,* Baltimore, University Park Press, 1981.

Murphy EA, Chase GA: *Principles of Genetic Counseling,* Chicago, Year Book Medical Publishers Inc, 1975.

President's Commission for the Study of Ethical Problems in Medicine and Behavioral Research: *Screening and Counseling for Genetic Conditions,* Washington, D.C., U.S. Government Printing Office, 1983.

Rowley PT: Genetic screening: Marvel or menace? *Science* 1984;225:138.

Thompson JS, Thompson MW: *Genetics in Medicine,* Philadelphia, WB Saunders, 1980.

Vogel F, Motulsky AG: *Human Genetics: Problems and Approaches,* New York, Springer-Verlag, 1979.

Chapter 3

Prenatal Diagnosis

Prenatal diagnosis, in offering prospective parents information about some of the serious risks of childbearing, is one of the redoubtable tools of genetic counseling. Though still over-shadowed by the total roster of catalogued genetic conditions and diseases, the number of disorders amenable to prenatal diagnosis is growing at an impressive rate. Table 3-1, provides a list of selected common genetic disorders for which prenatal diagnosis has been reported.

Several basic procedures lead to the diagnosis of hundreds of disorders, both genetic and nongenetic. The application of each is usually limited to specific types of conditions. Some are now routinely available, although others are still done only in specialized laboratories. In view of the rapid advances in genetic research, it is worthwhile to inquire into current availability of prenatal diagnosis, whenever a pregnancy is at risk for a potentially diagnosable disorder.

Prenatal detection can change probability (e.g., 25%) of producing an infant with an incurable or lethal disorder to the certainty that the fetus is either free of the specific disease or is almost surely affected. The parents can then make decisions on the basis of facts, rather than on risk calculations. Of course, no test can show that the baby is ''normal'', that is, free of all possible disorders.

The *benefits* of prenatal diagnosis vary, depending on the findings and on parental decisions. In most cases the parents can be reassured that their unborn child is free of the condition for which it was at increased risk. They then revert to the general population background risk—approximately 2% to 3%—for having a child with a significant birth defect.

If a fetus is found to have the disorder for which the prenatal diagnosis was conducted, parents may have several options. Rarely, prenatal treatment may be available, but usually the disorder is sufficiently severe that possible termination of the pregnancy was part of the original reason for the prenatal diagnosis. For couples who elect to terminate, a future pregnancy still offers the opportunity to bear a child free of that disease.

When a couple chooses to carry an affected fetus to term, knowledge that an infant will be born with a specific defect or disorder can significantly improve prognosis by readying therapeutic measures for the critical neonatal period. For example, surgical facilities can be prepared for the neonate known to have a neural-tube defect; dietary needs can be immediately available to the infant affected with galactosemia. In addition, the parents have time to prepare psychologically for the birth of a child with special needs.

Like all medical procedures, prenatal diagnosis has attendant *risks* and inherent *limitations,* which depend on the specific techniques used. They should be fully explained in counseling (see below, Prenatal diagnostic procedures and analytic techniques, and Psychologic considerations in prenatal diagnosis).

TABLE 3-1
Selected Common Genetic Disorders for Which Early Prenatal Diagnosis Is Available or Has Been Reported

DISORDER	INHERITANCE	DIAGNOSTIC METHOD
Chromosomal		
Ataxia-telangiectasia	AR	Chromosome studies on cultured fetal cells
Bloom's syndrome	AR	Chromosome studies on cultured fetal cells
Fanconi's anemia	AR	Chromosome studies on cultured fetal cells
Fragile X syndrome	XR	Fragile-X analysis of cultured fetal cells
Numerical or structural abnormalities	CH	Karyotype analysis of cultured fetal cells*
Xeroderma pigmentosum	AR	Chromosome studies on cultured fetal cells
Congenital malformations		
Achondroplasia	AD	Ultrasonography for abnormal head-to-body ratio
Anencephaly	MF	Alpha-fetoprotein determination on maternal serum and/or amniotic fluid; ultrasound; amniography
Camptomelic dysplasia	AR	Ultrasonography for femoral length
Diastrophic dysplasia	AR	Ultrasonography
Ectrodactyly	AD	Ultrasonography, fetoscopy
Hydrocephalus	Varied	Ultrasonography, radiography
Infantile polycystic kidneys	AR	Ultrasonography
Meckel's syndrome	AR	Ultrasonography, amniography; for encephalocele, polycystic kidneys, oligohydramnios
Potter's syndrome	?MF	Ultrasonography for absent kidneys and bladder, oligohydramnios
Robert's syndrome	AR	Ultrasonography for short or absent long bones, polycystic kidneys, oligohydramnios; chromosome analysis for centromere abnormality
Spina bifida	MF	Alpha-fetoprotein determination on maternal serum and/or amniotic fluid; ultrasound; amniography
TAR (thrombocytopenia/absent radii) syndrome	AR	Radiography
Metabolic		
Adrenogenital syndrome (21-hydroxylase deficiency)	AR	Hormone assay of amniotic fluid
Albinism—tyrosinase negative	AR	Fetoscopy—analysis of fetal hair bulbs
Apha$_1$-antitrypsin deficiency	AR	Recombinant DNA analysis of cultured fetal cells

TABLE 3-1 *Continued*

DISORDER	INHERITANCE	DIAGNOSTIC METHOD
Fabry's disease	XR	Enzyme assay of cultured fetal cells
Galactosemia	AR	Enzyme assay of cultured fetal cells
Gaucher's disease	AR	Enzyme assay of cultured fetal cells
Hunter's syndrome (mucopolysaccharidosis II)	XR	Enzyme assay of cultured fetal cells
Hurler's syndrome (mucopolysaccharidosis I-H)	AR	Enzyme assay of cultured fetal cells
Hypophosphatasia—congenital	AR	Ultrasonography—deficient bone mineralization; enzyme assay of cultured fetal cells
Krabbe's disease (globoid cell leukodystrophy)	AR	Enzyme assay of cultured fetal cells
Lesch-Nyhan syndrome	XR	Enzyme assay of cultured fetal cells
Maple syrup urine disease	AR	Enzyme assay of cultured fetal cells
Metachromatic leukodystrophy	AR	Enzyme assay of cultured fetal cells
Methylmalonic acidemia (Vitamin B_{12} responsive)	AR	Enzyme assay of cultured fetal cells
Phenylketonuria	AR	Recombinant DNA family study and analysis of cultured fetal cells
Pompe's disease (glycogen storage disease, type II)	AR	Enzyme assay of cultured fetal cells
Porphyria—congenital erythropoietic	AR	Amniotic fluid porphyrin assay
Severe combined immune deficiency due to adenosine deaminase deficiency	AR	Enzyme assay of cultured fetal cells
Tay-Sachs disease	AR	Enzyme assay of cultured fetal cells
Miscellaneous		
Cardiac anomalies	Varied	Ultrasonography
Congenital lamellar ichthyosis	AR	Fetal skin biopsy
Epidermolysis bullosa letalis	AR	Fetal skin biopsy
Familial hypercholesterolemia (homozygous)	AR	LDL uptake of cultured fetal cells
Hemophilia A and B	XR	Fetoscopy for a male fetus, for fetal blood studies of coagulation factor antigens; recombinant DNA family studies and analysis of cultured fetal cells in informative families (experimental)
Menkes' disease	XR	Copper uptake of cultured fetal cells
Muscular dystrophy (Duchenne's)	XR	Recombinant DNA family studies and analysis of cultured fetal cells
Myotonic dystrophy	AD	Family secretor linkage study and secretor status of amniotic fluid
Osteogenesis imperfecta congenita	AR	Ultrasonography; radiography; procollagen assay in cultured fetal cells
Sickle cell anemia	AR	Recombinant DNA analysis of cultured fetal cells
Thalassemia (α and β)	AR	Recombinant DNA family studies and analysis of cultured fetal cells; fetoscopy for fetal blood studies

Key: AD = autosomal dominant; AR = autosomal recessive; XR = X-linked recessive; CH = chromosomal; MF = multifactorial

*Cultured fetal cells were obtained by amniocentesis in the second trimester of pregnancy. Cells obtained by chorionic villus sampling in the first trimester can be cultured for the same purposes and, in some cases (e.g., karyotyping), may be analyzed directly without further culturing.

INDICATIONS FOR PRENATAL DIAGNOSIS

Prenatal diagnosis is an option in the following situations:

Advanced maternal age. The risk for conceiving a child with a chromosomal trisomy, particularly Down's syndrome, increases with advancing maternal age (see Chapter 6 for age-specific risks at birth). Prenatal diagnosis is generally offered to pregnant women with an estimated date of confinement at age 35, or over, when the risk starts to rise steeply. However, depending on the availability of services, the test may be an option for anxious parents at a younger age.

Previous conception with a chromosome anomaly. For mothers under 35 years, the birth of one child with a chromosome anomaly usually confers an additional risk of at least 1% for a subsequent pregnancy (however, recurrence risk for trisomies to the woman over 35 is still age-related). Similarly, spontaneous abortions with known karyotypic abnormalities can serve as an indication for prenatal diagnosis for later pregnancies (see Chapters 6 and 52).

A parent with a known chromosomal anomaly. A balanced translocation, chromosomal mosaicism, or fragile X carrier status, for example, increases the risk for chromosomally abnormal offspring (see Chapters 6 and 52).

A previous child with a metabolic or other genetic disorder for which prenatal diagnosis is available: Well over 100 genetic conditions have been diagnosed prenatally by various methods, and, increasingly, reports of new prenatal diagnoses appear in journals. While many of these techniques are still experimental, it is worthwhile to investigate the current state of the art for any serious genetic condition diagnosed in a family. Table 3-1 gives a representative list of common genetic disorders that have been diagnosed prenatally (many experimentally, in special laboratories). Other chapters provide information on intrauterine diagnosis of specific disorders.

Known or suspected carrier parents. Although carrier status of parents is generally inferred from the previous birth of an affected child (especially for errors of metabolism, most of which are autosomal recessive), for some conditions parents may be identified as carriers before such a tragedy has occurred. This is most likely for the ethnically related diseases for which effective screening protocols have been developed (see Chapters 17, 18, and 34). Rarely, mild clinical manifestations may identify the heterozygous carrier of a recessive disorder.

Prenatal diagnosis may also be an option when one parent is affected with an autosomal dominant disorder, particularly when the child may be more severely affected than the parent. Prenatal diagnosis for late-onset dominant conditions (e.g., Huntington's disease) may become an option for selected families. However, it is fraught with additional psychologic complications, because the parent may still be in the "at risk" category and unwilling to face the incidental presymptomatic diagnosis of a devastating illness that a positive finding in the fetus would generate.

Carrier status for X-linked recessive disorders may be identified or suspected from the family history or through partial expression. Although many X-linked disorders cannot be diagnosed prenatally, parents may be reassured that the fetus is unaffected when it is determined to be female. If it is male (and at 50% risk to be affected with a severe, prenatally undiagnosable disorder, such as muscular dystrophy), the option of termination may be considered.

Family history of neural-tube defects. The risk for occurrence increases not only when there is a previous affected child, but also when a parent, parent's sibling, or other close relative is affected. (See Chapter 7 for recurrence risks.)

A confirmed positive finding in a routine maternal serum alpha-fetoprotein screen is described under "Prenatal diagnostic procedures" below.

Prenatal diagnosis is also an option in several situations which are most appropriately evaluated individually, because their significance varies:

Multiple abortions. In the absence of a known obstetric cause, a series of first trimester spontaneous abortions may be related to chromosomal anomalies in parents (e.g., translocation chromosome). Peripheral blood karyotyping of the parents may provide an indication for chromosomal studies of the fetus (see Chapter 52).

A previous child with undiagnosed multiple anomalies. In the event that chromosome studies were not performed on the affected child, a chromosome anomaly cannot be ruled out. If chromosome study of the child is not feasible, parental chromosome studies could indicate the presence of a heritable anomaly.

Extended family history of nondisjunction: Rare families have been reported in which there is an apparent tendency to nondisjunction, that is an inordinate number of offspring with chromosomal trisomies unexplained by maternal age or carriers of structurally abnormal chromosomes.

PRENATAL DIAGNOSTIC PROCEDURES AND ANALYTIC TECHNIQUES

MATERNAL SERUM ALPHA-FETOPROTEIN SCREENING

Elevated alpha-fetoprotein levels in the serum of a pregnant woman may be an indication of an open neural-tube defect in the fetus, thus, an assay done be-

TABLE 3-2
Conditions Associated With Elevated Amniotic Fluid AFP

Multiple gestation
Underestimate of gestational age
Open neural-tube defects
Duodenal atresia
Esophageal atresia
Fetal bleeding
Fetal demise
Omphalocele
Gastroschisis
Exstrophy of the bladder
Turner's syndrome
Triploidy

tween the 14th and 22nd week of gestation can be used as a gross screening test. Positive findings must be confirmed, since they may reflect a multiple gestation, an underestimate of gestational age, or a number of other abnormalities (see Table 3-2). Two abnormal results, one week apart, followed by ultrasonography may constitute an indication for amniocentesis. Already widely used, this simple test is expected to become routine for all pregnancies in the near future.

Recent findings suggest that low maternal serum alpha-fetoprotein levels may be indicative of an increased risk for fetal trisomy, especially Down's syndrome. A routine screening of maternal serum AFP may thus serve the additional purpose of identifying pregnancies at increased risk for Down's syndrome in women of any age.

AMNIOCENTESIS

Transabdominal amniocentesis is currently the most frequently used technology to obtain a fetal specimen for prenatal diagnosis. Its applications for genetic diagnosis have been well established since 1967, when cytogenetic analysis of amniotic fluid cells successfully demonstrated a balanced fetal translocation. Since that year, indications for its use and the number of women to whom it has been made available have steadily grown. Nevertheless, the increasingly available procedure of chorionic villus sampling (see below) may replace amniocentesis as the most common procedure for prenatal diagnosis, because it is less invasive and is performed much earlier in the gestation.

Amniocentesis is optimally performed during the 16th to 17th week of gestation, as determined from the first day of the last menstrual period. Fetal safety and practical consideration dictate this approximate time. The volume of amniotic fluid is generally sufficient at this time to allow easy withdrawal, and the fluid contains ample cells for successful culture. The removal of 15 ml to 30 ml of fluid, which is rapidly replaced, is readily tolerated. Also, this period is early enough in the pregnancy to allow for laboratory procedures requiring two to five weeks (depending on the disorder), to be completed in time for parents to exercise all their options.

Ultrasonography as an adjunct procedure is recommended to confirm gestational age, provide information on multiple gestation, gross structural defects in the fetus (e.g. neural-tube defects), fetal demise, and obstetric complications, and to ensure an additional measure of safety for the fetus. An ultrasound scan reduces the likelihood of a tap that pierces the placenta producing a bloody fluid, and it is invaluable to forecast potential problems: a thick anterior placenta, oligohydramnios, or a fetal position which precludes a safe tap. Such situations can be reassessed after a postponement of one or two weeks, as can other contraindications for amniocentesis, such as an active infection on the surface of the maternal abdomen, a severe upper respiratory infection, or vaginal bleeding.

Given the invasive nature of amniocentesis, fetal and maternal risk cannot be altogether ruled out. Potential adverse effects include amniotic fluid leakage, maternal infection, fetal puncture, maternal or fetal hemorrhage, and spontaneous abortion. However, performed by an experienced practitioner under aseptic conditions with ultrasonography, amniocentesis carries a total morbidity risk of less than 1%. The Rh negative mother is a special consideration; immunization against Rh positive fetal blood type is generally offered at the time of amniocentesis to forstall possible complications in the current or future pregnancies.

Limitations of amniocentesis for prenatal diagnosis are inherent in the nature of the specimens available in this way. It can provide information only on disorders expressed in fetal cells sloughed off in fluid, in fetal chromosomes, or in the fluid itself. Although a growing number of biochemical assays are becoming possible, amniocentesis is indicated only when a pregnancy is known to be at increased risk for a particular disorder.

Technical problems can attend successive steps in the diagnostic evaluations, even after uncomplicated withdrawal of fluid. Fetal cells occasionally defy even the most reliable incentives to growth, and laboratory contamination is a possibility with any procedure dependent on cell culture. Time permitting, repeat taps may be offered under these circumstances.

Amniotic fluid contains fetal cells' which are primarily of epithelial or fibroblastic origin. The fluid and cells can be utilized for the following *analytic techniques:*

Chromosome analysis. Any known chromosomal anomaly can be demonstrated by cytogenetic analysis of cultured fetal cells. Cultured and stained by a variety of recently developed methods which more clearly characterize each chromosome or special abnormality, the fetal chromosome complement will reveal the presence of an abnormal number, structural defects, or functional aberrations which could signify clinical abnormality (see Chapter 5). Fetal sex is incidentally obtained from the karyotype, but may become important when prenatal diagnosis is conducted for an X-linked disorder.

Alpha-fetoprotein (AFP) determination. AFP levels are generally determined routinely, whenever amniocentesis is performed for any indication, and particularly for a family history of neural-tube defects or a confirmed finding of evaluated maternal serum AFP. AFP in amniotic fluid varies with gestational age, underlining the need for ultrasound determination of fetal age at the time of amniocentesis. An elevated level of AFP in amniotic fluid is an indication for further diagnostic procedures, (e.g., acetylcholinesterase assay or further ultrasonography), since a variety of conditions other than open neural-tube defects can be responsible for the finding. Table 3-2 provides a list of conditions reported to have been associated with elevated amniotic fluid AFP.

Biochemical assay. Cultured amniotic fluid cells are the basis for diagnosis of those inborn errors of metabolism expressed as enzyme deficiencies in these particular cells. Although a limited number of determinations can be obtained directly from the fluid (e.g., hexoseaminidase A deficiency in Tay-Sachs' disease), accuracy is improved by analysis of cultured cells. When an abnormal or missing enzyme has not been characterized or is not produced by obtainable cells, biochemical prenatal diagnosis is not possible.

Linkage analysis. As the human gene map is more precisely delineated, linked loci (genes close together on the same chromosome) may provide diagnostic information in instances where the gene defect is itself unknown or cannot be diagnosed prenatally. Previous determination, through family studies, that a deleterious gene is closely linked to a gene producing a measurable biochemical marker is thus the basis for prenatal diagnosis of specified disorders—such as myotonic dystrophy, which is linked to the secretor gene. The family studies should be done prior to, or very early in the pregnancy. The diagnosis is feasible only for informative families. The level of precision, though generally quite high, cannot provide complete certainty.

Recombinant DNA techniques. Restriction enzymes, which cleave DNA at specific nucleotide sequences, provide the basis for another kind of linkage analysis. Mutations closely linked to particular DNA fragments thus created, can be diagnosed prenatally, with the fragment serving as a marker. Briefly, the technique extracts fetal DNA from cultured fetal cells and compares it by molecular hybridization to a DNA probe of known composition.

The first disorders for which such diagnostic strategies became possible were the hemoglobinopathies, which until then, could be diagnosed only from fetal blood obtained through fetoscopy. As long as restriction-enzyme analysis is based on linkage data, family studies, prior to or very early in the pregnancy, are necessary and, not all families are informative. More sophisticated is site-specific retriction-enzyme analysis. This is possible when a restriction enzyme is discovered that cleaves DNA directly at the site of a mutation. Precise diagnosis can then be achieved without the need for family studies. This has been accomplished for sickle cell anemia.

Technology is still highly specialized and available only at selected centers. However, it is a rapidly growing field with exciting prospects for prenatal diagnosis. Recent developments include reports of recombinant DNA probes with potential application for some families with muscular dystrophy, $alpha_1$-antitrypsin deficiency, phenylketonuria, Huntington's disease, and hemophilia.

Viral studies. A suspected maternal viral infection can be confirmed by recovering and culturing viruses from amniotic fluid or cells. The value of this technology is underscored by increasing recognition of viruses as teratogens. Viruses, such as cytomegalovirus and rubella, have been recovered from amniotic fluid.

CHORIONIC VILLUS SAMPLING (CVS)

Still experimental, but very promising, chorionic villus sampling (biopsy by transcervical aspiration) is performed to obtain fetal cells. These cells can be used for all the analyses done now on amniocytes (but not on fluid). Since they are more viable than the sloughed off cells in the amniotic fluid, some determinations, such as chromosome analysis, can be done immediately, without the need for added time to culture them. The principal advantage of CVS over amniocentesis is that it is done much earlier in pregnancy (generally between eight and 10 weeks' gestation), permitting an early abortion, if termination is elected subsequent to a positive finding. However, although it is expected to be safe for routine use, not enough information has been collected on CVS to establish accurate estimates of associated risks (at this writing a 2% to 3% risk has been quoted for all complications, including spontaneous abortion). Eventually, it may replace amniocentesis as the most widely used procedure for prenatal diagnosis. At this writing, it is already offered in several centers in the United States.

SONOGRAPHY

The diagnostic value of ultrasonography is based on the ability of sound waves to be reflected back from tissues of varying densities. Once an adjunct to amniocentesis, ultrasound has been refined and increasingly has become a prenatal diagnostic modality in its own right.

Pulsed waves are beamed into the uterus through a transducer placed on the mother's abdomen, and the reflected waves are visualized on a screen and printed. The experienced diagnostician can determine the size of fetal structures and the distance between them. Ultrasound need not necessarily be the sole method of diagnosis for a given entity. It can, for example, sometimes confirm and locate a neural-tube defect suggested by elevated AFP levels in amniotic fluid.

Standardized atlases of fetal bone parameters have been compiled to assist in diagnostic evaluation. Performing the procedure in centers where such standards of reference are available offers evident advantage.

The usefulness of ultrasound as a diagnostic aid is limited to disorders whose effects are gross rather than subtle, structural rather than metabolic, and which are, moreover, apparent early in fetal development. Renal agenesis, for example, would be apparent on an ultrasound scan (see Table 3-1 for others).

There is no evidence to date that diagnostic ultrasound has had untoward effects on the fetus or neonate, and it can, in fact, sometimes be used to avoid the greater risks of other prenatal diagnostic procedures.

FETOSCOPY AND PLACENTAL ASPIRATION

Fetoscopy employs a small-gauge endoscope, inserted into a cannula pushed through the abdominal wall, to visualize directly the fetus and placenta in the pregnant uterus. Ultrasound is critical in this procedure to determine the position of the fetus, placenta, and umbilical cord. A local anesthetic, with or without fetal or maternal sedation, is administered.

The procedure carries a higher risk than amniocentesis, reflecting its more invasive quality—risk of miscarriage is approximately 3% to 5% even in experienced hands, and the risk of prematurity is also increased. Safety is heavily dependent on the experience of the obstetrical team performing the procedure, which is available in a few clinical centers on an applied research basis.

The advantages of fetoscopy lie in its access to fetal tissues—blood and skin—and to direct visualization of fetal surface morphology. It has been particularly useful for the diagnosis of disorders expressed in blood, but not in amniocytes (e.g., sickle cell anemia, hemophilia). However, with the advent of recombinant DNA techniques, which can be used with any nucleated cell, the need for this high-risk procedure may decline considerably.

Placental aspiration is an ultrasound-directed approach to fetal blood vessels on the placenta's inner surface. The blood sample obtained through placental aspiration is more likely to be contaminated with maternal blood than samples obtained via fetoscopy. Like fetoscopy, the procedure carries a higher morbidity than amniocentesis. However, with an anterior placenta, placental aspiration is sometimes the method of choice for obtaining fetal blood.

In general, fetal blood sampling, by whichever method, is delayed somewhat longer than amniocentesis—to approximately 18-20 weeks'—to ensure adequacy of the fetal blood supply.

FETAL RADIOGRAPHIC VISUALIZATION

Skeletal development is sufficient by 17 to 19 weeks' gestation for observation, using standard radiography, of long bones and ribs, and indeed, most tubular bones. Specialized contrast radiography can provide a much more precise visualization.

In amniography a a water-soluble dye is introduced into the amniotic fluid. This technique is informative about the fetus' ability to swallow, which, when abnormal, can be a diagnostic indication of gastrointestinal malformation or neurologic deficit.

Fetography, another variation, employs an oil-soluble dye. The fetal outline appears when the dye adheres to the vernix caseosa coating the skin. It is useful in outlining cystic spinal lesions.

The unquantified but small risks of irradiation—carcinogenic or mutagenic—must be balanced against the known risks of carrying a fetus with a severe developmental disorder. There may be an additional risk in employing contrast materials, which in several reports have had a transitory effect on thyroid function in the fetus. Finally, the techniques do not always clearly reveal specific bones even at 20 weeks' gestation, and may not be able to distinguish between the slow end of a normal maturation range in ossification and a pathologic developmental process.

PSYCHOLOGIC CONSIDERATIONS IN PRENATAL DIAGNOSIS

The couple undergoing prenatal diagnosis, which now generally entails a two- to four-week wait for news of such considerable emotional import, can expect to experience emotional stress. Ideally, empathetic counseling precedes prenatal diagnosis, preferably well in advance of the actual procedure.

Pregnancy has unique psychologic dimensions anyway, hence, it is important for those involved with genetic aspects of care to keep in mind that any suggestion of potential risks can create heightened apprehensions. Implicit in the idea of prenatal diagnosis for whatever indication is the realization that a pregnancy can have an abnormal and undesired outcome. Anxiety is almost inseparable from any procedure, and sensitive counseling can do much to minimize it. Diagnostic procedures should be explained fully, including risks, benefits, limitations, and time constraints (Figure 3-1 shows a sample consent form, which outlines the major points of preamniocentesis counseling). Relevant components of parents' family histories should be discussed in terms of recurrence risk and genetic mechanisms involved.

The prenatal diagnostic procedure under consideration should be viewed by the family as an option, without judgment on the part of the counselor. It is desirable for the parents to be counseled together, if possible, to be given the time and freedom to ask whatever questions they wish, and encouraged to air any concerns and conflicts they may have, so they can make a well-informed decision.

The period between procedure and results can be characterized by sleep disturbances, marital strain, and severe anxiety. Leaving the door open for further contact following the procedure, should the couple feel the need for it, can do much to lessen stress.

For the parents, particularly the mother, whose diagnostic results are positive, this period of stress is only a prelude. By the time amniocentesis results are available, gestational time is usually sufficiently advanced for the mother to have ex-

FIGURE 3-1 Amniocentesis Consent Form

This is to state that I request and authorize ______________________________,
Name of physician

or his designated associate, to perform amniocentesis for the purpose of prenatal diagnosis of the following condition(s):

____________ Chromosomal abnormality in the fetus

____________ Neural-tube defect in the fetus

____________ Biochemical abnormality in the fetus ______________________
Name of condition

____________ Other conditions ______________________________
Name of condition

The procedure has been fully explained to me; I understand that it involves withdrawal of fluid surrounding the fetus by means of a needle passed through the wall of the abdomen and into the uterus. I have also been informed that the procedure can involve the following risks or complications:

1. A small risk to mother or fetus can include infection, bleeding, leakage of amniotic fluid, and miscarriage.
2. There is a possibility culturing (growing) fetal cells may be unsuccessful. This may require a repeat amniocentesis. I understand that failure of fetal cells to grow does not indicate any abnormality in the fetus.
3. There is a possibility chromosomal studies or biochemical studies may be unsuccessful. This may require a repeat amniocentesis.
4. Normal results of tests indicated above do not eliminate the possibility that the fetus can have other abnormalities or disorders.

Signature ______________________________

Witness ______________________________

Date ______________________________

perienced quickening, and for bonds between mother and fetus to have begun forming. When feelings about terminating an affected pregnancy are ambivalent, a difficult decision under severe time pressure must be faced. With chorionic villus sampling, the procedure is done between eight and ten weeks and the period between it and the results may be much shorter, allowing termination before quickening and before the pregnancy has become obvious to outsiders. While this timing may lighten some of the burdens related to a termination, it by no means obviates the psychologic sequelae. Even for those who have little trouble deciding to terminate an affected pregnancy, its loss can be attended with almost the same grief and mourning as that of a desired infant carried to term. Feelings of grief are frequently compounded by the stigma and loss of self-esteem inherent in the conception of a child deemed genetically defective.

Supportive counseling cannot eliminate grief, nor is it intended to do so. It can, however, help the family adjust to its loss, and perhaps look forward to another pregnancy with a more successful outcome.

SUGGESTED READING

Antenatal diagnosis. U.S. Department of Health, Education and Welfare, NIH Publication No 79-1973, 1979.

Baumgarten A, Schoenfeld M, Mahoney MJ, et al: Prospective screening for Down syndrome using maternal serum AFP. *Lancet* 1985;1(8440):1280.

Blumberg B, Golbus M, Hanson K: The psychological sequelae of abortion performed for a genetic indication. *Am J Obstet Gynecol* 1975;122:799.

Chang JC, Kan YW: A sensitive new prenatal test for sickle cell anemia. *N Engl J Med* 1982;307:30.

Golbus MS, Loughman WD, Epstein CJ, et al: Prenatal genetic diagnosis in 3,000 amniocenteses. *N Engl J Med* 1979;300:157.

Golbus MS: Antenatal diagnosis of hemoglobinopathies, hemophilia and hemolytic anemias. *Clin Obstet Gynecol* 1981;24:1055.

Goossens M, Dumez Y, Kaplan L, et al: Prenantal diagnosis of sickle cell anemia in the first trimester of pregnancy. *N Engl J Med* 1983;309:831.

Hobbins JC, Grannum PA, Berkowitz RL, et al: Ultrasound in the diagnosis of congenital anomalies. *Am J Obstet Gynecol* 1979; 134:331.

Holbrook K, (ed): Prenatal diagnosis of inherited skin disease, *Seminars in Dermatology*. New York, Thieme-Stratton, 1984.

Kidd VJ, Golbus MS, Wallace RB, et al: Prenatal diagnosis of $alpha_1$-antitrypsin deficiency by direct analysis of the mutation site in the gene. *N Engl J Med* 1984;310:639.

Kolata G: Huntington's disease gene located. *Science* 1983;222:913.

Macri JN, Baker DA, Baim RS: Diagnosis of neural-tube defects by evaluation of amniotic fluid. *Clin Obstet Gynecol* 1981;24:1089.

Macri JN, Weiss RR: Prenatal serum α-fetoprotein screening for neural-tube defects. *Obstet Gynecol* 1982;59:633.

Miles JH, Kaback MM: Prenatal diagnosis of hereditary disorders. *Pediatr Clin North Am* 1978;25:593.

Milunsky A: *Genetic Disorders and the Fetus: Diagnosis, Prevention, and Treatment.* New York, Plenum Press, 1979.

Oberle I, Camerino G, Heilig, R, et al: Genetic screening for hemophilia A (classic hemophilia) with a polymorphic DNA probe. *N Engl J Med* 1985;312:682.

Orkin SH, Little PFR, Kazazian HH, Jr, et al: Improved detection of the sickle mutation by DNA analysis. *N Engl J Med* 1982;307:32.

Powledge TM, Fletcher J: Ethics of prenatal diagnosis. *N Engl J Med* 1979;300:168.

Robinson J, Tennes K, Robinson A: Amniocentesis: Its impact on mothers and infants. A one-year follow-up study. *Clin Genet* 1975;8:97.

Schrott HG, Omenn GS: Myotonic dystrophy: Opportunities for prenatal prediction. *Neurology* 1975;25:789.

Simoni G, Brambati B, Danesino C, et al: Efficient direct chromosome analyses and enzyme determinations from chorionic villi samples in the first trimester of pregnancy. *Hum Genet* 1983;63:349.

Stephenson SR, Weaver DD: Prenatal diagnosis: A compilation of diagnosed conditions. *Am J Obstet Gynecol* 1981;114:319.

Ward RHT, Modell B, Petrou M, et al: Method of sampling chorionic villi in the first trimester of pregnancy under guidance of real-time ultrasound. *Br Med J* 1983;286:1542.

Woo SLC, Lidsky AS, Güttler F, et al: Prenatal diagnosis of classical phenylketonuria by gene mapping. *JAMA* 1984;251:1998.

Section Two

Disorders in Clinical Genetics

Chapter 4

Mental Retardation

The term, mental retardation, encompasses three areas of impairment: maturation, learning, and social adjustment. It has been defined as: "significantly subaverage general intellectual function, existing concurrently with deficits in adaptive behavior and manifested during the developmental period."

While the definition may include some individuals who fall into the low end of the normal distribution of intelligence in the population, it describes primarily a wide array of heterogeneous conditions ranging from very mild to extremely severe and from isolated mental deficiency to disorders with multiple associated anomalies.

Depending on methods of classification and varied estimates, mental retardation as a whole is thought to affect from 1% to 3% of the total population.

Many changes have come about over the years in defining who is mentally retarded and to what degree. Table 4-1 shows currently used scales, based on psychometric testing, that were adopted by the American Association of Mental Deficiency in 1977.

Approximately 85% to 90% of all retarded persons in the United States fall within the mild or educable range, live at home or in the community, and are partially self-sufficient. However, those with genetic disorders are likely to be more seriously affected.

This chapter provides an overview of all types of mental retardation, with special emphasis on those with known or suspected genetic etiologic components. Some of the genetic forms are discussed in greater detail in chapters that deal with the underlying causes (e.g., chromosomal syndromes, Down's syndrome, phenylketonuria).

ETIOLOGY AND PATHOGENESIS

Although the etiology of mental retardation in the majority of retarded individuals still remains unknown despite extensive studies, many causes of mental retardation are now recognized and new diagnostic techniques promise to identify others.

Recognized etiologies of mental retardation run the gamut from well-defined genetic abnormalities to conditions of purely nongenetic origin, and include multifactorial disorders thought to be caused by both genetic and environmental components. Genetic etiology can also be indirect—via the intrauterine environment for instance, in the mental retardation of biochemically unaffected children born to mothers who themselves have phenylketonuria (PKU). Beyond the known reasons for mental deficiency remain the idiopathic cases for which no cause has been identified. Many of these may, in fact, have unidentified genetic components. The fragile X chromosome in some retarded males is an example of a newly revealed important causative factor.

Severe mental retardation is more likely to have a genetic etiology than do the milder forms. It has been estimated that about a third of the cases of severe mental deficit are clearly genetic and that perhaps half of the remainder also have genetic components. Except for those family members who have inherited the same deficit, families of severely retarded individuals generally are not intellectually impaired, indicating the likelihood of a single major cause. Mild or moderate mental retardation is often, though not always, thought to have a multifactorial etiology. Among borderline or mildly retarded individuals familial clustering frequently occurs, and socioeconomic causes are often invoked; but the clustering may indicate some genetic influence as well.

In an effort to establish a diagnosis, all genetic and nongenetic etiologies must be considered. Table 4-2 presents an etiologic classification of identified genetic and acquired mental retardation, with examples of the nongenetic categories. Table 4-3 gives a classification, notes examples of the more common genetic forms, and lists the modes of inheritance.

GENETIC CHARACTERISTICS AND MODE OF INHERITANCE

Genetic forms of mental retardation may be inherited as autosomal recessive, autosomal dominant, X-linked, or multifactorial traits, or may be caused by chromosomal aberrations. Most are associated with other abnormalities. About 200 of the 1,005 birth defects listed in the current Birth Defects Compendium are associated with mental retardation.

For characteristic examples of genetically determined forms of mental retardation, see Table 4-3. Recurrence risks depend on the mode of inheritance. Table 4-4 provides empiric recurrence risks for siblings of children with unexplained

TABLE 4-1
Levels of Mental Retardation

LEVELS	OBTAINED INTELLIGENCE QUOTIENT BY TEST STANFORD-BINET AND CATTELL (SD 16)	WECHSLER SCALES (SD 15)
Mild	67-52	69-55
Moderate	51-36	54-40
Severe	35-20	39-25
Profound	19 and below	24 and below
Children above the mild level, but under an IQ of 80, are classified as borderline.		

Source: Adapted from Grossman HJ (ed): *Manual on Terminology and Classification in Mental Retardation,* American Association on Mental Deficiency, Washington, 1977.

mental retardation of various types. Recurrence risks for other genetic disorders associated with mental retardation are given in relevant chapters. If an environmental cause has been identified, recurrence is, of course, unlikely, unless the harmful agent is still operating.

CLINICAL NOTES

The establishment of an accurate diagnosis and recurrence risks is of paramount importance, not only for counseling, but for a management protocol, since some forms of mental retardation can now be avoided or treated successfully.

TABLE 4-2
Common Causes of Mental Retardation

<u>Genetic</u>
Chromosome disorders
Inborn errors of metabolism
Hereditary degenerative disorders
Hormonal deficiencies
Primary CNS defects
Malformation syndromes
Sporadic syndromes with unidentified etiology of possibly genetic origin
Familial idiopathic mental retardation
Low end of normal distribution
<u>Acquired</u>
Prenatal
Infection (syphilis, rubella, toxoplasmosis, cytomegalic inclusion disease . . .)
Fetal irradiation
Toxins (fetal alcohol syndrome, lead poisoning, mercury poisoning, fetal hydantoin syndrome . . .)
Maternal metabolic problems (maternal PKU . . .)
Perinatal
Prematurity
Asphyxia (abruptio placentae, cord prolapse, meconium aspiration . . .)
Infection (meningitis, encephalitis, TORCH agents, syphilis, herpes simplex . . .)
Trauma (breech delivery, intracerebral hemorrhage . . .)
Hypoglycemia
Kernicterus
Postnatal
Brain injury (trauma, drowning, lightning . . .)
Poisoning (lead, carbon monoxide . . .)
Cerebrovascular accidents
Postimmunization encephalopathy (pertussis, rabies . . .)
Infection (meningitis, encephalitis, abscess . . .)
Early severe malnutrition
Hormonal deficiency
Psychosocial deprivation, abuse or neglect

TABLE 4-3
Selected Genetic Forms of Mental Retardation

TYPE	DISORDER	MODE OF INHERITANCE
Chromosomal abnormalities	Down's syndrome	CH
	Other autosomal trisomies	CH
	Unbalanced translocations	CH
	Fragile X syndrome	XR
Inborn errors of metabolism		
Aminoacidurias	Phenylketonuria	AR
	Maple syrup urine disease	AR
	Homocystinuria	AR
	Hartnup disease	AR
Mucopolysaccharidoses	Hunter's syndrome	XR
	Hurler's syndrome	AR
	Sanfilippo's syndrome	AR
Lipidoses	Tay-Sachs disease	AR
	I-cell disease	AR
	Pseudo-Hurler polydystrophy	AR
Carbohydrate metabolism	Fucosidosis	AR
	Galactosemia	AR
Purine metabolism	Lesch-Nyhan syndrome	XR
Hereditary degenerative disorders	Myotonic dystrophy	AD
Hormone deficiencies and dysfunction	Congenital hypothyroidism	AR or V
Primary CNS defects	Primary microcephaly	AR or V
	Congenital hydrocephalus	V
	Craniofacial dysostosis	AD
	Spina bifida	MF
Neuroectodermatoses	Tuberous sclerosis	AD
	Neurofibromatosis	AD
Miscellaneous syndromes	Menkes steely-hair disease	XR
	Smith-Lemli-Opitz syndrome	AR
	Seckel's syndrome	AR
	Laurence-Moon-Bardet-Biedl syndrome	? AR
	Noonan's syndrome	? AD
	Oculocerebralrenal syndrome of Lowe	XR
	Apert's syndrome	AD
	Prader-Willi syndrome	? CH
Disorders not proven genetic*	de Lange's syndrome	?
	Rubinstein-Taybi syndrome	?
	Cerebral gigantism	?
	Autism	?

Key: CH = chromosomal; AR = autosomal recessive; AD = autosomal dominant; XR = X-linked recessive; MF = multifactorial; V = varied; ? = unknown

*Most cases are sporadic with unknown etiology, but familial incidence and affected identical twin pairs have been reported. An empiric recurrence risk for siblings has been estimated at 2% for de Lange's syndrome and 1% to 2% for autism.

Among the treatable or potentially treatable genetic defects that cause mental retardation are the inborn errors of metabolism, notably PKU (low phenylalanine diet) and galactosemia (lactose-free diet). Another easily preventable cause is congenital hypothyroidism, when hormone treatment is begun early in life. Neonatal and heterozygote screening programs and prenatal-diagnosis protocols have been instituted to facilitate the early detection and treatment of some disorders associated with mental retardation, or to identify at-risk parents.

Treatment may involve such strategies as putting a woman with never-treated or no-longer-treated PKU on a low phenylalanine diet (preferably before conception) to prevent risk to the fetus conferred by the maternal metabolic imbalance. Surgical procedures, such as shunts for hydrocephalus, are also used to prevent mental manifestations. Reports of experimental prenatal shunting for hydrocephalus hold a promise of future preventive intervention.

Diagnostic evaluation depends on information derived from the family history, the prenatal, birth, and postnatal histories; the physical examination, with special emphasis on patterns of associated malformations, both major and minor; and appropriate laboratory studies. Sometimes examination of other family members may be helpful.

The *timing of onset* of disease manifestations provides a useful framework for systematic diagnostic evaluation. This is clearly evident when mental retardation has an environmental cause operating at a specific time of development (see Table 4-2). Although all genetic forms of mental retardation originate prenatally, the onset of symptoms can provide a clue to the cause.

Prenatal-onset mental retardation can often be predicted at birth on the basis of a primary CNS defect or a characteristic pattern of associated malformations. Multiple minor and major non-CNS anomalies and neurologic signs help to identify teratogenic agents, possible chromosomal abnormalities, and known malformation syndromes. A complete physical examination, careful gestational history, and neonatal screening will provide indications for additional diagnostic procedures.

Nongenetic *perinatal insult* to the brain, documented more often in premature infants than in others, is the most common cause of mental retardation originating during this period. However, caution is recommended before assuming that a potential perinatal factor is the primary cause of the deficit, since an underlying genetic etiology can be the reason for the prematurity or birth complications. Family and gestational histories, the clinical course, and detailed evaluation for any anomalies may help to identify or rule out potential causes.

Some may immediately show abnormalities, but generally children with *postnatal onset* of mental retardation appear normal at birth. Slowed development and/or mental deterioration set in after a variable period of normal function. These may be accompanied by other neurologic manifestations or somatic symptomatology. Among the types of mental retardation presenting in this way are the untreated or untreatable inborn errors of metabolism. Familial or multifactorial mild retardation is also usually diagnosed after infancy, as are those cases with a postnatal environmental etiology. The latter may, of course, often be recognized by a temporal relationship between a potential cause and actual onset of symptoms.

In addition, mental retardation may present with or without other CNS dysfunction (e.g., spasticity, hypotonia, seizures), without other evident causal factors, or at an unpredictable stage of development. The same can be true of mental retardation due to prenatal infectious disease, the manifestations of hypothyroidism, and mild retardation associated with some sex-chromosome anomalies.

As new diagnostic techniques are introduced, patients remaining undiagnosed after the initial workup may benefit from periodic re-evaluation in light of recent advances.

PROCEDURES FOR DIAGNOSTIC CONFIRMATION

Pregnancy history
Medical and developmental history
Pedigree analysis

TABLE 4-4
Empiric Risks for Recurrence of Mental Retardation of Unknown Etiology in Sibs and Children

AFFECTED	INDIVIDUAL AT RISK	RISK
I.Q. below 70		
Isolated case	Sib	1 in 10
I.Q. below 50		
Isolated case	Sib (both sexes)	1 in 35
	Male sib	1 in 25
	Female sib	1 in 50
Two or more affected males	Male sib	1 in 2.5
	Female sib	1 in 10
Two sibs regardless of sex	Sib	1 in 4
Isolated case (parents consanguineous)	Sib	1 in 7
One affected parent	Child	1 in 10
Two affected parents	Child	1 in 2
Severe retardation with associated malformations		
Isolated case with microcephaly	Sib	1 in 10
Isolated case with cranial contour abnormalities	Sib	1 in 16
Isolated case with multiple congenital anomalies	Sib	1 in 25

Source: Adapted from Harper PS: *Practical Genetic Counseling,* Baltimore, University Park Press, 1981; and Cox DR, Epstein CJ, Golbus MS, et al: *Medical Genetics Syllabus,* University of California, San Francisco, 1982.

Physical examination

Radiographic studies, including skull and skeletal films, CT scan, bone age

Chromosome studies of peripheral blood or other tissue

Fragile X chromosomal preparation

Blood and urine tests for amino acids; urine for organic acids and mucopolysaccharides; urine for reducing substances

Serum electrolytes, ammonia, CO_2, blood glucose

Serum uric acid

Serum creatine phosphokinase

Lysosomal enzyme analyses in blood, tears, platelets

Muscle or tissue biopsy for histologic studies, including electron-microscopy and/or biochemical analysis

Thyroid function tests

Developmental and psychologic testing

Assessment of hearing and vision

Electroencephalograms in awake and sleep states

TORCH antibody titers (toxoplasmosis, rubella, cytomegalovirus, herpes)

VDRL

CSF analysis, including cultures

Consultation with appropriate specialists

CONSIDERATIONS IN MANAGEMENT

Genetic
- Establish etiology and assess recurrence risk
- Establish carrier status of other family members, if indicated and available
- Risk counseling
- Consideration of reproductive options, as indicated by diagnosis
- Discussion about management of future pregnancies, if maternal prenatal factors are involved in etiology (e.g., maternal PKU, fetal alcohol syndrome)
- Consideration of prenatal diagnosis in future pregnancies
- Discussion of sharing information with other family members to encourage evaluation, available carrier testing, and counseling

Psychosocial, educational, familial
- Family counseling; support for acceptance of limited intellectual development
- Consideration of referral for psychosocial therapy and/or support
- Family education for effective home care
- Assessment of family/community resources, such as support groups, infant stimulation programs, special education facilities, recreational resources, respite care, group homes, day-hospital programs, residential settings, foster care and adoption agencies

Medical
- Dependent on diagnosis

SUGGESTED READING

Carter CH, (ed): *Medical Aspects of Mental Retardation*. Springfield, Charles C Thomas, 1978.

Milunsky A, (ed): *Prevention of Genetic Disease and Mental Retardation*. Philadelphia, WB Saunders Co, 1975.

Smith DW: *Recognizable Patterns of Human Malformation*. Philadelphia, WB Saunders Co, 1982.

Taft LT: Mental Retardation: Overview. *Pediatr Ann* 1973;2:10.

Chapter 5

Chromosomal Syndromes

Abnormalities in chromosome number or structure have been implicated in a variety of recognized syndromes. Down's syndrome, the most common, is discussed in detail in Chapter 6, and some abnormalities of sexual differentiation associated with chromosomal changes are covered in Chapter 27. This chapter provides an overview of the growing number of genetic syndromes in which chromosomal aberrations can be detected with current cytogenetic methods. The conditions are all characterized by a chromosomal abnormality present at conception or shortly thereafter. They are all subject to prenatal diagnosis, since the abnormality appears in cells of almost all tissues.*

About five in every 1,000 *liveborn infants* have a detectable chromosomal abnormality: In about half of these the chromosome abnormality is accompanied by congenital anomalies and/or mental retardation or by phenotypic changes that manifest later in life. In the other cases of chromosome abnormality there is either a rearrangement which does not result in chromosomal imbalance, or a sex chromosome abnormality not associated with phenotypic effects (e.g., some mosaics, some cases of XXX and some cases of XYY).

The highest proportion of chromosome anomalies occurs in pregnancies lost by *early spontaneous abortion;* in about 50% of such miscarriages, the fetus has abnormal chromosomes. About 7% of *stillbirths* and *perinatal deaths* have chromosomal anomalies; 3% are accounted for by Down's syndrome. It is thus important to obtain chromosome studies in such cases; these can be performed on specimens obtained at autopsy, or on fetal tissues from the placenta.

Surveys of seriously mentally retarded children have shown that about 12% have a chromosomal problem. When selection is limited to those with multiple congenital anomalies as well, or with low birthweight, the frequency increases to 23%, of which about half have Down's syndrome. These figures do not include cases of the fragile X syndrome, now thought to be almost as common as Down's syndrome. Chromosome analysis is thus essential for any child with mental retardation of uncertain etiology, particularly if accompanied by dysmorphic features, congenital abnormalities, and/or low birthweight.

ETIOLOGY AND GENETIC CHARACTERISTICS

The normal human chromosome constitution comprises 46 chromosomes—23 pairs, one of each pair from the mother, via the ovum, and one from the father,

*Chromosome abnormalities associated with somatic changes in malignancies are not discussed here.

via the sperm. The full complement is the diploid set, the half complement is called haploid. The 23 pairs include 22 pairs of autosomes, which are numbered according to size, from 1 to 22. The other pair is the set of sex chromosomes, termed X and Y (female = XX, male = XY).

For analysis, photomicrographs of chromosomes are arranged by pairs according to size, shape, and banding patterns (karyotype). Figures 5-1 and 5-2 show examples of normal human male and female karyotypes.

An abnormal chromosome constitution can be inherited or have its origin prior to conception in a parental gamete. It can also be the result of an error in chromosomal separation during an early embryonic cell division. Although agents known to promote chromosomal breakage (e.g., ionizing radiation, chemotherapeutic drugs) are thought to have a potential for producing chromosomally abnormal gametes, determining the cause of any particular chromosome aberration is usually impossible.

Chromosomal disorders may be classified according to the type of chromosomal change and the particular chromosome number involved in the abnormality: An internationally standardized formula is used to describe the karyotype of an individual. Examples of the nomenclature for normal chromosomes and for some of the more common disorders are given in Table 5-1.

Heritable chromosome aberrations can be transmitted in a balanced or unbalanced form from one generation to another. As a rule, this transmission does not follow mendelian patterns of inheritance, and recurrence risks are based on empiric data.

Most individuals with gross chromosomal abnormalities do not reproduce, either because of the associated disabilities or because they are not fertile. Thus, the risks listed below apply primarily to the recurrence of an abnormality in a family that has already had one affected child or is at risk on the basis of family history. The risks depend on the type of abnormality:

Abnormal number of chromosome sets

- Triploidy: three copies of each chromosome (total chromosome complement:69), instead of the normal two copies(total chromosome complement:46)—can result from complete failure of disjunction at meiosis, resulting in a diploid gamete instead of a haploid gamete, or from fertilization of one egg by two sperm (about 85% of cases).
- Tetraploidy: four copies of each chromosome(total chromosome complement:92),—apparently occurs because of a failure of mitotic division, in an originally chromosomally normal fertilized egg.

Both of these abnormalities are extremely rare in liveborn infants and are generally not compatible with survival past the neonatal period. In both instances, parental chromosomes are expected to be normal, and the recurrence risk for the condition is not expected to be increased above population values.

Abnormal number of chromosomes

- Monosomy: only one copy of one particular chromosome(total chromosome complement:45)

- Trisomy: three copies of one particular chromosome(total chromosome complement:47)

Either condition may be complete or mosaic. In a mosaic, some cells have the abnormality and some do not.

Monosomy for all but the X chromosome (Turner's syndrome) is extremely rare, and has been reliably described only for chromosome 21 in full-term pregnancies. Trisomy for most autosomal chromosomes is lethal at an early embryonic stage, resulting in miscarriage. Only a few kinds have been described in full-term births: for chromosomes 21 (Down's syndrome), 18, 13, and, more rarely, 8 and 9. Trisomy for the X chromosome, in females an extra X (XXY) or an extra Y (XYY) in males, are each found in about 0.1% of live births (higher numbers of X chromosomes have been reported in both men and women).

Complete trisomy results from abnormal segregation at meiosis leading to a gamete with an extra chromosome. This can be maternal or paternal in origin; evidence from trisomy 21 suggests it is maternal about 75% of the time. Monosomy also can result from the same kind of abnormality in meiosis, but concep-

FIGURE 5-1 Normal Male Karyotype

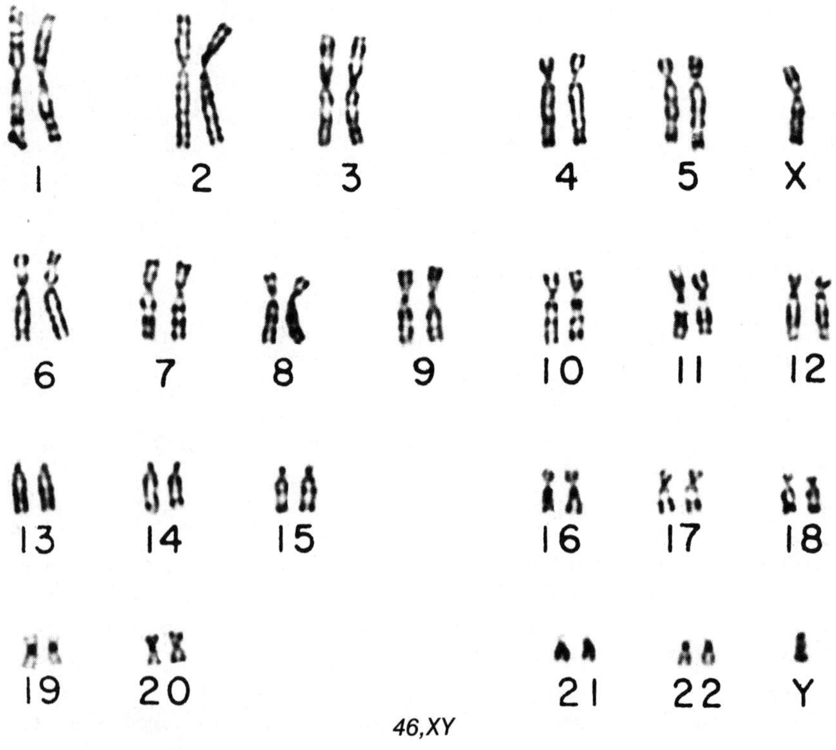

Courtesy of Lillian Hsu, M.D. and Sara Kaffe, M.D. of the Prenatal Diagnosis Laboratory of New York City.

tions with this condition do not usually survive to the stage of recognizable pregnancy, with the exception of monosomy X.

In cases of trisomy or monosomy, parental chromosomes are expected to be normal, except in very rare instances where a parent is a mosaic with some trisomic cells in the gonad. Mosaicism is demonstrated in less than 1% of trisomy cases.

All trisomies, including XXX and XXY, increase in frequency with the age of the mother, and can be detected by chromosome studies of amniotic fluid (for detailed age-related risks for trisomy 21 and age-related risks for any chromosome abnormality, see Chapter 6). In a woman over 40, about one third of the clinically significant chromosomal anomalies detected at amniocentesis will be trisomies other than 21.

When a young woman has had one trisomic pregnancy she is at an increased risk of about 1% for having another.

Women with trisomy X and men with an extra Y chromosome are fertile, producing mostly normal offspring. They are at an increased, but very low, risk for producing children who also have a sex chromosome abnormality.

FIGURE 5-2 Normal Female Karyotype

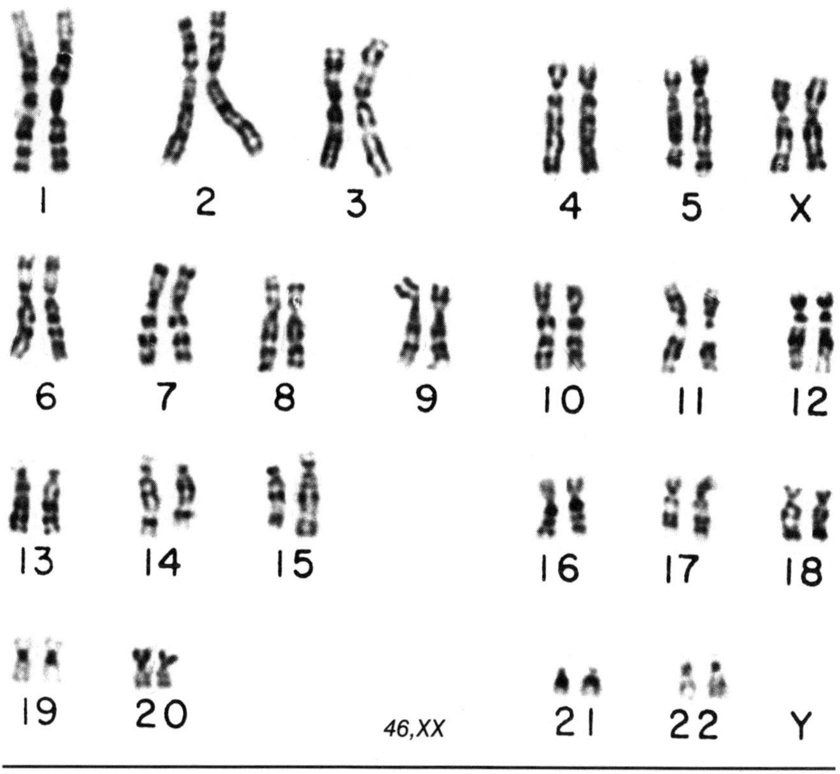

Courtesy of Lillian Hsu, M.D. and Sara Kaffe, M.D. of the Prenatal Diagnosis Laboratory of New York City.

Mosaic trisomy and monosomy result from an abnormal segregation of chromosomes during a mitotic division of the embryo. Different tissues of the body may have different proportions of normal and abnormal cells. Mosaicism is particularly common for sex chromosome abnormalities. Recurrence risks for mosaics have not been studied extensively. In theory, they are expected to be very low.

Chromosomal rearrangements leading to partial deletions or duplications of one or more chromosomes

- Deletion: a piece of one chromosome is missing (sometimes called partial monosomy)
- Duplication: a piece of one chromosome is extra (sometimes called partial trisomy)
- Translocation: the transfer of a segment from one chromosome to another
- Inversion: a chromosomal rearrangement where a broken piece is inserted in reverse direction
- Ring chromosome: breakage at both ends of a chromosome resulting in fusion of the two broken ends to form a ring

Chromosomal rearrangements initially result from breakage which is not correctly repaired. Translocations and inversions can be carried in a "balanced" form, where chromosomal material has not been lost or gained, but merely rear-

TABLE 5-1
Examples of Karyotype Nomenclature

DESCRIPTION	KARYOTYPE FORMULA*
Normal male	46,XY
Normal female	46,XX
Trisomy 21, male (Down's syndrome)	47,XY + 21
Translocation Down's syndrome, female (unbalanced karyotype with extra 21; robertsonian translocation)	46,XX, – 14, + t(14q21q)
Balanced robertsonian translocation carrier, male (father of translocation Down's syndrome patient)	45,XY, t(14q21q)
Trisomy 18, male (Edward's syndrome)	47,XY,+18
Trisomy 13, female (Patau's syndrome)	47,XX,+13
Mosaic with normal cells and triple-X cells	46,XX/47,XXX
Female with deletion of short arm of chromosome 5 (cri-du-chat syndrome)	46,XX,5p – or del (5p)
Male with deletion of long arm of chromosome 18	46,XY,18q – or del (18q)
Male with balanced translocation between chromosome 8 and chromosome 14	45,XY,t(8;14)
Female with ring chromosome 13	46,XX,r(13)
Female with monosomy X (Turner's syndrome)	45,X

*Number of chromosomes, sex chromosomes, abnormalities
Key: p = short arm of chromosome; q = long arm

ranged. Individuals carrying such a balanced rearrangement will be phenotypically normal. However, at gametogenesis, they are at risk to produce gametes with "unbalanced" chromosome constitutions. They can also produce gametes which can lead to chromosomally normal offspring, or balanced carriers like themselves.

Unbalanced rearrangements will result in partial deletions and duplications of chromosomal material, which can produce congenital malformations and mental retardation in a liveborn infant, or lead to miscarriage or stillbirth.

Any chromosome can be involved in a rearrangement, and the breaks involved can be anywhere in the chromosome. While, in general, breaks seem to occur at random, there are a few rearrangements which are more common than others. Some deletions and duplications have been described often enough that a recognizable clinical syndrome can be said to be associated with that particular chromosomal change, for example, 5p- or cri-du-chat syndrome, and translocation Down's syndrome. However, for many unbalanced rearrangements one cannot predict the exact nature of the abnormalities expected.

Recently, small deletions detectable only with high resolution techniques have been found in some patients with retinoblastoma, del(13)(q14); Prader-Willi syndrome, del(11)(p13); and aniridia-Wilms' tumor syndrome, del(11)(p13). More such associations will likely be discovered.

When a chromosomal rearrangement is detected in a child, parental chromosomes should be studied to determine if either is a carrier of a balanced rearrangement. About half the time, a chromosome change will be found in one of the parents. If the parents have normal karyotypes, then the rearrangement is said to have occurred "de novo" in either the egg or the sperm. Such couples are not at increased risk for having another child with a chromosomal anomaly.

If a parent is a carrier of a balanced rearrangement, there is a substantial risk of recurrence of a chromosomal defect. However, the exact risk cannot usually be stated, since this will vary with the type of rearrangement and the chromosomal breakpoints involved. Many of the unbalanced products of conception can be expected to abort spontaneously; the risks for abnormal full-term births may vary from 2% to 3% to 100%, depending upon the nature of the rearrangement. Prenatal diagnosis can detect any unbalanced offspring that survive to the time of the study.

When a parent is found to be a carrier of a balanced translocation or inversion, other phenotypically normal family members may carry the same chromosome rearrangement and thus be at the same risk for offspring with an unbalanced chromosome defect. Chromosome studies of the carrier's parents, siblings, and unaffected children can reveal the extent to which the aberration exists in the family.

Fragile sites. In some individuals frequent breakage is seen in a particular chromosomal region. Such "fragile sites" have been described on several chromosomes, often without any apparent clinical consequences. However, one fragile site, seen as a constriction of the long arm of the X chromosome at bands Xq27 to Xq28, is associated with *fragile X syndrome,* characterized by moderate mental retardation. This appears to be a relatively common anomaly (estimated at about 1/2,000 males).

Fragile X syndrome is inherited as an X-linked recessive trait, the X chromosome with the fragile site being passed on by carrier females. The full-blown syndrome affects males who have the fragile X chromosome because they have no normal X chromosome. However, roughly 30% of carrier females have mild mental retardation. Current thinking is that the mother of a sporadic affected male is an obligate carrier.

- The risk is 50% for sons of a carrier female to be affected and for daughters to be carriers.
- If the male reproduces, all his daughters are obligate carriers and all his sons are unaffected and will not transmit the disorder.

Characteristics of fragile X syndrome, in addition to the moderate mental retardation, are large ears, dysmorphic facial features, enlarged testes (in 80% of postpubertal fragile-X positive males), mitral valve prolapse, perseverative speech, and autistic symptoms. It has been reported that occasional males will be fragile-X positive and be phenotypically and mentally normal. They will, however, pass the fragile X on to all their daughters. Since mental retardation due to fragile X can also occur in the absence of other phenotypic stigmata, it is important to rule out fragile X before entertaining a diagnosis of another X-linked mental retardation, such as Renpenning syndrome—microcephaly, small stature, and normal or small testes).

Syndromes associated with increased spontaneous chromosome breakage. Higher-than-normal rates of chromosomal breakage are found in cultures from patients with certain recessively inherited diseases that often have associated growth failure, congenital anomalies, immunologic deficiencies, and high rates of cancer. These include Fanconi's anemia, Bloom's syndrome, ataxia telangiectasia, and xeroderma pigmentosum.

- These syndromes are inherited as autosomal recessive traits with an occurrrence risk of 25% when both parents are carriers.

CLINICAL NOTES

A large number of clinical features are associated with all the syndromes resulting from chromosomal abnormalities.

Multiple congenital anomalies and mental retardation are characteristic of autosomal trisomies. Of these, individuals with Down's syndrome (trisomy 21), and those with trisomy 8 are most likely to reach adulthood. Other trisomies are more severe and survival past infancy is less likely.

While mosaic cases may be more mildly affected than nonmosaic ones, this is not always so. Experience indicates that the degree of a defect cannot be predicted by the proportion of abnormal cells in any one tissue.

The mildest clinical consequences are associated with some of the sex chromosome anomalies, which may well go undiagnosed because they may not show any obvious phenotypic expression at all. This is particularly true of women with

Table 5-2
Phenotypic Features: Common Chromosomal Syndromes

Trisomy 18 (Edward's syndrome)

Growth and development — low birthweight, failure to thrive, severe mental retardation
CNS — hypertonia
Craniofacial — prominent occiput, low-set malformed ears, micrognathia, cleft lip and/or cleft palate (15%)
Extremities — flexion deformities of fingers, overlapping fingers, clubfoot, short dorsiflexed thumbs, and halluces
Cardiac — usual: VSD, PDA, and ASD (most common)
Abdominal — Meckel's diverticulum
Renal — variety of malformations
Skin and hair — single crease on digits
Genital — cryptorchidism
Remarks — 10% survive past 1 year

Trisomy 13 (Patau's syndrome)

Growth and development — severe mental retardation
CNS — deafness, seizures, apneic spells, holoprosencephaly, hypotonia
Craniofacial — microcephaly, microphthalmia, cleft lip and/or cleft palate (75%), micrognathia, colobomata, low-set ears, hypotelorism
Extremities — flexion deformity of fingers, polydactyly, clubfoot
Cardiac — usual: VSD, PDA, coarctation
Abdominal — omphalocele or umbilical hernia
Renal — polycystic kidneys and other abnormalities
Skin and hair — scalp defect, hemangiomata
Genital — cryptorchidism, bicornuate uterus
Remarks — $<$20% survive past 1 year

Trisomy 8

Growth and development — mild to moderate mental retardation
Craniofacial — prominent forehead, prominent ears, plump nose, micrognathia
Extremities — joint abnormalities
Cardiac — defects sometimes present
Renal — abnormalities common
Skin and hair — deep flexion creases in palms and soles
Remarks — about $^3/_4$ cases are mosaic with normal cells

Trisomy 9p

Growth and development — severe to moderate mental retardation
CNS — microcephaly
Craniofacial — small deep-set eyes, prominent ears, down-turned mouth
Extremities — hypoplasia of middle phalanx, syndactyly
Cardiac — defects rare
Remarks — normal life expectancy

Trisomy 3q

Growth and development — severe mental retardation, failure to thrive, short stature
Craniofacial — oblique palpebral fissures, cloudy corneas, depressed nasal bridge, cleft palate
Extremities — clubfeet
Cardiac — defects common
Abdominal — omphalocele and umbilical hernia
Renal — defects common
Genital — cryptorchidism
Remarks — early death usual

TABLE 5-2 *Continued*

Triploidy (may be 69,XXX or 69,XXY)

Growth and development — low birthweight
CNS — hypotonia, hydrocephalus or meningomyelocele
Craniofacial — asymmetry, microphthalmia, colobomata, low-set malformed ears
Extremities — syndactyly
Cardiac — defects common
Renal — cystic changes
Genital — ambiguous in XXY cases
Remarks — some diploid/triploid mosaics reported to have survived infancy; most are aborted spontaneously; large cystic placenta; nonmosaics all died in infancy

XXY (Klinefelter's syndrome)

Growth and development — mild to moderate mental retardation in a small proportion, large stature with eunuchoid build
Skin and hair — scant facial and pubic hair
Genital — small testes, azoospermia, small penis, increased levels of gonadotropins

Monosomy X (Turner's syndrome)

Growth and development — short stature, usually normal psychomotor development
CNS — defective vision and hearing
Craniofacial — webbed neck, short neck, low hair line, epicanthal folds
Extremities — short metacarpals, cubitus valgus
Cardiac — sometimes defects present, usually coarctation of aorta
Renal — defects common
Skin and hair — lymphedema in infancy; increased pigmented nevi
Genital — sexual infantilism, primary amenorrhea, scanty axillary and pubic hair
Remarks — about $\frac{1}{3}$ of liveborn cases are mosaics with additional normal or abnormal cell lines; 45,X is common among spontaneous abortions

5p- (cri-du-chat syndrome)

Growth and development — profound mental retardation, low birthweight, failure to thrive, short stature
CNS — high-pitched catlike cry in infancy
Craniofacial — microcephaly, hypertelorism, strabismus, downward slanting palpebral fissures, epicanthal folds
Cardiac — defects sometimes present
Skin and hair — premature graying of hair
Remarks — frequent survival to adulthood

4p-

Growth and development — profound mental retardation, low birthweight, failure to thrive
CNS — seizures
Craniofacial — microcephaly, prominent glabella, beaked nose, low-set "simple" ears, preauricular tags and dimples, down-turned mouth, cleft lip and/or cleft palate, colobomata, epicanthal folds, hypotelorism
Extremities — clubfoot
Cardiac — defects common
Skin and hair — scalp defect, hypoplastic dermal ridges
Genital — hypospadias

18q-

Growth and development — low birthweight, short stature, severe mental retardation
CNS — hypotonia
Craniofacial — mid-face hypoplasia, prominent deformed ears, cleft lip and/or cleft palate

(Continued)

Fragile X

Growth and development— delayed and perseverative speech, moderate mental retardation (IQ about 35-55), autism, occasional self-mutilation
Craniofacial— large protruding ears, prognathism
Genital— macro-orchidism

Key: VSD = ventricular septal defect; PDA = patent ductus arteriosus; ASD = atrial septal defect

a 47,XXX constitution, or men with 47,XYY (the original suggestion of a "criminal personality" in this syndrome has not been borne out, but some behavioral problems and mental retardation may be associated with the presence of the extra Y chromosome). Men with Klinefelter's syndrome (47,XXY) may also be minimally affected and have often been diagnosed only in the course of an infertility workup. However, further additional X chromosomes are associated with mental retardation, as well as with infertility.

Often, an unbalanced rearrangement may lead to a combination of chromosomal deletion and duplication not previously described; and in such cases precise prognosis is difficult or impossible. However, one can say with assurance that extra and missing pieces of chromosomal material large enough to be seen with standard techniques almost always result in some degree of mental retardation, often fairly severe, and some associated dysmorphic features.

A few of the more common or phenotypically distinctive characteristics for some defined conditions are outlined in Table 5-2.

Prenatal diagnosis is feasible for chromosomal abnormalities via cytogenetic studies of fetal cells. Currently amniotic fluid cells obtained at 16 to 18 weeks' gestation are generally used for this analysis. Chorionic villus sampling, just recently becoming available for routine use, will permit first-trimester studies (see Chapter 3).

PROCEDURES FOR DIAGNOSTIC CONFIRMATION

Routine chromosome analysis using G, R or Q banding for chromosomes from:
- PHA stimulated peripheral blood lymphocytes
- Other cultured tissue (skin biopsy, autopsy, fetal)
- Bone marrow (recommended in exceptional cases needing an immediate result)

High-resolution chromosome analysis for detection of very small abnormalities (available in special laboratories)

Fragile X analysis, requiring special cytogenetic techniques

Chromosome breakage studies

CONSIDERATIONS IN MANAGEMENT

Genetic
- Chromosome analysis to establish diagnosis
- Risk counseling
- Establish carrier status of other family members, if indicated
- Consideration of prenatal diagnosis for pregnancies at risk

Discussion of sharing information with other relatives to encourage evaluation and counseling, if indicated by diagnosis

Psychosocial, educational, familial

Family counseling

Assessment of community resources for appropriate services and/or support

Medical

Dependent on diagnosis

SUGGESTED READING

deGrouchy J, Turleau C: *Clinical Atlas of Human Chromosomes,* New York, John Wiley & Sons, 1977.

Hagerman RJ, McBogg PM, (eds): *The Fragile X Syndrome,* Dillon, Colo., Spectra Publishing, 1983.

Summitt RL: Cytogenetic disorders, in Jackson LG, Schimke RN, (eds): *Clinical Genetics, A Source Book for Physicians,* New York, John Wiley & Sons, 1980.

Warburton D: Current techniques in chromosome analysis. *Pediatr Clin North Am* 1980;27:753.

Yunis JJ: *New Chromosomal Syndromes,* New York, Academic Press, 1979.

Chapter 6

Down's Syndrome

Down's syndrome, a leading cause of mental retardation, is the most common chromosome disorder. Overall incidence has been estimated at 1/650 to 1/1,000 and is dependent upon the maternal-age structure of the population. Risk for affected offspring increases with advancing maternal age and rises significantly for pregnant women 35 and over whose pregnancies are thus considered high risk. Based on recent findings of low levels of maternal serum alpha-fetoprotein in Down's syndrome pregnancies, high-risk pregnancies at any maternal age may become subject to identification. Retardation ranges from moderate to severe (IQ generally between 35 and 50). Prenatal diagnosis is now routinely available for high-risk pregnancies.

ETIOLOGY, GENETIC CHARACTERISTICS AND MODE OF TRANSMISSION

Down's syndrome is a chromosomal disorder caused by the presence of an extra No. 21 chromosome (see Figure 6-1 for karyotype examples). The excess 21 chromosomal material can result from any of the following mechanisms:

Trisomy 21 with 47 chromosomes (92% to 95% of liveborn affected children) occurs as a result of a cell-division error (nondisjunction) during gamete formation (meiosis), whereby two copies of a No. 21 chromosome are included in the ovum (75% to 80% of the time) or sperm (20% to 25%). Fertilization adds another No. 21 chromosome, bringing the total to three, instead of the normal two. About 25% of these conceptions survive to term.

As women age, they are at increased risk for errors during meiosis and, therefore, for offspring with this type of Down's syndrome. Table 6-1 illustrates the age-related incidence of live births of infants with all types of chromosomal abnormalities and of those with Down's syndrome specifically. Incidence at amniocentesis is higher, because some 20% of the cases so ascertained would miscarry before term. The possibility of a paternal-age effect has been investigated, but results are inconclusive. If there is any such effect, it is likely to be weak and apply only to men over age 55.

If one conception with either trisomy 21 or another trisomy has occurred, the risk increases for future offspring with trisomy 21. Table 6-2 shows occurrence and recurrence risks for Down's syndrome under various conditions.

Translocation Down's Syndrome with 46 chromosomes (3% to 5% of all cases, about 9% of those when maternal age is under 30), is caused by a rear-

FIGURE 6-1 Down's Syndrome Karyotypes

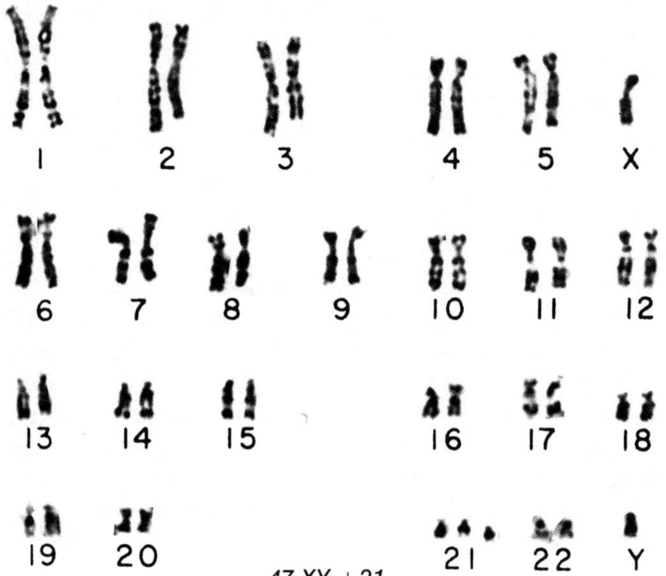

47,XY, +21

Trisomy 21 results from meiotic nondisjunction

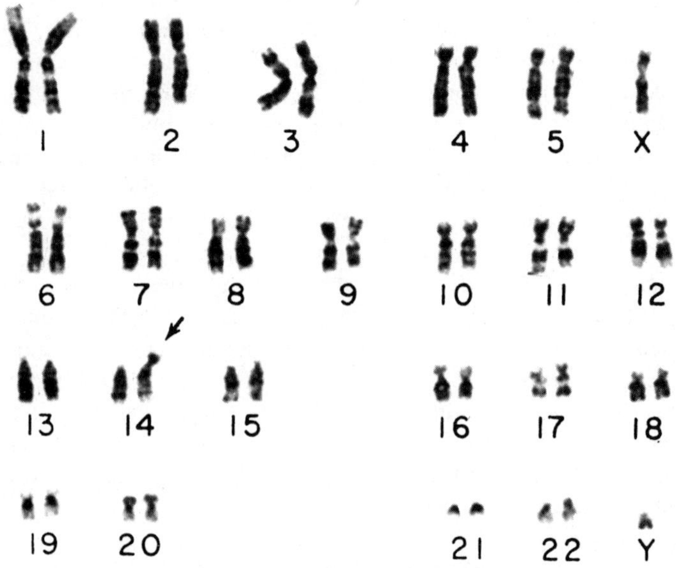

46,XY, – 14, + t(14q21q)

In this translocation trisomy, the trisomic 21 is attached to a No. 14 chromosome (arrow).

Courtesy of Lillian Hsu, M.D. and Sara Kaffe, M.D. of the Prenatal Diagnosis Laboratory of New York City.

rangement of chromosomal material. This is the result of breakage of a No. 21 chromosome and subsequent reattachment of the broken portion to another chromosome (usually No. 14, rarely 13 or 15, sometimes 22 or the other 21). This type of rearrangement can be familial if one of the parents has a "balanced" translocation. The balanced translocation carrier has only 45 chromosomes (including just one unattached No. 21); the total amount of chromosomal material is normal with the attached 21 chromosome functioning normally. Empiric risks of Down's syndrome in progeny of such a carrier vary (see Table 6-2). Translocation Down's syndrome can also arise de novo in the child, with neither parent having a balanced rearrangement. In this instance, the risk of recurrence for siblings is not increased. Thus, determination of parental karyotype is essential for accurate risk assessment when translocation trisomy 21 has occurred.

Mosaic Trisomy with 47 chromosomes in some cells and 46 in others (2% to 3% of cases), can occur as a result of a parental meiotic error, with the subsequent loss of a chromosome No. 21 in an early zygotic division, or as a result of an error in early mitotic division in the zygote itself, that adds the extra chromosome to some cells. The mechanism cannot always be determined in a given case, but risk for recurrence depends on it.

TABLE 6-1
Estimated Incidence of Chromosome Abnormalities

	LIVEBORN INFANTS			
MATERNAL AGE	DOWN'S SYNDROME		ALL CHROMOSOME ABNORMALITIES	
		%		%
21	1/1,500	(0.07)	1/500	(0.20)
27	1/1000	(0.10)	1/450	(0.22)
33	1/600	(0.17)	1/300	(0.33)
34	1/450	(0.22)	1/250	(0.40)
35	1/400	(0.25)	1/200	(0.50)
36	1/300	(0.33)	1/150	(0.66)
37	1/220	(0.45)	1/125	(0.80)
38	1/175	(0.57)	1/100	(1.00)
39	1/140	(0.71)	1/80	(1.25)
40	1/100	(1.00)	1/60	(1.66)
41	1/80	(1.25)	1/50	(2.00)
42	1/60	(1.67)	1/40	(2.50)
43	1/50	(2.00)	1/30	(3.33)
44	1/40	(2.50)	1/25	(4.00)
45	1/30	(3.33)	1/20	(5.00)
46	1/25	(4.00)	1/15	(6.66)
47	1/20	(5.00)	1/10	(10.00)
48	1/15	(6.70)	1/9	(11.11)
49	1/10	(10.00)	1/7	(14.28)

Source: Adapted from Hook EB, Cross PK: Interpretation of recent data pertinent to genetic counseling for Down syndrome, in Willey AM, Carter TP, Kelly S, et al (eds): *Clinical Genetics: Problems in Diagnosis and Counseling*, New York, Academic Press, 1982.

Schreinmachers DM, Cross PK, and Hook EB: Rates of trisomies 21, 18, 13 and other chromosome abnormalities in about 20,000 prenatal studies compared with estimated rates in live births, *Human Genet* 61:318, 1982.

CLINICAL NOTES

Aside from the hallmarks of Down's syndrome (mental retardation, flat facies, oblique palpebral fissures, epicanthal folds, short broad hands, flattened occiput), the following may be observed: cardiac abnormalities; duodenal obstruction; hypotonia; absent moro reflex; Brushfield's spots of the iris; abnormal dermatoglyphics, including simian crease; seizures; increased susceptibility to respiratory infections; increased incidence of leukemia; and premature aging.

Clinically, Down's syndrome is not as obvious in the neonate as it becomes later. Neither false-negative nor false-positive diagnoses are unusual, underscoring the importance of cytogenetic confirmation.

While some of the associated conditions may be life threatening, especially in early infancy, advances in management have increased the life expectancy of Down's syndrome patients to a current mean of 20 years. Affected individuals are generally sociable. If they receive extensive stimulation and intensive education they are able to function on a higher level than was thought possible some years ago. However, considerable mental retardation cannot be prevented.

Prenatal diagnosis with amniocentesis and cytogenetic analysis of cultured amniocytes is currently offered at 16 to 18 weeks after the last menstrual period to women with increased risk for Down's syndrome conceptions (see Table 6-2). Chorionic villus sampling, performed at eight to 10 weeks' gestation, is expected to become more widely available and to provide prenatal diagnosis in the first trimester, reducing the traumatic aspects accompanying a potential option of termination.

TABLE 6-2
Empiric Risks* for Down's Syndrome Live Birth

SOURCE	RISK (%)
Young mother with previous trisomy 21 (or other trisomy) live birth, miscarriage, or stillbirth	About 1
Occurrence of trisomy 21 in a 2nd- or 3rd-degree relative; maternal age below 35	Somewhat increased but still less than 1
Rare families with two or more cases of trisomy 21	Risk markedly increased for 1st- and 2nd-degree relatives
Mother with a 21/13, 21/14, or 21/15 translocation	About 15
Father with a 21/13, 21/14, or 21/15 translocation	About 5
Mother with a 21/22 translocation	About 10
Father with a 21/22 translocation	About 2
Either parent with a 21/21 translocation	100
Decreased levels of maternal serum alpha-fetoprotein	Recent studies indicate increased risk

*Maternal-age specific risks not included

PROCEDURES FOR DIAGNOSTIC CONFIRMATION

Chromosome analysis of peripheral blood, amniocytes, or other tissues

CONSIDERATIONS IN MANAGEMENT

Genetic
- Establish karyotypic information for affected individual
- Determination of parental karyotype, if translocation is present in the affected individual
- Chromosome analysis of other close relatives, if a balanced translocation is identified in the family
- Risk counseling
- Consideration of prenatal diagnosis for the following:
  - Pregnancy at age 35 or older
  - Previous conception with trisomy 21 or another trisomy
  - Down's syndrome in a close relative
  - Known translocation in either parent
  - Family history suggestive of a familial tendency to nondisjunction
  - Reduced alpha-fetoprotein levels in routine maternal serum alpha-fetoprotein screen
- Discussion of sharing information with other family members, if a familial problem is present, to encourage evaluation and counseling

Psychosocial, educational, familial
- Family counseling; support for acceptance of limited intellectual development
- Assessment of family/community resources, such as infant- stimulation programs, support groups, special-education facilities, recreational resources, day programs, residential settings, placement options

Medical
- Surveillance for associated conditions
- Supportive therapy

SUGGESTED READING

Baumgarten A, Schoenfeld M, Mahoney MJ, et al: Prospective screening for Down syndrome using maternal serum AFP. *Lancet* 1985;1(8440l):1280.

Hook EB, Cross PK: Interpretation of recent data pertinent to genetic counseling for Down syndrome, in Willey AM, Carter TP, Kelly S, et al: *Clinical Genetics: Problems in Diagnosis and Counseling*. New York, Academic Press, 1982.

Horrobin JM, Rynders JE: *To Give an Edge—A Guide for New Parents of Down's Syndrome (Mongoloid) Children*. Minneapolis, The Colwell Press Inc, 1974.

Smith DW, Wilson AC: *The Child with Down's Syndrome*. Philadelphia, WB Saunders Co, 1973.

Smith G, Berg JM: *Down's Anomaly,* Livingston, New York, Churchill Press, 1975.

Tamaren J, Spuhler K, Sujansky E: Risk of Down syndrome among second- and third-degree relatives of a proband with trisomy 21, *Am J Med Genet* 1983;15:393.

Chapter 7

Neural-tube Defects: Anencephaly/Spina Bifida

Neural-tube defects (NTDs) are among the more common birth defects. Incidence varies worldwide; overall in the United States it is approximately one to two in 1,000 births. Most cases represent the first occurrence in a family.

ETIOLOGY AND PATHOGENESIS

The neural tube usually closes by the fourth week of embryonic life. Failure to close in the anterior end results in anencephaly, and in the caudal section, in spina bifida. Both genetic and environmental factors have been implicated in the etiology of isolated NTDs, but the specific cause usually is not known. Recent research in the United Kingdom suggests that one possible factor could be a vitamin deficiency. A significant reduction in recurrence rates was documented for women who had had children with NTDs and were then given vitamin supplements prior to the next pregnancy and for six or more weeks after conception.

TABLE 7-1
Common Syndromes That May Include Neural-Tube Defects

DISORDER	OTHER CHARACTERISTICS	MODE OF INHERITANCE
Meckel's syndrome	Microcephaly, posterior or dorsal encephalocele, cleft palate, polydactyly, polycystic kidneys, cryptorchidism, incomplete genital development in males	Autosomal recessive
Trisomy 18 syndrome	Developmental and mental retardation, prominent occiput, low-set ears, micrognathia, clenched hands, rocker-bottom feet, cardiac defects	Chromosomal
Oculoauriculovertebral dysplasia (Goldenhar's syndrome)	Anomalies usually asymmetric; unilateral facial hypoplasia with microtia, dermoids of the eye, coloboma of upper eyelid	Usually sporadic
Amniotic band complex	Ringlike constrictions of limbs, facial clefts, encephalocele; probable cause: early rupture of amnion with formation of constricting amnion and/or chorion bands	Usually sporadic

Similar protocols are planned to investigate the effect of such supplementation in the United States. NTDs generally are isolated defects, but they may also occur as a part of a syndrome (see Table 7-1).

GENETIC CHARACTERISTICS AND MODE OF INHERITANCE

Unless a syndrome (with its own associated risk) is identified, inheritance is considered to be multifactorial since anencephaly and spina bifida each confer a recurrence risk; for both, statistics reflect a risk for NTDs in general. Empiric recurrence risks are given in Table 7-2.

In addition, an increased rate of spontaneous abortion has been noted in families with NTDs.

TABLE 7-2
Empiric Recurrence Risks For Neural-Tube Defects

AFFECTED RELATIVE(S)	RECURRENCE RISK (%)
1 affected child	3 to 5
Affected parent	3 to 5
2 or more affected children	6 to 9
NTD in other family member	Somewhat higher than population risk, but unlikely to exceed 0.5

CLINICAL NOTES

Clinical severity of NTDs is related to the extent and location of the lesion. Common NTDs include:

Anencephaly —partial or complete absence of the calvarium and cranial vault with the cerebral hemispheres missing or greatly reduced. This defect is incompatible with protracted postnatal life.

Encephalocele —herniation of brain and meninges through a defect in the skull. If only meninges herniate, the prognosis may be better than with herniation of brain tissue.

Spina bifida —defect in a portion of the spinal column with a saclike protrusion through the opening. Further classification is based on tissues contained in the sac.

Myelomeningocele —most common of the clinically significant types of spina bifida; usually an open lesion. Herniation includes meninges, spinal cord structure, and nerve roots. Hydrocephalus caused by Arnold-Chiari malformation accompanies about 80% of cases. Clinical findings include weakness or flaccid paralysis in the lower extremities, lack of sphincter control with rectal prolapse or urinary incontinence, and clubfeet.

Meningocele —about 10% to 15% of cases. The skin-covered sac includes the meninges only. Prognosis is better than for open lesions.

Spina Bifida Occulta —closed defect; estimated to occur commonly in the general population. It is usually detected incidentally in the course of x-ray evaluation and often has no clinical significance. When spinal dysraphism is present (often associated with a pigmented area or hairy tuft), the risk for overt neural-tube defects in offspring may be higher than when only vertebral arches are absent. In the latter group there is no evidence of increased risk for overt neural-tube defects in relatives.

Prenatal diagnosis for neural-tube defects by amniotic fluid assay is indicated for pregnancies at increased risk for such a defect; it is usually done routinely when amniocentesis is performed for prenatal diagnosis for other reasons. The amniotic fluid assay for elevated alpha-fetoprotein (AFP) levels effectively identifies nearly all open NTDs, but rarely diagnoses a closed lesion. AFP levels may also be elevated with twins, esophageal and duodenal atresia, omphalocele, threatened or missed abortion, fetal blood, etc. Therefore, positive findings must be further evaluated by such methods as acetylcholinesterase determination and ultrasonography.

Since increased maternal serum AFP levels have also been documented in pregnancies involving an NTD, they can be used to screen for NTDs. This test is not as reliable as the amniotic fluid assay, because it is subject to even more sources of error that cause both false negatives and false positives. However, it is useful as a simple noninvasive preliminary evaluation for identifying at-risk pregnancies and may become routine for all pregnancies. Positive findings, if confirmed by a second maternal serum assay and by ultrasonic verification of gestational age and number of fetuses, must be followed by an amniotic fluid assay and possibly further studies. Five percent to 10% of mothers with a confirmed elevated serum AFP level have been reported to have a fetus with a neural-tube defect.

PROCEDURES FOR DIAGNOSTIC CONFIRMATION

- Physical examination with careful neurologic examination
- Transillumination of the skull in the neonate
- CT scan of the skull to rule out hydrocephalus

CONSIDERATIONS IN MANAGEMENT

- Genetic
  - Risk counseling
  - Consideration of prenatal testing:
    - Maternal serum AFP determination, if available in qualified hands; followed by careful ultrasonography, if positive
    - Amniocentesis for AFP determination for increased-risk group, with further studies, if indicated
      - Acetylcholinesterase assay
    - Ultrasonography
- Psychosocial, educational, familial
  - Family counseling
  - Education and guidance in management of handicapped individual
  - Survey of community facilities including schools for handicapped
  - Survey of available public funds for the handicapped

Medical

Consider preconception review of nutritional status, and potential vitamin supplementaion before and during pregnancy, particularly for women who are planning pregnancies and are at high risk for offspring with NTDs (see Etiology and Pathogenesis).

Surgical repair when indicated

Control of hydrocephalus

Control of intercurrent infections, especially of CNS and genitourinary tract

SUGGESTED READING

Crandall BF: Alpha-fetoprotein: The diagnosis of neural-tube defects. *Pediatr Ann* 1981;10:38.

Lorber J: Selective treatment of myelomeningocele: To treat or not to treat. *Pediatrics* 1974;53:307.

Macri JN, Haddow JE, Weiss RR: Screening for neural-tube defects in the United States: A summary of the Scarborough Conference, *Am J Obstet Gynecol* 1979;133:119.

Macri JN, Weiss RR: Prenatal serum alpha-fetoprotein screening for neural-tube defects. *Obstet Gynecol* 1982;59:633.

Milunsky A: Prenatal detection of neural-tube defects, VI. Experience with 20,000 pregnancies. *JAMA* 1980;244:2731, 1980.

Second Report of the UK collaborative study on alpha-fetoprotein in relation to neural-tube defects: Amniotic fluid alpha-fetoprotein measurement in antenatal diagnosis of anencephaly and open spina bifida in early pregnancy. *Lancet* 1979;II:651.

Smithells RW, Nevin NC, Seller MJ, et al: Further experience of vitamin supplementation for the prevention of neural-tube defect recurrences. *Lancet* 1983;1(8332):1027.

Toriello HV, Higgins JV: Occurrence of neural-tube defects among first-, second-, and third-degree relatives of probands: Results of a United States study. *Am J Med Genet* 1983;15:601.

Chapter 8

Hydrocephalus

Hydrocephalus is a condition characterized by an excess of fluid volume in the ventricles of the brain, resulting from an imbalance in the amount of cerebrospinal fluid (CSF) secreted and the amount absorbed. It is usually accompanied by an increase in intracranial pressure. The forms discussed here have their onset prenatally (75% to 90%) or in infancy, and may or may not be genetic in origin. (Hydrocephalus ex vacuo, which refers to an increase in CSF under normal pressure, secondary to atrophy in the brain, will not be discussed in this chapter).

Hydrocephalus can occur in isolation, as part of a sporadic syndrome, or in association with other defects. Incidence is estimated at two per 1,000 births, with figures varying from 0.3 to 4.2/1,000. The disparity may be due to inconsistency in reporting the isolated condition in stillborns, lack of uniformity in reporting when associated with other defects (primarily neural-tube), and failure to report cases not apparent at birth.

ETIOLOGY AND PATHOGENESIS

Hydrocephalus almost always develops as a result of an obstruction in CSF circulation. It can also be caused by interference with the CSF absorption mechanism or, more rarely, by an oversecretion of CSF.

The underlying etiology of the condition is diverse, including a variety of embryologic and perinatal mishaps, as well as genetic origins. Often the cause is not determined. The major reasons for hydrocephalus are post-traumatic or postinflammatory lesions and congenital malformations. Table 8-1 illustrates pathogenic mechanisms associated with common forms of congenital hydrocephalus.

Table 8-2, lists a number of genetic syndromes that may include hydrocephalus, giving some phenotypic features and the mode of inheritance.

GENETIC CHARACTERISTICS AND MODE OF INHERITANCE

Since hydrocephalus can result from a variety of pathologic mechanisms, specific diagnosis is necessary for a precise assessment of any risks of recurrence that may apply. Pedigree analysis may be informative.

- In the absence of additional information, the empiric recurrence risk for a sporadic case of uncomplicated hydrocephalus is about 1%. It must be kept in mind that this includes unrecognized cases of forms with higher recurrence risks, such as X-linked recessive aqueductal stenosis, which is thought to constitute 2% of all uncomplicated hydrocephalus.

- Sporadic uncomplicated aqueductal stenosis, in a male or female, has an empiric recurrence risk of about 4.5%, including the X-linked cases, which, it has been suggested, may account for as many as 25% of the males in this category.
- A sporadic case of aqueductal stenosis in a male carries an empiric recurrence risk of 12% for male siblings.
- When X-linked aqueductal stenosis is diagnosed, a carrier female (usually identified by the fact that two sons or one son and another male maternal relative are affected) has a 50% risk of an affected son with each male fetus.
- Daughters of carrier females have a 50% risk of being carriers.
- Autosomal recessive forms—e.g., Dandy-Walker syndrome—carry a recurrence risk of 25% for siblings of the affected child.
- When multifactorial inheritance is assumed, (e.g., association with a neural-tube defect), the empiric recurrence risk for siblings is 3% to 5%.
- It is not always possible to distinguish between the various forms of transmission, therefore, a range of figures may be quoted.
- When hydrocephalus is due to a known nongenetic cause, the recurrence risk for siblings is not increased, provided the etiologic factor does not recur.
- When associated with a specific syndrome, the risk of recurrence will be that assigned to the syndrome (see Table 8-2 for common syndromes and mode of inheritance).

TABLE 8-1
Pathogenesis for Some Common Forms of Hydrocephalus

PRIMARY PATHOLOGY	ETIOLOGY
Stenosis of aqueduct of Sylvius	Posthemorrhagic Postinflammatory Genetic (XR)
Dandy-Walker syndrome (occlusion of 4th ventricle at foramina of Magendie and Luschka)	Genetic (AR)
Arnold-Chiari malformation (downward displacement of hindbrain through foramen magnum)	Associated with neural-tube defects
Obstruction of basilar cisterns and arachnoid pathways	Post-traumatic Postinflammatory (sequela of prenatal teratogenesis: meningitis, toxoplasmosis, cytomegalovirus infection . . .)
Papilloma of choroid plexus	Unknown

Key: XR = X-linked recessive; AR = autosomal recessive

CLINICAL NOTES

In about 30% of cases an enlarged head is evident at birth. For the rest, the diagnosis is generally made in the first few months of life, when a disproportionate increase in head size is observable. Eighty percent of cases are diagnosed by the first year.

TABLE 8-2
Genetic Syndromes That May Include Hydrocephalus

SYNDROME	OTHER CHARACTERISTICS	MODE OF INHERITANCE
Albers-Schönberg (recessive osteopetrosis)	Thick brittle bones, cranial nerve compression, progressive deafness and blindness; death in infancy or childhood	AR
Achondroplasia	Short limbed dwarfism, caudal narrowing of spinal cord	AD
Acrodysostosis	Peripheral dysostosis, short stature, short hands, stubby fingers, mental retardation	?Sporadic
Meckel's	Microcephaly, encephalocele, cleft palate, polydactyly, polycystic kidneys	AR
Orofacial digital I	Oral frenula and clefts, digital abnormalities, mental retardation	XD
Osteogenesis imperfecta congenita	Short bones, multiple fractures, blue sclerae	AR
Riley-Day (familial dysautonomia)	Dysautonomia, failure to thrive, lack of tearing, paroxysmal hypertension	AR
Robert's	Hypomelia, hypotrichosis, facial hemangioma, cleft lip/palate; usually death in early infancy	AR
Chromosomal abnormalities (triploidy, trisomy 13, trisomy 18)	Multiple congenital malformations	

Key: AR = autosomal recessive; AD = autosomal dominant; XD = X-linked dominant

Characteristic features may include: macrocephaly; frontal bossing; widely separated cranial sutures; an enlarging anterior fontanelle; distended scalp veins; and, in severe cases, downward displacement of the eyes, cranial nerve paralysis, and mental retardation.

The prognosis for isolated hydrocephalus is least favorable when the condition is present from birth: most do not survive; of those who do, only 11% to 14% are expected to have normal mental development without intervention. While some children with untreated hydrocephalus have been noted to have spontaneous arrest and normal function, surgical shunts are generally used to drain fluid from the head before the excess pressure has led to further cranial enlargement and other pathologic conditions. This treatment has increased survival past infancy, with normal or educable mental function to be expected in over half of the cases.

In acquired hydrocephalus, surgical removal of the obstruction may be an option.

Prenatal diagnosis by ultrasound has become widely available for pregnancies with a known risk for hydrocephalus or when hydrocephalus is suspected (e.g., increased uterine size for the date, inactive fetus, abnormal pelvic examination). Prenatal sex determination is an option for pregnancies at risk for X-linked aqueductal stenosis.

In-utero shunting of prenatally diagnosed hydrocephalus has been utilized on an experimental basis. New postnatal therapeutic endeavors are based on the promotion of CSF absorption or inhibition of CSF production.

PROCEDURES FOR DIAGNOSTIC CONFIRMATION

- Serial head circumference
- Skull x-ray
- Brain CT scan
- Air contrast studies
- Chromosome analysis, in selected cases
- Antibody screens to rule out bacterial and viral etiology

CONSIDERATIONS IN MANAGEMENT

- Genetic
  - Determination of etiology
  - Risk counseling
  - Prenatal testing for at-risk group
    - Prenatal serial ultrasonography
    - Maternal serum AFP
    - Consideration in selected cases for:
      - Amniotic fluid AFP determination
      - Fetal chromosome analysis
      - Fetal sex determination for X-linked hydrocephalus
  - Discussion with other family members at risk to encourage counseling and evaluation
- Psychosocial, educational, familial
  - Family counseling; crisis counseling for loss of child or to promote acceptance of handicap
  - Education and guidance in management of handicapped individuals
  - Survey of community facilities for the handicapped
- Medical
  - Control of hydrocephalus (surgical shunts)
  - Control of intercurrent infections, especially of CNS tract
  - Consideration of fetal surgery for severe cases diagnosed prenatally (may be available in selected centers)

SUGGESTED READING

Bay C, Kerzin L, Hall BD: *Recurrence Risk in Hydrocephalus,* vol 15, no. 5C. (Birth Defects: Original Article Series.) White Plains, N.Y., March-of-Dimes Birth Defects Foundation, 1979, p 95.

Burton BK: *Empiric Recurrence Risks for Congenital Hydrocephalus,* vol 15, no. 5C. (Birth Defects: Original Article Series.) White Plains, N.Y., March-of-Dimes Birth Defects Foundation, 1979, p 107.

Donn SM: Fetal Hydrocephalus, A Perinatal Dilemma. *J Reprod Med* 1982;27:589.

Freiherr G: Fetal Surgery: Saving the Unborn. *Research Resources Reporter* 1983;7:1.

Harrison MR, Golbus MS, Filly RA: Management of the fetus with a correctable congenital defect. *JAMA* 1981;246:635.

Warkany J: *Congenital Malformations: Notes and Comments,* Chicago, Year Book Medical Publishers, 1971.

Chapter 9

Epilepsy

An etiologically heterogeneous disorder, epilepsy is estimated to affect 1% to 2% of the general population. Cumulative risk for any individual to have recurrent afebrile seizures is age-related, and is estimated to be 1% by age 20, 2% by 40, and 3% by 80. This includes seizures that are associated with other disorders. A person is considered to have idiopathic epilepsy only if he or she has had two or more seizures in the absence of known precipitating factors (e.g.,fever, infection, trauma, metabolic disease, or neurotoxic illness).

Seizures are generally classified into two types: partial, which include focal or local seizures, and generalized, which include absence seizures (petit mal); myoclonus; infantile spasms; clonic; tonic; tonic-clonic (grand mal); atonic seizures; and akinetic seizures. Genetic factors undeniably play a role in predisposition to seizure disorders; the extent to which they influence development of seizures is not always clear.

ETIOLOGY AND PATHOGENESIS

About 80% of epilepsy is idiopathic. The remaining 20% is symptomatic, or secondary to a host of conditions, including perinatal injury, infections, trauma, tumor, underlying hereditary disease, etc. Exact pathophysiology remains unknown; it may involve some combination of altered membrane structure and function, neuron reaction, and neurotransmitter function.

GENETIC CHARACTERISTICS AND MODE OF INHERITANCE

Epilepsy may occur alone or as part of a mendelian or chromosomal syndrome. Although over 130 mendelian traits are associated with epilepsy or seizures, they account for only a small proportion of all seizure disorders. Nonetheless, a precise diagnosis is important in counseling for epilepsy, as recurrence risks may be greatly affected. Table 9-1 summarizes genetic disorders commonly associated with epilepsy.

Suspicion of a genetic disorder underlying epilepsy is raised by a family history of seizures, or by one or more associated abnormalities, including cutaneous lesions, mental retardation, neurologic manifestations, metabolic disturbance, peculiar physical characteristics, or multiple congenital anomalies.

Two relatively common single-gene disorders that often present as seizure disorders are tuberous sclerosis and neurofibromatosis. Both are inherited as autosomal dominant traits, both exhibit neurocutaneous manifestations with a

TABLE 9-1
Mendelian Disorders Associated with Epilepsy

DISORDER	CHARACTERISTIC FEATURES	MODE OF INHERITANCE
Centralopathic epilepsy (centrencephalic EEG)	Bilaterally synchronous EEG (spike-wave trait), petit mal, centrencephalic epilepsy, automatisms, akinetic or atonic seizures, grand mal, myoclonic petit mal	AD, low penetrance at birth, almost complete penetrance by age 4 to 16
Cerebrohepatorenal syndrome (Zellweger's syndrome)	Seizures, paroxysmal discharges, severe muscular hypotonia, absent reflexes, craniofacial dysmorphology, death in first two years	AR
Ceroid lipofuscinoses	Seizures, myoclonic jerks, dementia, neuronal inclusions	AR
Congenital adrenal hypoplasia	Neonatal onset seizures, adrenal insufficiency, normal external genitalia and gonadal function, hypoglycemia, electrolyte disturbance	Sporadic, AR, XR
Craniodiaphyseal dysplasia	Cranial nerve compression, mental retardation, hyperostosis and sclerosis of cranial bones, tubular-bone anomalies, enlarged head	AR
Epilepsy, photogenic	Light-induced seizures	AD, reduced penetrance
Epilepsy, reading	Adolescent-onset seizures, jaw jerks triggered by reading	? AD
Galactosemia	Hepatomegaly, jaundice, anorexia, cataracts, milk intolerance, variable growth and mental retardation	AR
Glycogen-storage diseases	Mental retardation	AR
Hereditary ataxias (Friedreich's ataxia, Charcot-Marie-Tooth disease, etc.)	Seizures, ataxia, muscle wasting	AR, AD
Holoprosencephaly	Bilateral cleft lip, absent philtrum, hypotelorism, nasal anomalies, eye anomalies, severe retardation	Heterogeneous (associated with trisomy 13)
Huntington's disease	Choreiform movements, dementia, clumsy gait, adult onset	AD
Hyperparathyroidism, neonatal familial	Hypercalcemia, low serum phosphate levels	AR
Hypoglycemia, familial neonatal	Hypercalcemia	AR, AD
Hypoparathyroidism, infantile	Hypocalcemia, tetany, dental anomalies	XR
Lesch-Nyhan syndrome	Progressive choreoathetosis, mental retardation, motor defect self-mutilation, gout, HGPRT deficiency	XR

(Continued)

DISORDER	CHARACTERISTIC FEATURES	MODE OF INHERITANCE
Lysosomal storage diseases (Tay-Sachs, Niemann-Pick, etc.)	Mental retardation	AR
Menkes (kinky hair) syndrome	Seizures; progressive spasticity and decerebration; abnormal hair, mental and growth retardation	XR
Mucolipidoses	Mental retardation	AR
Myoclonic epilepsy, Hartung type	Myoclonic epilepsy; no Lafora bodies found at autopsy, diffuse atrophy	AD
Myoclonus, hereditary essential	Sudden brief muscle contractions, no epilepsy, affects proximal muscles of extremities	AD
Myoclonus, progressive hereditary essential (Lafora-Unverricht-Lundborg)	Myoclonus, grand mal, childhood onset, cerebellar ataxia, dementia	AR
Neurofibromatosis	Seizures, café-au-lait spots, neurofibromas, skeletal anomalies, CNS tumors, variable mental retardation	AR
Pelizaeus-Merzbacher disease	Jerky arm and limb movements, rolling nystagmus, delayed milestones and motor development, slow progression	XR
Phenylketonuria and other amino acid disorders	Mental retardation, seizures	AR
Porphyrias: Acute intermittent Variegate Coproporphyria	Acute abdominal pain, psychiatric disturbances, photosensitivity	AD
Rickets, vitamin-D-dependent	Convulsions and tetany, growth failure, rickets, hypotonia, motor retardation, enamel hypoplasia	AR
Sturge-Weber syndrome	Seizures, facial angioma, intracranial anomaly, variable mental retardation, glaucoma	? AD or sporadic
Tuberous sclerosis	Infantile spasms with hypsarrhythmic EEG, generalized or focal seizures with onset in first or second decade, adenoma sebaceum (angiofibromas), mental retardation, white nevi, shagreen patches	AD

Key: AD = autosomal dominant; AR = autosomal recessive; XR = X-linked recessive wide range of expression, even within families. In tuberous sclerosis, these include multiple small firm cutaneous nodules (angiofibromas) in a butterfly pattern around the nose and cheeks; multiple depigmented or white nevi, especially in infants; and shagreen skin patches. Seizure types include infantile spasms and myoclonic astatic epilepsy. In neurofibromatosis, symptoms include multiple café-au-lait spots and cutaneous neurofibromas. When either disease is suspected, examination of family members may reveal previously undiagnosed subtle manifestations.

When no mendelian or chromosomal disorder is identified, epilepsy is generally counseled as multifactorial, with recurrence risks for different types of seizures based on empiric data. Age and sex are important modifying features for the expression of seizures in relatives. Females and those with early-onset epilepsy (before age 5) are more likely to exhibit a familial pattern. Recurrence risks in offspring are reported to be higher for a woman with epilepsy than for a man. Risks to offspring are greater if parents have idiopathic rather than symptomatic seizures. A familial tendency has been identied in some seizures secondary to trauma or infection, although this tendency is not as strong as that for idiopathic epilepsy or febrile seizures. Recurrence risks for siblings and offspring of affected individuals are generally thought to be of the same magnitude.

Familial EEG patterns exist; many are inherited multifactorially, some as autosomal dominant traits, and others in an unknown way. EEG abnormalities can exist with or without overt clinical symptoms, although they tend to predispose to seizures.

Summarized below are estimated recurrence risks for various types of seizure disorders. Recurrence is not always of the same seizure type.

Generalized Seizure Disorders (bilaterally symmetric)

- Grand mal epilepsy, or major motor epilepsy, is the most common type. Cumulative risks to siblings have been reported to be 7.5% by age 20 if the proband's age at onset of seizures was before 4, or 4.3% by age 20, if age at onset of seizures in the proband was between 4 and 15.
- Petit mal (absence seizures) carries a general recurrence risk for offspring and siblings of about 2% to 5%; it may be as high as 10% for a daughter of an epileptic mother. Mode of inheritance is most likely polygenic, with greater heritability and a lower threshold for expression in females.
- Infantile spasms (hypsarrythmic EEG pattern) occur in early infancy and childhood and can have multiple causes. Some are idiopathic and may be accounted for by polygenic factors, with general recurrence risks for siblings around 5% to 10%. An excess of males has been noted among those with a secondary or symptomatic cause. Some spasms are associated with genetic syndromes, such as PKU (autosomal recessive), tuberous sclerosis (autosomal dominant), and there is an X-linked form. It has been reported that 25% of infants with infantile spasms develop tuberous sclerosis, but in general, a familial incidence is low, with about 5% of all cases exhibiting a positive family history. The Lennox-Gastaut syndrome (myoclonic astatic epilepsy) may develop from infantile spasms, or can newly develop, with complex absence seizures the most commonly observed seizure type. This syndrome is rarely observed in siblings, although a few cases have been reported, suggesting a possible autosomal recessive inheritance in some families.

- Febrile seizures occur in 3% to 5% of all children. Several studies have found an 8% to 11% risk (approximately twofold) for siblings to also have them.Some reports show 5% to 6% risks to develop epilepsy following the febrile seizures if a parent or sibling also had them.Still other studies found increased risks to develop epilepsy after febrile seizures only if there was a family history of epilepsy. Observed recurrence risks may be explained by multifactorial inheritnce patterns or autosomal dominant inheritance with reduced penetrance. Febrile seizure tendency may also be inherited differently in different families.

Partial Seizure Disorders (local onset)

- Partial epilepsy usually carries general recurrence risks for siblings of about 1% to 4%. The risks may be as high as 30% if the proband has seizures associated with centrotemporal spikes.
- Benign rolandic epilepsy is strongly familial. Expression is age-limited; with onset after age 3, the epilepsy disappears during adolescence in half of all cases.

Familial EEG Abnormalities

- Centrencephalic EEG (generalized spike-wave) is a presumably mendelian EEG trait that may lead to development of a variety of seizure types (e.g., petit mal, grand mal). Inheritance of this EEG pattern appears to be autosomal dominant, with reduced penetrance. If the trait followed a strictly dominant inheritance pattern, one would expect an average of 50% of siblings of an affected individual to also have the EEG trait. The proportion observed is on the order of 37%, with only one-fourth of siblings who have the trait actually developing epilepsy. Therefore, overall risk for a sibling to develop epilepsy is about 8%. Penetrance is low at birth, and almost complete by ages 4 to 16 years. Similar risks exist for offspring of an affected individual.
- Photosensitive EEG patterns are also found in relatives of individuals who have them, especially if the proband is female. Although the inheritance pattern remains unclear, it is known to be influenced by age and sex. Photic-induced epilepsy develops only in a small proportion of those with photosensitive EEG patterns, but when it does occur, it appears to be highly inheritable.

CLINICAL NOTES

Precise diagnosis of epilepsy often includes differentiating breath-holding attacks in children, syncope, heart block, transient ischemic attacks, hypoglycemia, and hysteria from true recurrent epileptic seizures.

The incidence of epilepsy is greater in males than in females.

Epilepsy with unknown origin generally has onset early in life, although it can begin at any age. Peak ages at onset are 2 years and puberty. Epilepsy caused by an identifiable lesion generally has its onset in the early years or in the third decade.

Mortality among individuals with epilepsy is estimated to be twice that of the general population. Sudden and accidental deaths, including suicides, have been observed with high frequency.

Pregnancy in a woman with epilepsy poses several risks. Offspring of epileptic women are known to be at increased risk for congenital malformations, such as heart defects and cleft palate and many anticonvulsant agents have been associated with such effects (see Table 9-2 for a summary of their effects during

TABLE 9-2
Effects of Anticonvulsant Agents Used During Pregnancy

DRUG	TERATOGENIC RISK FACTOR CATEGORY*	EFFECTS ON FETUS
Hydantoins		
Phenytoin (Dilantin)	D	Fetal hydantoin syndrome—characteristic facial features, protruding teeth, thick lips, flattened nose and forehead, nail dysplasias, hypoplasia and ossification of distal phalanges; risks to fetus may be dose related; neonatal hemorrhage risk
Mephenytoin (Mesantoin)	C	No evidence of fetal effects found in limited reports
Ethotoin (Peganone)	D	Toxicity thought to be lower than phenytoin; 1/3 of reports in literature showed associations with heart defects and cleft palate
Diones		
Trimethadione (Tridione)	X	Fetal trimethadione syndrome—increased risk of fetal loss; 60% risk cleft lip and/or palate; 20% risk congenital heart defects; 20% risk microcephaly/mental retardation
Paramethadione (Paradione)	X	Tetralogy of Fallot, mental retardation, failure to thrive, increased incidence of spontaneous abortions
Barbiturates		
Phenobarbital	?B	Although it was formerly considered the least likely anticonvulsant to be teratogenic, recent observations question the safety of phenobarbital during pregnancy
Primidone (Mysoline)	D	Risks may exceed benefits
Mephobarbital (Mebaral)	B	No reports linking use with congenital defects
Succinimides		
Ethosuximide (Zarontin)	C	Neonatal hemorrhage risk
Methsuximide (Celontin)	C	No evidence of adverse effects (observed in five pregnancies)
Phensuximide (Milontin)	C	
Benzodiazepines		
Diazepam (Valium)	D	Some studies have shown increased risk for cleft lip and/or palate; first-trimester use has been associated with inguinal hernia, pyloric stenosis, and cardiac defects; second-trimester use has been associated with hemangiomas and cardiac or circulatory problems Risks may be increased if mother is a smoker
Clonazepam (Clonopin)	C	No reports available

(*Continued*)

DRUG	TERATOGENIC RISK FACTOR CATEGORY*	EFFECTS ON FETUS
Valproic acid		
Depakene	D	Similar to fetal hydantoin syndrome; increased risk for neural-tube defects
Miscellaneous		
Carbamazepine (Tegretol)	D	Conclusions not possible
Phenacemide (Phenurone)		No information available

*According to FDA definitions, categories as assigned in: Briggs GG, Bodendorfer TW, Freeman RK, et al: *Drugs in Pregnancy and Lactation: A Reference Guide to Fetal and Neonatal Risk*, Baltimore, Williams & Wilkins, 1983.

A: Possibility of fetal harm appears remote

B: No confirmed evidence of a risk

C: Insufficient information. Drugs should be given only if the potential benefit justifies the potential risk to the fetus.

D: Positive evidence of human fetal risk; benefits from use in pregnant women may be acceptable despite the risk (e.g., if the drug is needed in a life-threatening situation or a safer drug cannot be used).

X: Known teratogen

pregnancy). Controversy exists as to whether increased risk for congenital malformations is due solely to medications or also to some other unknown maternal factor, such as an altered metabolism due to the maternal epileptic state, or to a combination of the two. Also, in some women, pregnancy can exacerbate occurrence of seizures, which may increase fetal risks. In others—for example, some women with myoclonic petit mal epilepsy—frequency of seizures may actually improve or they may disappear during pregnancy. In general, the risk for fetal malformations seems to be lower for women who are both seizure-free and unmedicated.

Evaluation of individual cases may enable some patients to go through pregnancy without medication. In others, treatment plans will depend on risk-versus-benefit considerations. If medication is necessary, newborn infants may be subject to bleeding tendencies due to exposure in utero. This can be successfully treated with vitamin K.

PROCEDURES FOR DIAGNOSTIC CONFIRMATION

History of a typical attack
History of prior trauma, infection, or toxic episode
CBC and urinalysis
Fasting blood sugar
Serum glucose, calcium, and phosphorous
Studies to exclude meningeal infections
Skull x-ray films
EEG, serial EEGs

Further evaluation, as indicated for suspected underlying disorders, for example:
- Lumbar puncture
- CT scan
- Angiogram or pneumoencephalogram
- Chromosome analysis
- Metabolic screening, e.g., urinary amino acids
- TORCH titer
- Ophthalmoscopy

CONSIDERATIONS IN MANAGEMENT

Genetic
- EEG evaluation of first-degree family members
- Establishment of diagnosis, pedigree analysis, recurrence risk assessment
- Risk counseling
- Consideration of prenatal diagnosis for pregnancies at risk for some specific syndromes which include epilepsy (e.g., chromosomal disorders, hereditary metabolic disorders)
- Evaluation of anticonvulsant therapy in pregnant epileptic women or for those considering pregnancy

Psychosocial, educational, familial
- Family counseling and support for the possibility of emotional disturbances, depression, adjustment problems, stigmatization
- Education to promote increased compliance with management
- Support in learning to live with epilepsy
- Family education for emergency procedures
- Counseling for avoidance of precipitating factors, e.g., alcohol, lack of sleep, stress, etc.

Medical
- Drug therapy
- Surgical treatment
- Consideration of discontinuing medication prior to pregnancy for epileptic women who may be able to manage without it
- Prophylactic administration of vitamin K, parenterally to treated epileptic mothers before delivery and to infant after birth, to prevent neonatal hemorrhagic sequelae of some anticonvulsants

SUGGESTED READING

Anderson VE, Hauser WA, Penry JK, et al, (eds): *Genetic Basis of the Epilepsies.* New York, Raven Press, 1982.

Dalessio DJ: Seizure disorders and pregnancy. *N Engl J Med* 1985;313:559.

Glaser GH: Convulsive Disorders (Epilepsy), in Merritt HH (ed): *Textbook of Neurology,* 6th ed. Philadelphia, Lea & Febiger, 1979.

Newmark ME, Penry JK: *Genetics of Epilepsy: A Review,* New York, Raven Press, 1980.

Sands H, (ed): *Epilepsy: A Handbook for the Mental Health Professional.* New York, Brunner/Mazel, 1982.

Chapter 10

Tuberous Sclerosis

Tuberous sclerosis is classically defined as the combination of epilepsy, adenoma sebaceum (actually angiofibromas), and mental retardation, usually in an individual with hypopigmented macules on the trunk and limbs. It has been listed as one of the conditions seen most frequently for genetic evaluation. Incidence has been estimated between 1/10,000 and 1/100,000. This wide span may reflect differences in ascertainment because of the extensive clinical variability of this disorder, which ranges from the fullblown syndrome, through partial expression (e.g., only mental retardation), to asymptomatic cases that are diagnosed only when more seriously affected relatives are born. Among severely retarded individuals, 0.5% are reported to have tuberous sclerosis, and among infants with infantile spasms, showing an EEG pattern of hypsarrhythmia, 25% later develop the overt disorder.

ETIOLOGY AND PATHOGENESIS

The basic defect in tuberous sclerosis is not known, nor is the pathogenic process. Gross examination of the brain reveals firm nodules on the surface, in the cortex, the basal ganglia, the lining of the lateral and third ventricles, as well as in the cerebellum, brain stem, and spinal cord. The subependymal nodules adjacent to the ventricles, as seen on a CT, are considered diagnostic in otherwise asymptomatic individuals. Skull radiography shows calcification, especially in the area of the basal ganglia. The characteristic facial lesions and areas of fibrosis originate from the terminal nerves in the subcutaneous region.

Clinical symptoms, particularly the seizures and mental retardation, are thought to be produced by the cerebral lesions, but a correlation between a positive CT scan or skull radiograph, and mental retardation or seizures, has not been documented.

GENETIC CHARACTERISTICS AND MODE OF INHERITANCE

Tuberous sclerosis is transmitted as an autosomal dominant disorder, with a recurrence risk of 50% for offspring of affected individuals.

Fifty percent to 85% of isolated cases have been estimated to be the result of new mutations (with no increased recurrence risk for siblings). However, recent findings suggest that these estimates may be too high since apparently healthy parents of an affected child may be asymptomatic carriers of the tuberous sclerosis gene more often than suspected. A careful examination of both parents (see

Procedures for diagnostic confirmation) is expected to identify most undiagnosed affected individuals.

CLINICAL NOTES

Clinical expression of tuberous sclerosis is extremely variable. Cases that are clinically significant tend to become progressively more severe.

Epilepsy (usually generalized or focal major motor seizures, or infantile spasms) and/or congenital ash-leaf-shaped white macules on the trunk and limbs are the two most consistent findings in diagnosed patients. However, the macules may not be apparent, especially in otherwise asymptomatic individuals, except under Wood's lamp examination.

The typical butterfly rash, consisting of angiofibromas (historically, adenoma sebaceum) in the malar and nasal region, is usually present by age 4. Other skin lesions include leathery shagreen patches, periungual or subungual fibromas, multiple skin tags of the neck and axillae, café-au-lait spots, hemangiomas, and gingival fibromas. Malignant degeneration is infrequent but may occur in the skin lesions or the cerebral nodules.

Mental retardation has been reported in roughly two thirds of overt cases and is thought to be more severe, the earlier the onset of seizures. However, it may also be the only clinical manifestation, thus suggesting consideration of tuberous sclerosis in the differential diagnosis of unexplained mental retardation.

Retinal phakomas are the most common ocular manifestation. Cystlike areas in the bones of the hands and feet are frequent features. Renal hamartomas and rhabdomyomas of the heart may be among the serious complications that often lead to death before age 25.

In mildly affected individuals, however, the disease is compatible with normal intelligence and life span.

PROCEDURES FOR DIAGNOSTIC CONFIRMATION

- Wood's lamp examination of skin for hypomelanotic macules
- Fluorescein angiography to rule out retinal phakomas
- Skull radiography to rule out califications
- CT scan for visualization of characteristic subendymal periventricular nodules
- Renal-function studies
- Ultrasonic studies of the kidneys to rule out typical cysts
- Biopsy of angiofibromas

CONSIDERATIONS IN MANAGEMENT

- Genetic
  - Evaluation of family members, especially CT scan of parents, for identification of asymptomatic gene carriers to determine whether a prior family history exists or whether disorder is more likely to be due to a new mutation
  - Risk counseling
  - Discussion of potential teratogenic risks conferred by medication, with affected women on anticonvulsants, who are or may become pregnant
- Psychosocial, educational, familial

Family counseling; support for acceptance of handicaps, dependent on the severity of the case
Education for management of seizures
Education and psychologic support for management of cosmetic disfigurement
Assessment of community resources for mentally retarded individuals when intellectual impairment is a feature, e.g., special education programs, vocational training, support groups

Medical
Anticonvulsants for control of seizures
Surgical removal of tumors or nodules, if indicated
Management of complications

SUGGESTED READING

Gomez MR, (ed): *Tuberous Sclerosis*. New York, Raven Press, 1979.

Katzman R: Tuberous sclerosis, in Merritt HH, (ed): *A Textbook of Neurology,* 6th ed. Philadelphia, Lea & Febiger, 1979.

Scott LN, Bartoletti SC, Rosenbaum A, et al: The value of CT in genetic counseling in tuberous sclerosis. *Pediatr Radiol* 1980;9:1.

Chapter 11

Neurofibromatosis

Neurofibromatosis (von Recklinghausen's disease) is a pleiotropic genetic disorder characterized by abnormal skin pigmentation, internal and external tumor formation, and growth abnormalities. It is one of the more common mendelian conditions, with an estimated incidence of 1/3,300 to 1/2,500. Heterogeneity of signs and symptoms and wide variability of expression are major features.

ETIOLOGY AND PATHOGENESIS

It is postulated that cells most vulnerable to tumor transformation are those derived from embryonic neural crest tissue. This includes cells of the posterior root ganglia, sensory ganglia or cranial nerves, autonomic ganglia, Schwann cells, suprarenal cells, and melanocytes. Pathogenesis is not clear. Current focus of research is on nerve-growth factor and other trophic factors.

GENETIC CHARACTERISTICS AND MODE OF INHERITANCE

Neurofibromatosis is transmitted as an autosomal dominant disorder.

- Risk of recurrence is 50% for offspring of affected individuals. The gene cannot be transmitted by unaffected individuals. However, minimally affected individuals are often overlooked, thereby obscuring risk of recurrence in some families: about 95% of mild cases are believed to escape detection but may have more severely affected offspring.
- Mutation rate is high; it is estimated that 50% of cases are new mutations, with no increased recurrence risk for siblings. But to establish a diagnosis of a new mutation, it is necessary to rule out minimal expression in both parents.

CLINICAL NOTES

Manifestations include: neurofibromas of skin, peripheral and central nervous systems, eye, stomach, liver, intestine, kidney, bladder, and larynx. Neurofibromas can resemble small skin tags and may be misdiagnosed, unless a biopsy is performed.

Café-au-lait spots are common and may be the sole finding. Axillary freckling may be present.

Other types of tumors associated with the disorder are: gliomas of the optic nerve and chiasma, acoustic neuromas, meningiomas, and pheochromocytomas (rare).

Sarcomatous degeneration of tumors may occur in 3% to 5% of cases.

Other possible findings are: congenital pseudoarthrosis, seizures, mental retardation (40% have some intellectual handicap; 2% to 5%—mental retardation), kyphoscoliosis, growth retardation, asymmetric growth, overgrowth (e.g., macrocephaly), other hamartomatous growths.

Intrafamilial variability is common regarding tumor sites, number of tumors, and disease severity.

Puberty and pregnancy may exacerbate the disorder.

A central form with acoustic neuroma, perhaps associated with other CNS tumors, but unaccompanied by peripheral manifestations, has been observed in some families; it is almost certainly a distinct entity, which is also inherited as an autosomal dominat condition. Malignant transformation has been reported at 30% to 50%.

PROCEDURES FOR DIAGNOSTIC CONFIRMATION

Diagnosis based on findings of neurofibromas, café-au-lait spots, and histologic examination of tumors

Any one of the following are highly suggestive, even without positive family history:
- Six or more café-au-lait spots 1.5 cm or greater in diameter for adults; or 0.5 cm or greater for children
- Three or more neurofibromas
- Five or more iris nodules

Bilateral acoustic neuromas always represent familial central neurofibromatosis

Significance of unilateral acoustic neuroma is not yet clear

CONSIDERATIONS IN MANAGEMENT

Genetic
- Evaluation of family members to determine if disorder has been transmitted or is the result of a mutation
- Risk counseling
- Consideration of biopsy for at-risk individuals of childbearing age with minimal signs

Psychosocial, educational, familial
- Dependent on severity of case
- Intellectual impairment, may require special schooling
- Support for individual and families when surgical intervention becomes necessary (particularly repeated instances)
- Psychologic management of cosmetic disfigurement
- Referral to support group, if desired

Medical
- Surveillance for elevated blood pressure, problems related to tumors, growth abnormalties
- Surgery when indicated (e.g., pain, impairment of function, suspicion of malignancy)
- Orthopedic management of associated problems (e.g., scoliosis, tibial bowing)

SUGGESTED READING

Riccardi VM: Von Recklinghausen neurofibromatosis. *N Engl J Med* 1981;305:1617.

Mulvihill JJ, Riccardi VM, (eds): *Neurofibromatosis (von Recklinghausen's disease): Genetics, Cell Biology and Biochemistry,* vol 29. (Advances in Neurology). New York, Raven Press, 1981.

Crowe FW, Schull J, Neel JW: *A Clinical Pathological, and Genetic Study of Multiple Neurofibromatosis*. Springfield, Ill., Charles C. Thomas, 1956.

Chapter 12

Alzheimer's Disease

Alzheimer's disease, which affects an estimated 2% to 3% of the population, accounts for 50% to 60% of adult dementia cases in the United States. Historically, Alzheimer' disease was referred to as dementia occurring before the age of 65. Increasing knowledge of the clinical and histopathologic similarities between the senile and presenile dementias, however, has blurred the age distinction. Alzheimer's disease may still refer to the presenile disorder, but senile dementia which meets the diagnostic criteria is now also referred to as senile dementia of the Alzheimer's type.

ETIOLOGY AND PATHOGENESIS

The etiology of Alzheimer's disease is unknown. Although the clinical course may be difficult to distinguish from a number of conditions of differing etiologies, the histopathology of the brain in Alzheimer's disease is characteristic. The microscopic hallmarks are neurofibrillary tangles within neuronal cell bodies, and extracellular neuritic (senile) plaques, which generally have a central core of amyloid. Both lesions affect the cerebral cortex and hippocampus, and the severity of the disease closely follows their concentration in these areas.

The cholinergic neurotransmitter system is affected in Alzheimer's disease; choline acetyl transferase (ChAT) activity is reduced by as much as 90%, and there is evidence for selective loss of cholinergic neurons. The reduction of ChAT activity correlates with both the number of neuritic plaques and the degree of intellectual impairment.

GENETIC CHARACTERISTICS AND MODE OF INHERITANCE

The contribution of genetic factors to this disease is unclear, and might best be approached via individual family histories.

- At least half the reported cases are sporadic, i.e., with no known affected family members.
- Overall, there is increased risk for first-degree relatives of index cases in these families of approximately four times that of the general population, that is 8% to 12%.
- Families have also been reported with an apparently autosomal pattern of inheritance (with variable penetrance) over several generations. In these families, recurrence risk to first-degree relatives of affected individuals is increased and may approach 50%.

Familial cases tend to have a younger age of onset, and the disease is clinically more severe than in nonfamilial cases. There are reports in the literature of various chromosomal anomalies associated with Alzheimer's disease: aneuploides, acentric fragments, and atypical chromatic configurations. A two-way association with Down's syndrome has also been observed: many Down's syndrome patients who live until adulthood show Alzheimer's histopathology, and, conversely, family studies of Alzheimer's patients have reported a higher than expected incidence of Down's syndrome in first-degree relatives. The significance of these associations is unclear at this time, but there has been conjecture that disordered microtubule function or synthesis would be consistent both with the Alzheimer's neurofibrillary tangles and with chromosomal anomalies due to nondisjunction.

CLINICAL NOTES

The onset of Alzheimer's disease is insidious; the first noticeable symptom is generally a problem with recent memory. Speech disturbances characteristically follow—for example, inability to name common objects, and difficulty with comprehension of oral or written speech. Personality disturbances accompany increasing loss of intellectual function. Seizures may occur in advanced stages of the disease. Death usually results from intercurrent infection.

Age of onset is variable, usually from 50 to 75. Extremes on either side of the range are seen, however. Cases have been reported at age 30 and into the 80s.

The disease is more severe in younger patients. Although it appears to run a more rapid course in the younger age-group (below 70), this tendency may be masked by the older patient's succumbing quickly because of frailty or complicating disorders.

PROCEDURES FOR DIAGNOSTIC CONFIRMATION

- Histologic evaluation of brain tissue at biopsy or autopsy
- CT scan to exclude focal brain disease
- Laboratory and clinical tests to rule out reversible causes of dementia, such as syphilis, brain tumor, stroke, hepatic/metabolic/thyroid disorders, depression, and the effects of medication

CONSIDERATIONS IN MANAGEMENT

- Genetic
  - Risk counseling on the basis of family history and empiric data
- Psychosocial, educational, familial
  - Structured environment with goal of maintaining function as long as possible, especially in home or other familiar environment
  - Measures to counter sensory deprivation and maintain orientation in time
  - Appropriate safety measures
  - Referral of family to support groups to alleviate stress and educate care givers
- Medical
  - Psychotropic agents for management of depression, agitation, anxiety, etc.
  - Custodial care may be required in terminal stages of the disease

SUGGESTED READING

Cook RH, Ward BE, Austen J: Studies in aging of the brain: IV. Familial Alzheimer disease: Relation to transmissible dementia, aneuploidy, and microtubular defects. *Neurology* 1979;29:1402.

DeBoni U, McLachian DR: Senile dementia and Alzheimer's disease: A current view. *Life Sci* 1980;27:1.

deLeon MJ, Ferris SH, George AE: Computed tomography and positron emission tomography, in Crook T, Ferris S, Bartus R (eds): *Assessment in Geriatric Psychopharmacology*. New Canaan, Powley Associates, 1983.

Heston L, White J: A family study of Alzheimer's disease and senile dementia: An interim report, in Cole JE, Barrett JE (eds): *Psychopathology in the Aged*. New York, Raven Press, 1980.

Katzman R: The prevalence and malignancy of Alzheimer disease. *Arch Neurol* 1976;33:217.

Steinberg G: Long term continuous support for family members of Alzheimer patients, in Reisberg B (ed): *Alzheimer's Disease*. New York, Free Press, 1983.

Terry RD, Davies P: Dementia of the Alzheimer type. *Ann Rev Neurosci* 1980;3:77.

Chapter 13

Huntington's Disease

Huntington's chorea or Huntington's disease (HD) is a hereditary disorder of adult onset, in which progressive degeneration of the basal ganglia and cerebral cortex result in choreiform movements and progressive mental deterioration. The rare juvenile form, referred to as the akinetic-rigid form, is characterized by seizures, ataxia, mental retardation, and by hypertonicity and bradykinesia rather than chorea.

ETIOLOGY AND PATHOGENESIS

Huntington's disease is caused by an autosomal dominant gene, which has recently been assigned to chromosome 4, via recombinant DNA marker studies in two large kindreds. The basic defect is not known. There are widespread degenerative changes in the brain. Neuronal loss and reactive gliosis exist, affecting primarily the caudate nucleus, putamen, and cerebral cortex. Glutamic acid decarboxylase activity is decreased in the basal ganglia, resulting in a deficiency of gamma-aminobutyric acid (GABA). In some cases, choline acetylase activity is decreased. Both these changes are associated with the degeneration of small interneurons.

GENETIC CHARACTERISTICS AND MODE OF INHERITANCE

HD is transmitted as an autosomal dominant condition with complete penetrance. Both sexes are affected and both can transmit the disease.

- Every child of a person affected with HD has a 50% risk of developing the disease.
- Cases resulting from a new mutation are rare.
- In the large majority, onset is in the fourth or fifth decade, but the range of age of onset is broad: from childhood to the eighth decade. Thus, until carrier testing is available and/or acceptable (see Clinical notes), most individuals at risk for inheriting the gene reproduce before they can know whether they have inherited it and whether or not they may pass it on to their children.

CLINICAL NOTES

The nature of presenting symptoms, clinical course, and response to treatment all vary. Onset is insidious, usually subtle, and may reflect emotional, congnitive, or motor disturbances. Mental impairment may be apparent long before motor signs appear. Premonitory symptoms may include restlessness, clumsiness, dysarthria, emotional lability, and lack of concentration; but these symptoms are frequently recognized only retrospectively.

The clinical course is variable, but the average life span after onset of symptoms is 15 years. Death usually results from intercurrent infections.

Differential diagnosis of dementing symptoms includes Alzheimer's or Pick's disease early in the course of the disease or when choreiform movements are minimal or absent. HD is often mislabeled "schizophrenia" or "affective disorder" when behavioral features predominate or when neurologic signs are not recognized. Senile chorea may be differentiated from HD by the (relative) lack of intellectual and emotional impairment, more benign course, and negative family history.

At this time there is no generally available way of predicting which individual at risk has inherited the Huntington gene, before the person becomes symptomatic. However, the recombinant DNA marker studies that localized the gene on chromosome 4 hold promise for routine availability of *presymptomatic indentification of gene carriers* and *prenatal diagnosis*. This breakthrough raises hopes of reducing the toll of this devastating disorder. However, it poses grave ethical and psychologic concerns on behalf of young persons at risk, electing testing or prenatal diagnosis. They may find that they carry the gene and will then have to live with the certainty that they will become demented.

PROCEDURES FOR DIAGNOSTIC CONFIRMATION

There is no laboratory test; recombinant DNA marker studies may provide one.

Diffuse EEG changes may be present but are not specific.

A pneumoencephalogram may reveal dilation of the anterior horn of the lateral ventricles caused by disappearance of the head of the caudate nucleus, but a normal pneumoencephalogram does not rule out a positive diagnosis.

Postmortem examination is essential because of the importance of diagnosis to other family members.

CONSIDERATIONS IN MANAGEMENT

Genetic

- Risk counseling
- Consideration of evaluation of individuals at risk; for early symptoms or for gene carrier status presymptomatically or prenatally, when techniques become available: selected large HD families are currently eligible for testing on an experimental basis and a national registry has been created to coordinate DNA banking for future family linkage studies, when currently affected individuals may no longer be living
- Consideration of reproductive options., e.g., artificial insemination by donor, adoption

Psychosocial, educational, familial

- Counseling and support for affected individuals and their families
- Patient may be cared for initially at home, but with progressive dementia and dyskinesis, institutionalization is necessary
- Sustained counseling and support for all at-risk family members
- Referral of family to support group, if desired

Medical

Chorea:

Drugs which block dopaminergic function, such as the anti-psychotic phenothiazine or butyrophenone, improve chorea but may produce restlessness, parkinsonism, dystonic reactions or tardive dyskinesia. Reserpine may result in hypotension or severe depression.

Affective disturbances:

Tricyclic antidepressants are preferred in treating depression.

In cases in which pyschosis occurs, symptoms generally respond favorably to antipsychotic drugs.

Efficacy of treatment is variable and individual, but is still palliative only.

SUGGESTED READING

Chase TN, Wexler N, Barbeau A, (eds): Huntington's disease, vol 23. *Advances in Neurology*. New York, Raven Press, 1979.

Conneally PM: Huntington disease: Genetics and epidemiology. *Am J Hum Genet* 1984;36:506.

Hayden M: *Huntington's Chorea*. New York, Springer-Verlag, 1981.

Kolata G: Huntington's disease gene located. *Science* 1983;222:913.

Yahr MD: Chronic progressive chorea (Huntington's Disease, Adult Chorea), in Merrit HH, (ed): *A Textbook of Neurology*. Philadelphia, Lea & Febiger, 1979, pp 493-498.

Chapter 14

Genetic Neuromuscular Disorders

Muscle tone and function are controlled at many different levels of the central and peripheral nervous systems. The effects of malfunction range from hypotonia, weakness, paralysis and atrophy, to inability to control muscular action and balance. Onset of symptoms may be congenital or later in life. Degeneration is typical. Many genetic disorders with other primary characteristics have neuromuscular manifestations; some of these are discussed in other chapters. Table 14-1 lists selected genetic disorders with muscular weakness. Table 14-2 lists those that include ataxia.

This chapter will review some of the more common neuromuscular diseases in the following categories: cerebral hypotonia, spinal muscular atrophy, hereditary polyneuropathy, hereditary ataxia, torsion dystonia, and familial incidence of generally sporadic neurologic disorders.

DESCRIPTION, GENETIC CHARACTERISTICS, AND MODE OF INHERITANCE

Cerebral Muscular Hypotonia. Cerebral hypotonia is a finding in many inherited disorders that are generally diagnosed on the basis of other symptoms, such as lipidoses, gangliosidoses, aminoacidurias, chromosome defects, and malformations of the brain (see Table 14-1 and relevant chapters).

Prader-Willi syndrome is an example of a disorder presenting in a newborn with severe cerebral hypotonia and failure to thrive. The infant later becomes more alert and develops a gargantuan appetite, which leads ultimately to obesity. Short stature, small hands and feet, and delayed sexual development are characteristic; hypogonadism and infertility are the rule. About 90% of affected children have an IQ below 75. Typically, an affectionate personality can give way to occasional severe temper tantrums.

- Most cases are sporadic, but rare familial occurrence has been reported. In almost half of those affected, an interstitial deletion or another abnormality in the long arm of chromosome 15 has been identified by high-resolution chromosome studies.

In neonates, Prader-Willi syndrome must be distinguished from other serious conditions with severe hypotonia, such as congenital muscular dystrophy (see Chapter 15), Werdnig-Hoffmann disease, and *benign congenital (essential) hypotonia* discussed next.

TABLE 14-1
Selected Genetic Disorders With Muscle Wasting and Weakness

DISORDER	OTHER CHARACTERISTIC FEATURES	MODE OF INHERITANCE
Chromosome Defects		
Down's syndrome	An extra chromosome 21; flattened facies, oblique palpebral fissures; flat occiput; abnormal dermatoglyphics; mental retardation	CH
Trisomy 18	An extra chromosome 18; failure to thrive, poor suck, congenital heart disease, overlapping fingers, multiple anomalies, developmental retardation	CH
Cri-du-chat syndrome	Deletion in short arm of chromosome 5; catlike weak cry in infancy; severe mental retardation	CH
Sphingolipidoses		
Tay-Sachs disease	Hexosaminidase A deficiency; mainly in Ashkenazi Jews; cherry-red spot in macula; psychomotor deterioration after age 4 to 6 months	AR
Fabry's disease	Alpha-galactosidase A deficiency; corneal opacities; acroparesthesia; skin lesions	XR
Krabbe's disease	Beta-galactosidase deficiency; irritability; developmental deterioration, onset 4 to 6 mo	AR
Farber's disease	Lipogranulomatosis; early-onset hoarseness; restriction of joint movements; lumpy masses over wrists and ankles; rapidly deteriorating course	AR
Infantile metachromatic leukodystrophy	Defective cerebroside sulfatase activity; progressive failure of motor development, regression, dysarthria	AR
Niemann-Pick disease	Sphingomyelinase deficiency; hepatosplenomegaly in infancy; pulmonary infiltration; developmental slowing	AR
Glycogen-Storage Diseases		
Pompe's disease	Glycogen storage disease, type II; acid maltase deficiency; cardiomegaly, dyspnea and cyanosis in infancy	AR
McArdle disease	Glycogen storage disease, type V; muscle phosphorylase deficiency; extensive cramping; weakness after mild or moderate exercise	?AR
Aminoacidurias		
Glutaric aciduria	Vomiting, metabolic aciduria in infancy; involuntary movements, mental retardation	AR
Muscular dystrophy	Weakness and wasting of diversified muscle groups	Varied
Myotonic dystrophy	Inability to relax muscles; wasting or weakness of face, neck, and distal limbs	AD
Congenital myopathy	Various disorders with abnormal muscle histopathology	?

(Continued)

DISORDER	OTHER CHARACTERISTIC FEATURES	MODE OF INHERITANCE
Spinal muscular atrophy		
Werdnig-Hoffmann disease	Hypotonia, weakness, absent or decreased deep-tendon reflexes; progressive deterioration	AR
Polyneuropathy		
Charcot-Marie-Tooth disease	Slowly progressive weakness and wasting first in legs, later hands; clubfeet	AR, AD, XR
Endocrine disease		
Congenital hypothyroidism	Goiter; impairment of intellect, somatic retardation; typical facies	AR
Thyroid dysgenesis	Neonatally; large posterior fontanelle; prolonged hyperbilirubinemia; mild myxedema of face and neck; hypothermia; later signs of hypothyroidism	?
Other		
Abetalipoproteinemia	Malabsorption; diffuse CNS signs; acanthocytes	AR
Spongy degeneration of the brain	Early optic atrophy; megacephaly; later hypertonicity; primarily in Ashkenazi Jews	AR
Zellweger's syndrome	Profound hypotonia; decreased or absent reflexes; hepatomegaly; typical craniofacial dysmorphology; glaucoma; cataracts; patellar and acetabular stippling; renal subcapsular cysts	AR
Diastematomyelia	Gait disturbance; weakness or atrophy of lower limb(s); bladder dysfunction	?
Leigh disease	Infantile necrotizing encephalopathy; pyruvate decarboxylase defect	AR
Moebius syndrome	Congenital paralysis of 6th and 7th cranial nerves; masklike face, neonatal difficulty in feeding, drooling	AD
Oculocerebrorenal (Lowe's) syndrome	Male child with bilateral congenital cataracts; mental retardation; renal tubular dysfunction	XR
Periodic paralysis	Episodes of paralysis; lowered or elevated serum potassium levels	AD
Prader-Willi syndrome	Infantile hypotonia; hypoplastic genitalia; later polyphagia, obesity, hypogonadism, mental retardation	? or CH

Key: CH = chromosomal; AR = autosomal recessive; AD = autosomal dominant; XR = X-linked recessive

Essential hypotonia. For essential hypotonia, familial incidence has been reported in 20% of cases. A self-limited condition, it is probably etiologically heterogeneous and is consistent with essentially normal development. Good muscle strength and an alert appearance may differentiate it from the severe disorders.

Motor development is only slightly delayed and the condition is generally outgrown by school age. The presence of other serious manifestations such as contractures or hip dislocations puts the diagnosis in doubt.

Spinal Muscular Atrophies. These atrophies are almost always genetic and share degeneration and loss of motor neurons in the anterior horns of the spinal cord and/or in the bulbar motor nuclei. They are characterized by progressive atrophy and weakness, loss of tendon reflexes, fasciculations, and contractures. Intelligence is not affected, and sensory modalities are intact. The spinal muscular atrophies differ in localization of weakness, age of onset, length of course, and mode of inheritance.

Carrier determination and prenatal diagnosis are not available for any spinal muscular atrophies.

Proximal spinal muscular atrophies comprise more than 80% of spinal muscular atrophies. They begin with weakness in the proximal muscles, which spreads to distal ones. Classification into subtypes is somewhat arbitrary, in view of intermediate cases and rare families with more than one type; but it is generally useful because in the majority of families only one subtype is seen.

Most cases of proximal spinal muscular atrophy are inherited as an autosomal recessive disorder.

- There is a 25% recurrence risk for each child of two asymptomatic carriers.
- Unaffected siblings of affected individuals have a 67% risk for being carriers, but their risk for having an affected child is very low, except in consanguineous matings.

The principal proximal atrophies include the following:

Werdnig-Hoffmann disease occurs in 1/25,000-1/20,000 newborns. General muscle weakness begins prenatally or in the first year of life and progresses relentlessly. There may be arthrogryposis, fasciculations of the tongue, and fine tremors. Tendon reflexes are decreased or absent, and contractures develop. Through all this, these infants socialize well. Although they usually die by one year of age, some have survived to 4-6 years.

In some children, the downhill course stops or slows down (*arrested Werdnig-Hoffman disease*). If respiratory function is preserved, such patients may lead useful, albeit nonambulatory, lives for decades.

Kugelberg-Welander disease, a variant of proximal spinal atrophy, manifests in childhood or adolescence. It may resemble Becker's or limb-girdle muscular dystrophy. Early clinical symptoms include a waddling gait and difficulties in climbing stairs. The weakness spreads to the shoulder girdle and to the extremities. Fasciculations and contractures may appear. Although some patients cease to walk during their late teens, others remain ambulatory for several decades.

Adult proximal spinal atrophy is characterized by muscle weakness that usually becomes noticeable between ages 18 and 50. Progression is slow.

Other spinal muscular atrophies include several rare disorders that differ from the proximal ones in localization of muscle weakness. They enter the differential diagnosis for more conditions. Among them are *juvenile progressive bul-*

bar palsy (childhood facial palsy), *facioscapulohumeral spinal muscular atrophy,* and *distal spinal muscular atrophy,*(early childhood weakness in the distal muscles of the legs).

Hereditary Polyneuropathies. Genetic disorders involving neuropathy include intermittent porphyria, analphalipoproteinemia, and Refsum's disease. Peripheral polyneuropathy also occurs as a symptom of various leukodystrophies, such as adrenoleukodystrophy, metachromatic leukodystrophy, and globoid leukodystrophy of Krabbe.

Among the more common hereditary disorders characterized primarily by polyneuropathy are the following:

Charcot-Marie-Tooth disease (peroneal muscular atrophy), encompasses a group of chronic disorders. Type I, the most common, is characterized by segmented demyelination of the nerve sheath; concentric proliferation of the Schwann cells with onion bulb formation; and reduced motor-nerve conduction velocity. Type II, another variant, is characterized by axonal degeneration and normal or near-normal motor-nerve conduction.

With onset generally in the first to third decade, early atrophy and weakness of the feet and lower legs lead to a slapping gait and typical "stork legs." Other findings are pes excavatus, talipes equinovarus, distal sensory deficit, eventual contractures (e.g., claw hand), scoliosis, optic atrophy and external ophthalmoplegia. Progression is slow, and prognosis for lifespan is favorable; but varying degrees of disability may result.

Most cases of Charcot-Marie-Tooth disease are inherited as autosomal dominant disorders.

- Children of an affected parent have a 50% risk for inheriting the mutant gene. Since expression is variable and not all gene carriers show clinical manifestations, parents and siblings of affected individuals should be carefully checked for evidence of subclinical manifestations to help distinguish a new mutation from a familial case.

Familial dysautonomia, (Riley-Day syndrome), is an autosomal recessive disorder that occurs predominantly in the Ashkenazi Jewish population, where carrier frequency is estimated at 2%. Studies reveal decreased activity of beta-hydroxylase, an increase in the beta fraction of nerve growth factor, and reduction of both peripheral and autonomic neurons.

Swallowing deficits, decreased pain sensation, absence of lacrimation, and absent fungiform papillae of the tongue are hallmarks of the disorder. Other features include vasomotor and emotional instability, GI disturbance, ataxia, seizures and secondary pneumonia. The multitude of manifestations distinguishes the disorder from *isolated congenital indifference to pain,* also autosomal recessive. Death may occur in infancy or as late as early adulthood. Carrier testing and prenatal diagnosis are not available.

- The risk of occurrence is 25% for each child when both parents are carriers.
- Unaffected siblings have a two-in-three risk of being carriers.

Hereditary Ataxias. Characterized by disturbances of muscular coordination and the resultant inability to control muscular action and balance, ataxia can result from impairment of diverse structures of the nervous system. These range from the cerebellum to the brain stem, the spinal cord, and even to the posterior root ganglia. In addition to the conditions in which ataxia is the primary manifestation, it may be a major feature in a variety of other genetic disorders (see Table 14-2).

A significant proportion of the ataxias includes the numerous forms of *spinocerebellar degeneration*, many of which are exremely rare individually. Age of onset ranges from infancy to adulthood, but once symptomatic, all show progressive deterioration leading to severe disability and a restricted lifespan.

Friedreich's ataxia, already considered the most common of the hereditary ataxias, with more than 7,000 known cases in the United States, is suspected to

TABLE 14-2
Genetic Disorders With Ataxia

DISORDER	OTHER CHARACTERISTIC FEATURES	MODE OF INHERITANCE
Abetalipoproteinemia	Malabsorption; acanthocytes; atypical retinal pigmentation	AR (AD)
Acoustic neuromata	Central neurofibromatosis; progressive auditory and vestibular dysfunction	AD
Ataxia-telangiectasia	Oculocutaneous telangiectasia; sinopulmonary infections; immune deficiencies	AR
Cerebellar ataxia and chorioretinal degeneration	Various disorders involving both ataxia and retinal degeneration	Varied
Late infantile GM_1 gangliosidosis, type II	Beta-galactosidase deficiency; mental and motor retardation; seizures; dementia	AR
Juvenile metachromatic leukodystrophy	Cerebroside sulfatase deficiency; mild mental retardation; polyneuropathy	AR
Olivopontocerebellar atrophies	Varying age of onset, which may have other manifestations, e.g., ophthalmoplegia, dysarthria, mental deterioration	Most: AD
Refsum's disease	Phytanic acid storage; retinitis pigmentosa; peripheral neuropathy	AR
von Hippel-Lindau disease	Retinal and CNS hemangioblastomas; vertigo; nystagmus; dysarthria	AD
Niemann-Pick disease	Sphingomyelinase deficiency; organomegaly; pulmonary infiltration; developmental slowing	AR
Usher's (Hallgren) syndrome	Congenital deafness; retinitis pigmentosa	AR

Key: AR = autosomal recessive; AD = autosomal dominant

be the cause of additional cases of undiagnosed degenerative illness. Carrier testing and prenatal diagnosis for Friedreich's ataxia are not available.

Pathologic findings reveal degeneration of neurons in the dorsal spinocerebellar and corticospinal tracts, as well as in the cerebellar nuclei. Biochemical studies suggest genetic heterogeneity.

Typical Friedreich's ataxia begins insidiously in the first or second decade of life with clumsiness involving the upper or lower extremities. Common findings are truncal and appendicular ataxia, a positive Romberg test, nystagmus, dysarthria, decreased position and vibration sense, extensor plantar response, areflexia, muscle weakness, muscle atrophy, and kyphoscoliosis. Mental deterioration is rare. Cardiomyopathies and ECG changes, notably an atrioventricular block, are frequent. Rarely, cardiomyopathy may precede neurologic manifestations. Diabetes mellitus occurs in 20% of patients. The neurological symptoms progress steadily, but patients may remain professionally active for years; the mean age at death is 37.

Inheritance is autosomal recessive (except in about 10% of cases where autosomal dominant inheritance is indicated by family history).

- There is a 25% risk of recurrence for each child of two clinically asymptomatic carriers.
- Siblings of affected patients, who have passed the age of risk for onset without developing symptoms, have a 67% risk for being carriers.
- All children of affected individuals are carriers.
- The risk for potential carriers to have affected offspring is negligible, except in consanguineous matings or marriage to a relative of another affected individual.

Ataxia-telangiectasia, a chromosomal fragility syndrome, is diagnosed in approximately 2/100,000 children. Deficient DNA repair is thought to be responsible for chromosomal breakage and rearrangements and, indirectly, for the clinical findings, which include progressive cerebellar ataxia, ocular apraxia, telangiectasias, humoral and cellular immune deficiency, susceptibility to lymphoproliferative neoplasms, muscular weakness and atrophy, endocrine malfunction, and intellectual decline. Death in childhood or young adulthood is frequent. Carriers are believed to be at increased risk for developing malignancies. Carrier detection is possible through studies of cellular DNA repair; and prenatal diagnosis is available, using fetal cells for chromosome breakage studies.

Ataxia-telangiectasia is inherited as an autosomal recessive disorder.

- Each offspring of two carrier parents has a 25% risk of occurrence.
- Unaffected siblings of affected individuals have a 67% risk of being carriers.

Other ataxias include such rare disorders as striatonigral degeneration (Joseph's or Azorean disease), various olivopontocerebellar ataxias, and familial spastic paraplegia.

Torsion Dystonia. Torsion dystonia, which reflects dysfunction within the basal ganglia, may be a symptom of a variety of cerebral disorders or a reaction to a pharmacologic agent. In the absence of a precipitating condition it may signal the presence of a genetic disorder.

Idiopathic torsion dystonia (dystonia musculorum deformans), describes the familial disorder. Involuntary movements of somatic muscles build up to intense sustained muscular contractions, and to twisting and turning the affected body part. Attacks are intermittent but may lead to constant contortion.

One of two hereditary forms is found predominantly in Ashkenazi Jewish populations. Onset is in childhood. Pharmacologic and surgical intervention have been used to ameliorate the condition. Inheritance is autosomal recessive, with a 25% risk of occurrence for each pregnancy of two carrier parents.

An autosomal dominant form of dystonia musculorum deformans is not associated with any particular ethnic group. Onset is later than in the recessive form, and the course is more benign. Elevated serum dopamine-beta-hydroxylase has been noted in patients with this form of the disorder.

Familial Incidence of Neuromuscular Disorders (generally not considered genetic). A number of relatively common conditions are seen mostly as sporadic cases, but occasionally show familial aggregation, suggesting a genetic etiology. Among them are the following.

Amyotrophic lateral sclerosis (ALS, Lou Gehrig's disease), is characterized by progressive atrophy and weakness of the muscles innervated by degenerating upper and lower motor units of the medulla oblongata and spinal cord. Although ALS can occur at any age, typical onset is from the fifth to the seventh decade of life. The incidence has been estimated at about 1/70,000 in the United States; from 5% to 10% of the cases are thought to be familial. Onset in the latter has been reported to be somewhat earlier than in sporadic cases.

Since early symptoms may mimic other neuromuscular disorders, precise diagnosis is necessary before recurrence risks can be estimated.

- If the diagnosis of ALS is confirmed, the familial nature is generally identified through a positive family history, which most often reveals autosomal dominant inheritance, with a 50% risk for siblings or children of an affected individual.

Myasthenia gravis may begin, typically in adulthood, with ptosis and facial weakness, resembling myotonic dystrophy. A defect in neuromuscular transmission, due to antibodies to the acetylcholine receptor, it is ultimately characterized by rapid onset of fatigue and decreasing muscle strength during exercise. A fairly normal life may be achieved with cholinergic drugs. Familial occurrence has been reported in 2% to 5% of cases—more likely the families with early onset (infancy or childhood). The mode of inheritance is not clear.

About 15% of babies born to mothers with myasthenia gravis are born with *transient myasthenia gravis of the newborn*. Caused by a transplacental maternal factor, this is not genetic. It disappears within a few weeks, but can be life threatening during that time, if untreated.

Multiple sclerosis (MS) varies considerably in geographic incidence. In the United States, estimates of 6/10,000 to 1/1,000 have been made. Other neuromuscular disorders involving weakness, parasthesias, and movement disturbance may lead to an erroneous diagnosis of MS. They should be ruled out, if possible, when familial incidence of MS is reported. While the cause of MS is not known and infectious factors have been suggested, an immunogenetic susceptibility, perhaps related to HLA-Dw2, has been proposed as a possible explanation for familial incidence. Empirically, a recurrence risk of 1% to 4% for first-degree relatives and 0.5% to 1% for second- and third-degree relatives has been estimated.

Parkinson's disease must be distinguished from parkinsonism—progressive tremor, rigidity, bradykinesia and loss of postural reflexes—which, with a prevalence of 1/1,000, is one of the most common neurologic conditions. Parkinsonism may occur secondary to another issue, such as encephalitis, toxic reaction, antipsychotic medication, or be part of another syndrome. Among these are genetic disorders like striatonigral degeneration, familial Alzheimer's disease, Wilson's disease and Huntington's disease. Reports of parkinsonism in a family may thus have to be evaluated for misdiagnosis or, perhaps, for denial of a more devastating diagnosis.

If primary Parkinson's disease is confirmed, multifactorial inheritance is postulated for the small proportion of familial cases. A recurrence risk of 12% for first-degree relatives has been estimated, particularly for early-onset cases (before the peak age of onset: 6th or 7th decade), but this may include families in which another etiology prompted the diagnosis. No presymptomatic tests have been developed.

Cerebral palsy is not a disorder in itself, but rather a condition generally found subsequent to perinatal damage to the nervous system. While most cases may well be due to a birth injury, genetic defects (e.g., chromosome disorders, metabolic errors) may also be the cause and are best ruled out before the nongenetic nature is accepted. This is especially important when two or more individuals in a family are affected or when an old diagnosis has not been re-evaluated for many years.

CLINICAL NOTES

Onset of neuromuscular conditions tends to be insidious and may share early signs and symptoms with many other disorders, often making the differential diagnosis a challenge, even to the specialist. Family-history analysis and careful examination for associated symptomatology or for minimal expression in relatives may greatly aid specific diagnosis.

PROCEDURES FOR DIAGNOSTIC CONFIRMATION

Serum CK
Urinary CK
Electromyography
Motor- and sensory-nerve conduction velocity
Muscle biopsy
High resolution chromosome analysis (Prader-Willi)
Response to neostigmine or edrophonium (myasthenia gravis)

Intradermal histamine injection (familial dysautonomia)
Methacholine chloride drip in conjunctival sac (familial dysautonomia)
EEG, ECG (Friedreich's ataxia)
CSF analysis (Friedreich's ataxia)
Chromosome-breakage studies (ataxia-telangiectasia)
Serum protein electrophoresis (ataxia-telangiectasia)
Myelin basic protein analysis (multiple sclerosis)
Spinal fluid gammaglobulins (multiple sclerosis)
Spinal fluid oligoclonal bands (multiple sclerosis)
Fractionated catecholamines (Parkinson's disease)
Serotonin levels (Parkinson's disease)
Appropriate tests for associated findings, for example, metabolic screens

CONSIDERATIONS IN MANAGEMENT

- Genetic
  - Establish diagnosis
  - Pedigree analysis for evidence of mode of inheritance or familial incidence
  - Evaluation of potential non-expressing gene carriers, as indicated by diagnosis
  - Risk counseling
  - Consideration of prenatal diagnosis, when indicated and available
  - Discussion of alternate reproductive options, as indicated (e.g., donor insemination or adoption)
  - Discussion about meaning of heterozygous state in ataxia-telangiectasia and possible need for increased surveillance for neoplasias
  - Discussion of sharing information with other family members to encourage evaluation and/or risk counseling
- Psychosocial, educational, familial
  - Family conseling: support for poor prognosis and prevention of guilt feelings; reassurance when indicated
  - Education for appropriate home care
  - Vocational consultation
  - Assessment of family/community resources: for example, supportive services and/or groups, special educational facilities, sheltered work options, recreational resources, group homes, institutional care
- Medical
  - As indicated by diagnosis

SUGGESTED READING

Baraitser M: *Genetics of Neurological Disorders,* New York, Oxford Press, 1982.

Bradley WG, Madrid R, Davis CJ: The peroneal muscular atrophy syndrome. Clinical, genetic and nerve biopsy studies. Part 3. Clinical, electrophysical and pathological correlations. *J Neurol Sci* 1977;32:123.

Brooke MH, Carroll JE, Ringel SP: Congenital hypotonia revisited. *Muscle & Nerve* 1979;2:84.

Brooks AP, Emery AEH: A family study of Charcot-Marie-Tooth disease. *J Med Genet* 1982;19:88.

Dubowitz V: *Muscle Disorders in Childhood,* vol. 16. (Major Problems in Clinical Pediatrics.) Philadelphia, WB Saunders, 1978.

Dubowitz V: *The Floppy Infant, Clinics in Developmental Medicine,* 2nd ed. Philadelphia, JB Lippincott, 1980.

Goldensohn ES, Appel SH: *Scientific Approaches to Clinical Neurology* 1975;2:1075.

Holm VA, Pipes PL: *Prader-Willi Syndrome.* Baltimore, University Park Press, 1981.

Ionasescu V, Zellweger H: Genetics of neuromuscular diseases, *Handbook of Clinical Neurology* 1979;41:405.

Merritt HH, (ed): *A Textbook of Neurology.* Philadelphia, Lea & Febiger, 1979.

Vinken PJ, Bruyn GW: *Handbook of Clinical Neurology,* New York, Elsevier, 1975.

Walton JN, (ed): *Disorders of Voluntary Muscle,* 4th ed. New York, Churchill Livingston, 1981.

Chapter 15

Genetic Muscle Disorders

Genetic disorders of muscles, as classified here, are primarily those in which symptoms and signs seem to relate directly to dysfunction of the muscles themselves, rather than being manifestations clearly secondary to a nervous system malfunction. Weakness and/or muscle wasting are the most common shared features, although these are also prominent in the neuromuscular disorders, which may enter the differential diagnosis. In addition, muscular weakness is a characteristic of a host of genetic disorders of assorted etiology (see Table 14-1, Chapter 14).

This chapter will review the major categories of primary myopathies, including the muscular dystrophies, myotonias, congenital non-progressive myopathies, and metabolic myopathies.

ETIOLOGY AND PATHOGENESIS

Although specific biochemical defects of the various primary muscle disorders are generally not known, the different myopathies have typical electrical, biochemical, and histologic features. The muscular dystrophies are distinguished by degeneration and regeneration of muscle fibers, which are eventually replaced by fat and fibrous tissue. Myotonic dystrophy, a disorder with manifestations in many other organ systems shows early selective muscle fiber atrophy, and an abundance of central nuclei, progressing to degeneration and regeneration of muscle fibers later in the course of the disease. The congenital nonprogressive myopathies are distinguished by anomalies evident in histochemical and electron-microscopic analysis. Recent progress in studies of metabolic myopathies has implicated abnormalities in mitochondrial lipid metabolism in the development of muscle weakness.

DESCRIPTION, GENETIC CHARACTERISTICS, AND MODE OF INHERITANCE

Most genetic disorders are inherited as single gene conditions with all types of transmission. In some, expressivity may vary greatly, making examination of family members necessary before recurrence risks can be established.

The major forms of primary genetic muscle disease can be classified according to clinical categories:

Muscular Dystrophies comprise a group of hereditary disorders which are all characterized by progressive muscle weakness and wasting, and are distinguished by typical clinical and genetic patterns.

Duchenne's muscular dystrophy. With an incidence estimated at 1/3,000 to 1/4,000 male births, Duchenne's dystrophy is by far the most common of the muscular dystrophies, and the most devastating. With X-linked recessive inheritance, the full-blown disorder affects primarily boys carrying the mutant gene on their one X chromosome. Although highly elevated serum creatine phosphokinase (CK) levels are detectable at birth, clinical onset is insidious, with early signs, such as hypotonia, often overlooked. Diagnosis is usually made at about age 3, when such manifestations as frequent stumbling, difficulty in climbing stairs, or getting up from the floor become apparent, and symptoms of pseudohypertrophy of the calf muscles may become obvious. Muscle weakness begins in the pelvic girdle and rapidly progresses to involve other skeletal muscles, going from proximal to distal. Contractures, kyphoscoliosis, and muscle atrophy follow. Patients lose the ability to walk between 5 and 15 years. About one third have mental retardation. Death occurs before age 20 for most patients.

- Women are carriers if they have inherited a mutant gene on one of their two X chromosomes. Most are asymptomatic, but about 5% have been reported to have some muscular weakness.
- Their sons have a 50% risk to inherit the mutant gene and be affected.
- Their daughters have a 50% risk to be carriers like the mother.
- Affected males almost never pass on the mutant gene, since nearly all are totally disabled before reaching reproductive age.
- The gene cannot be transmitted through unaffected males.
- Women are considered obligate carriers by family history, if they have one affected son and one affected brother or maternal uncle, more than one affected son, or an affected son and a carrier daughter.
- Women are at risk for being carriers if they have one affected son, one affected brother, one maternal uncle, or a sister with an affected son.
- Most women who have only one affected son or one carrier daughter are carriers, although a new mutation may be responsible for the mutant gene in their one affected or carrier offspring. Their theoretical risk is 67%, but practical experience at several centers has suggested that about 85% are, indeed, carriers.

Definitive carrier testing for mothers of sporadic cases, sisters of affected males, and other female relatives is yet to become a routine procedure. CK determinations identify 60% to 70% of all carriers, when an increased level has been shown on three different days and when confounding variables, such as pregnancy, nonspecific elevation of serum CK, other diseases, and age at testing are taken into account. Assay of a combination of four different serum enzymes has also been suggested. Test results considered together with the family history, and with tests in other female relatives, can be used to evaluate the risk for carrier status more closely than any one method. Recent recombinant DNA studies, however, promise carrier detection of much greater accuracy, at least for informative families (see below—prenatal diagnosis).

Prenatal diagnosis for Duchenne's muscular dystrophy is not yet generally available. However, the gene has been localized to the short arm of the X chromosome, and recombinant DNA techniques have already been used to achieve prenatal diagnosis. Such studies, though still experimental, are now offered in selected centers to members of large families with appropriate restriction enzyme polymorphisms. More routinely, the option of prenatal diagnosis for fetal sexing can be offered to high-risk families who elect to avoid the lethal illness by having only girls.

Becker's muscular dystrophy. About one tenth as common as Duchenne's muscular dystrophy, Becker's resembles it in many ways. It is also X-linked recessive, but thought to be caused by mutation on a different gene or allele. Clinically, it is a slow-motion copy of Duchenne's muscular dystrophy, eventually leading to comparable debility. Onset can be in the first or second decade; progression is slower, but similar (with pseudohypertrophy of the calves and other shared features). The range of clinical expression is wider. Patients are often able to walk until age 25 or later, and survival past 35 is common. Prenatal sex determination is available for high-risk pregnancies.

- Inheritance generally is the same as in Duchenne's muscular dystrophy, and approximately the same proportion (5%) of carriers have some muscle weakness. Only about half of them, however, have elevated CK, making carrier testing less reliable.
- One difference in inheritance patterns is that mildly affected males do reproduce. Since they do not pass on their X chromosome to their sons, none of their sons are affected. All of their daughters inherit the X chromosome with the mutant gene and are obligate carriers.

Scapulohumeral dystrophy (Emery-Dreifuss). Notable for early contractures, particularly of the elbows, it is among the very rare X-linked muscular dystrophies.

Limb-girdle muscular dystrophy. This is thought to be genetically heterogeneous. Onset can occur from childhood through early adulthood, beginning with weakness in the shoulder girdle and/or the pelvic girdle, or in selected muscle groups. Progression tends to be slow, with incapacitation only after 20 years or more. Life expectancy is shortened somewhat. Diagnosis is complicated by the fact that the disorder may be difficult to distinguish from Becker's muscular dystrophy, chronic spinal muscular atrophy, and facioscapulohumeral syndrome.

In most cases inheritance is autosomal recessive (although individual families with dominant inheritance have been reported). No carrier testing is available, nor is prenatal diagnosis.

- There is a 25% risk of recurrence for each child, (male or female) when both parents are carriers.
- Unaffected siblings of an affected individual have a 67% risk for being carriers.
- All children of an affected person are carriers. However, the risk for occurrence of the disorder in the next generation is negligible, except in consanguineous matings.

Congenital muscular dystrophy. Also a very rare muscular dystrophy, it is distinguished by prenatal onset, which may come to the mother's attention through decreased fetal movements in the latter part of pregnancy. A floppy infant is born, perhaps with arthrogryposis, and unable to suck or swallow. Many die in infancy, but a few have been known to survive to reproductive age. Inheritance is autosomal recessive with a 25% recurrence risk for each pregnancy of the carrier parents.

Facioscapulohumeral muscular syndrome. This is another extremely rare condition. Characteristically, facial muscles are involved. Inability to pucker the lips is common. By adolescence, weakness in the shoulder girdle is usually evident. Variable pathologic findings suggest that a number of different biochemical defects may be responsible for the disorder. However, expression within the same family is also variable, with some affected individuals suffering no disability and others being subjected to early incapacitation. Inheritance is autosomal dominant with a 50% risk for offspring of an affected individual. Penetrance is very high, in spite of the variable expression, allowing diagnosis by signs in otherwise asymptomatic affected individuals. Thus family studies may be necessary to identify all affected individuals and establish recurrence risks. A minimally affected person can have a severely affected child.

Ophthalmoplegic muscular dystrophy is one of various disorders with ptosis and external ophthalmoplegia. Three variants all have autosomal dominant inheritance.

Distal muscular dystrophy. Very rare in the United States, it is a slowly progressive autosomal dominant condition, most notable for the fact that it may resemble Charcot-Marie-Tooth disease and distal spinal muscular atrophy.

Myotonias comprise a special subgroup among the most frequently encountered disorders of the muscle in adults. They are characterized by delayed relaxation of contracted muscle (especially in cold temperatures), as in the delayed opening of a clenched fist or a handshake. Myotonic manifestations may also be found in hyperkalemic and hypokalemic periodic paralysis.

Myotonic dystrophy. A hereditary multisystem disease, it has very variable expression. Prevalence estimates have ranged from 1/20,000 to 1/7,000, but may be even higher, since mildly affected patients may not seek medical help, thus escaping clinical diagnosis.

Predominant among clinical characteristics are myotonia, weakness and atrophy of facial muscles (ptosis, masklike face), neck muscles, and the distal muscles of the extremities. More than 90% of patients have typical lenticular opacities when examined by slit lamp, and pigmentary retinal alterations are common. Other features include gonadal atrophy, menstrual disturbances, cardiomyopathy, GI and gallbladder disturbance, male frontoparietal baldness, nasal speech or dysarthria, infertility, and low plasma immunoglobulin IgG levels, due to accelerated catabolism. Mental subnormality and asocial behavior are found at higher frequency than in the general population. Symptoms are temporarily exacerbated by pregnancy.

The symptomatology is extremely variable. Many individuals do not realize that they are affected, having learned to live with some myotonia without paying attention to it. Others may report gallbladder troubles or GI disturbances or may have cataracts, without raising anyone's suspicion that these represent manifestations of myotonic dystrophy.

While myotonic dystrophy may be diagnosed at any age, the average age of onset has been reported to be in the third decade in index cases, but earlier in their offspring, a phenomenon that may be related to the awareness that the disorder is in the family.

Myotonic dystrophy is inherited as an autosomal dominant disorder.

- There is a 50% risk for children of an affected parent to inherit the gene.
- When the mother is the affected parent and a child has inherited the gene, a chance exists that it may be born severely affected, even when the mother is clinically asymptomatic (see congenital myotonic dystrophy, below).
- Clinically asymptomatic daughters of affected patients can be carefully evaluated for carrier status before reproduction.

Carrier detection (though not 100% reliable) has been conducted by use of electromyography, slit-lamp examination for lens opacities, and electroretinography. A normal electroretinogram at the age of 20 has been reported to exclude carrier status with high probability.

Prenatal diagnosis is expected to become available by restriction enzyme analysis, and has been reported by analysis of linkage to the ABH secretor gene for families with informative genotypes.

Congenital myotonic dystrophy. The special variant of the disorder that occurs in some of the offspring who have inherited the mutant gene for myotonic dystrophy from their mother, it is thought to be the result of the combination of the abnormal intrauterine environment and the fetal abnormality. Expression may be noted prenatally through maternal hydramnios, breech position, and reduced fetal movement. The infant is born with severe neonatal and infantile hypotonia and may have club feet and diaphragmatic flaccidity. Sucking and swallowing reflexes are absent; myotonia appears later. Many children with congenital myotonic dystrophy die in infancy. If they survive, they show delayed psychomotor development and more pronounced mental retardation than expected for myotonic dystrophy. Severe speech difficulties and dysarthria are common.

Another consideration in pregnancies of women with myotonic dystrophy is the potential teratogenicity of some of the medications used to treat the disorder (e.g., diphenylhydantoin).

Rare myotonias have been reported. Among the rare myotonias are myotonia congenita Thomsen (autosomal dominant), which improves with advancing age; an autosomal recessive variant of myotonia congenita; paramyotonia congenita Eulenberg (autosomal dominant), which may be symptomatic only in a cold environment; and chondrodystrophic myotonia (Schwartz-Jampel syndrome, autosomal recessive), in which myotonia is one of many manifestations, including short stature.

Congenital Nonprogressive Myopathies are usually benign. A number of them have been described within the last two decades, since new histochemical and electron microscopic techniques have led to a more profound understanding of muscle pathology. The muscle weakness, as a rule appearing at birth, may be accompanied by skeletal abnormalities, such as dislocated hips or scoliosis. The course may be non-progressive; the hypotonia may improve or may become more severe. Life expectancy is generally within the normal range, although death in infancy or childhood may occur in those cases with compromised respiratory musculature.

Some of these disorders, named according to the pathologic findings are: *central core disease, multicore or minicore disease, myotubular (centronuclear) myopathy, congenital fiber type disproportion, and nemaline or rod myopathy.*

Various patterns of inheritance have been suggested for each of the different types, but genetic counseling cannot be given until a mode of inheritance has been established in a particular family. Since histologic evidence of myopathy without any clinical signs is found occasionally in relatives of clinically affected individuals, muscle biopsies of parents and/or siblings may be informative.

Metabolic Myopathies have been recognized through progress in understanding various metabolic functions. Among them are a number of glycogen storage diseases, including *glycogenosis IIa* (Pompe's disease, autosomal recessive, which may resemble Werdnig-Hoffmann disease) and familial hypokalemic, normokalemic, and hyperkalemic *periodic paralysis* (autosomal dominant conditions with transient attacks of weakness).

Identification of abnormalities in lipid metabolism has led to an improved understanding of some myopathies, particularly those that originate in defects involving mitochondrial function. Among them are *myopathic carnitine deficiency* (childhood onset weakness, mostly in pelvic- and shoulder-girdle musculature), *carnitine-palmitoyl transferase deficiency* (recurrent attacks of weakness and muscle pains), and *systemic carnitiner deficiency* (congenital, slowly progressive muscle weakness with encephalopathic attacks). Autosomal recessive inheritance has been suggested for all of these disorders.

CLINICAL NOTES

The majority of muscle disorders have a genetic basis. Thus, except in the case of a clear inflammatory myopathy, especially in children, the diagnosis of a primary muscle problem suggests an investigation of the family history.

A high index of suspicion for genetic etiology is also raised by the birth of a floppy baby. This is especially so when there are associated anomalies (see Table 14-1 in Chapter 14). When a child has difficulty in swallowing and/or club feet, examination of the mother may lead to the diagnosis of congenital myotonic dystrophy for the child.

Technology in the field of muscle disease is advancing rapidly, therefore, referral to a specialist may best provide the most current benefits.

PROCEDURES FOR DIAGNOSTIC CONFIRMATION

Serum CK
Electromyography

Muscle biopsy for histology

Additional procedures, useful for myotonic dystrophy
- Ophthalmologic studies
- GI motility studies
- Serum immunoglobulins
- ECG

Diagnosis of congenital myotonic dystrophy may depend on eliciting myotonic symptoms in the mother

CONSIDERATIONS IN MANAGEMENT

Genetic
- Establish specific diagnosis of proband
- Pedigree analysis
- Evaluation of asymptomatic at-risk family members, as indicated by diagnosis
- Establish probability of carrier status in female relatives via family history and carrier testing, in X-linked disorders
- Risk counseling
- Investigation of current status of carrier testing and/or prenatal diagnosis by restriction enzyme analysis and recombinant DNA technology, prior to or very early in a pregnancy
- Consideration of prenatal diagnosis for fetal sex determination in X-linked disorders
- Discussion of sharing information with other relatives to encourage carrier studies and risk counseling

Psychosocial, educational, familial
- Family counseling: support for acceptance of poor prognosis; prevention of parental guilt feelings and overprotection; support and guidance for dealing with mental retardation and asocial behavior
- Education for effective home care
- Assessment of family/community resources, e.g., contact with other families with similarly affected children; support groups; special education programs; special therapy; recreational facilities for the handicapped; sheltered workshops; financial assistance; respite care; institutional facilities for advanced cases

Medical
- No specific treatment for the weakness of any form of muscular dystrophy is available
- Quinine, diphenylhydantoin, procainamide for marked myotonic symptoms
- Physical therapy, orthopedic measures, corrective surgery, as indicated
- Treatment of associated symptoms
- Consideration of increased surgical risk for malignant hyperthermia in myotonic dystrophy

SUGGESTED READING

Bakker E, Goor N, Wrogemann K, et al: Prenatal diagnosis and carrier detection of Duchenne muscular dystrophy with closely linked RFLPs. *Lancet* 1985;1(8430):655.

Baraitser, M: *The Genetics of Neurological Disorders,* New York, Oxford University Press, 1982.

Dubowitz V: Muscle disorders in childhood. *Major Problems in Clinical Pediatrics,* Philadelphia, WB Saunders, 1978.

Ionasescu V, Zellweger H: Genetics of neuromuscular diseases, *Handbook G Clinic Neurol* 1979;41:405.

Ionasescu V, Zellweger H: *Genetics of Neurology*, New York, Raven Press, 1983.

Percy ME, Andrews DF, Thompson MW: Duchenmne muscular dystrophy carrier detection using logistic discrimination: serum creatine kinase, hemopexin, pyruvate kinase, and lactate dehydrogenase in combination. *Am J Med Genet* 1982;13:27.

Schortt HG, Omenn GS: Myotonic dystrophy: opportunities for prenatal prediction. *Neurology* 1975;25:789.

Walton Sir JA: *Disorders of Voluntary Muscle,* 4th ed. Edinburgh and London, Churchill Livingstone, 1981.

Chapter 16

Heritable Clotting Disorders

Genetic abnormalities have been delineated at every stage in the coagulation process (Table 16-1). With rare exceptions, heritable clotting disorders cause deficiency or dysfunction of a single clotting factor. Multiple factor anomalies, in contrast, are likely to be acquired (e.g., the result of vitamin K deficiency) and more frequent than congenital forms. In general, an inherited factor deficiency produces spontaneous and prolonged hemorrhages into soft tissue and weight-bearing joints; the bleeding responds less to local therapy than that due to platelet or vascular defects. Symptoms appear early in life, often at birth or in the first decade.

The intrinsic clotting mechanism is activated by vascular damage, one of two convergent pathways ending in clot formation. A ''cascade'' of enzyme-mediated events follows in stepwise succession, each factor interdependent with others (see Figure 16-1). The four disorders discussed below reflect defects of specific factors required in the intrinsic coagulation pathway.

HEMOPHILIA A (FACTOR VIII DEFICIENCY)

ETIOLOGY AND PATHOGENESIS

The term hemophilia describes several hemorrhagic coagulation disorders. Hemophilia A, the classic form, is the most common—incidence is 1/10,000 white male births—and most frequently studied. It is the result of a functionally abnormal or missing factor VIII glycoprotein molecule. Several mutations occur, with different levels of residual clotting activity (VIII:c) and with varying specific activity (clotting activity/immunoreactivity). Affected males generally have a low level of coagulant activity, with normal or elevated levels of factor-VIII-related antigen. The platelet aggregation function is unaffected in hemophiliacs.

GENETIC CHARACTERISTICS AND MODE OF INHERITANCE

Hemophilia A is inherited as an X-linked recessive disorder, affecting males, who have only one X chromosome, almost exclusively.

- Women, having two X chromosomes, are carriers if they have one copy of the gene on one X chromosome.

TABLE 16-1 Summary of Heritable Clotting Disorders

NUMBER	FACTOR	COAGULATION PATHWAY*	DISORDER	MODE OF INHERITANCE	CLINICAL CHARACTERISTICS
I	Fibrinogen	I and E	Afibrinogenemia	AR	Variable: bleeding from umbilicus, gastrointestinal bleeding, severe postsurgical/traumatic bleeding
			Dysfibrinogenemia	AD	Mild: epistaxis, wound dehiscence
II	Prothrombin	I and E	Hypoprothrombinemia Dysprothrombinemia	AR AD	Excessive bruising, bleeding from mucous membranes, posttraumatic bleeding
III	Tissue thromboplastin Tissue factor	E			
IV	Calcium	I and E			
V	Proaccelerin Labile factor	I and E	Labile-factor deficiency parahemophilia	AR	Excessive bruising, bleeding from mucous membranes, menorrhagia, postsurgical bleeding
VII	Proconvertin	E	Hypoproconvertinemia	AR	Bleeding from umbilicus, wide range of hemorrhagic manifestations, variable in severity
VIII	Antihemophilic factor	I	Hemophilia A (classic)	XR	Hemorrhage following circumcision, excessive bruising on ambulation, spontaneous, recurrent bleeding, hemarthroses
			von Willebrand's disease	AD	Epistaxis, excessive bruising
IX	Plasma thromboplastin component Christmas factor	I	Hemophilia B, Christmas disease, PTC deficiency	XR	Clinically indistinguishable from hemophilia A
X	Stuart-Prower factor	I and E	Stuart-Prower deficiency	AR	Bleeding from umbilicus, mucous membranes; excessive bruising
XI	Plasma thromboplastin	I	PTA deficiency	AR	Mild to moderate: epistaxis, posttraumatic bleeding, menorrhagia
XII	Hageman factor, antihemophilic factor D	I	Hageman trait, Hageman factor deficiency	AR, AD	Asymptomatic despite abnormal laboratory findings
XIII	Fibrin stabilizing factor, fibrinase	I and E	Fibrin stabilizing factor deficiency, fibrinase deficiency	AR, XR	Bleeding from umbilicus, GI bleeding, intracranial bleeding, prolonged bleeding following trauma

*I denotes that factor is an intrinsic component, E denotes extrinsic. Key: AR = autosomal recessive; AD = autosomal dominant; XR = X-linked recessive

- The sons of carriers have a 50% risk to inherit the mutant gene and be affected.
- The daughters of carriers have a 50% risk to be carriers like the mother.
- Women are at risk for being carriers if they have one affected son, brother, or maternal uncle, or a sister with an affected son.
- All the daughters of a hemophiliac will be carriers (obligate carrier status).

FIGURE 16-1 Clotting Cascade

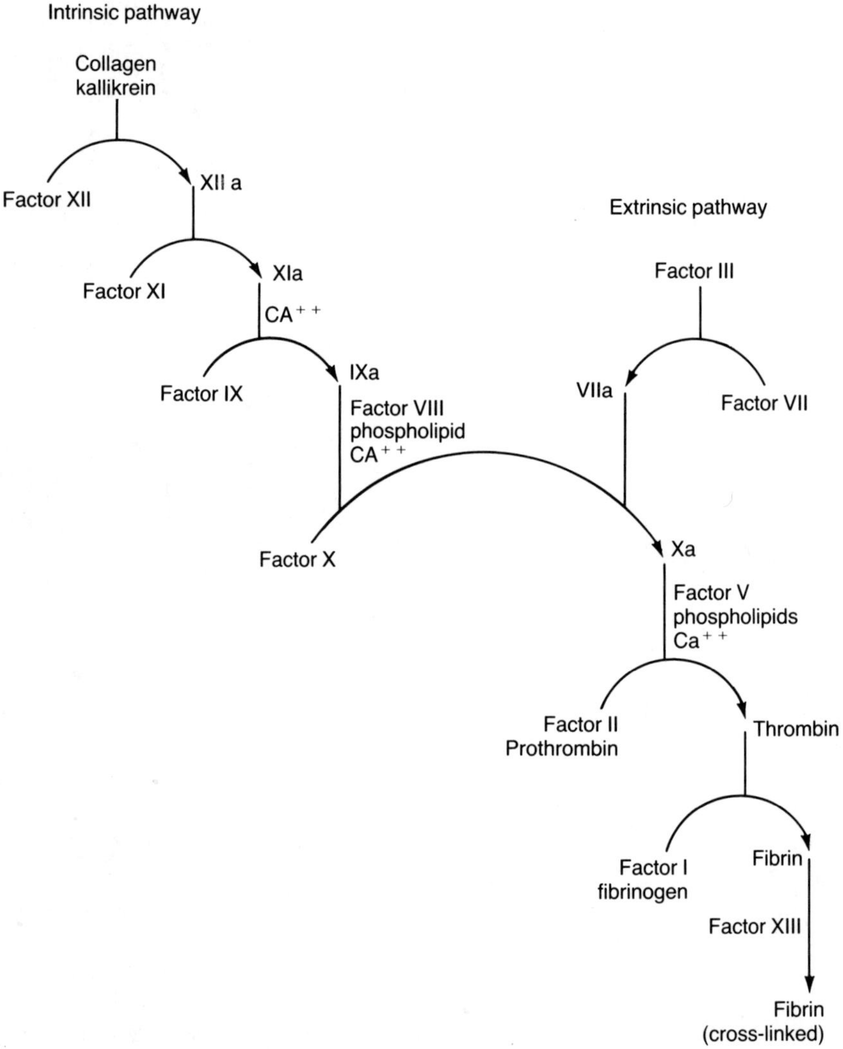

Reproduced, with permission, from the Annu Rev *Biochem*, Vol. 44:799 © 1975; Annual Reviews Inc.

- All the sons of a hemophiliac will be unaffected and cannot transmit the gene.
- Daughters of hemophiliacs will, in general, have a level of factor VIII coagulant activity halfway between the level in the affected father and the level in the normal mother.
- Mildly affected carrier females have been reported. The full-blown disease can occur only in a female who is the daughter of an affected father and a carrier mother, or who has 45, XO karyotype (Turner's syndrome).
- Approximately 50% to 70% of cases have a positive family history; the remainder represent de novo mutations or families in which the disease has passed unexpressed through the female line in preceding generations.

CLINICAL NOTES

The severity and prognosis of the disease is dependent on the level of factor VIII coagulant activity in plasma:

Less than 1%—severely affected
1% to 4% of normal level—moderately severly affected
5% or more of normal level—mildly affected

The level of factor VIII activity is consistent among affected members within a family, and remains constant in an individual patient.

Carrier females may have bleeding tendencies. Factor VIII activity falls rapidly after infusion of factor VIII concentrate (half-life: eight to 12 hours).

About 7% to 12% of hemophiliacs develop a circulating antibody (an IgG) directed against factor VIII. The appearance of this inhibitor is more likely in a severely deficient patient, though it is not precluded in mild cases. The tendency to develop an inhibitor may not be consistent within a family. Transfusions of factor VIII may stimulate antibody formation, but there is no direct relationship between the number of transfusions and the initial formation of antibody.

Other complications of therapy have long included the risk for contracting hepatitis and have become more ominous with the risk for contracting AIDS. However, heat treatment of the factor VIII products is now expected to prevent future infections with the AIDS virus.

Carrier testing, based on factor VIII activity, has not proved to be reliable enough for specific assignment of carrier status. Recent reports, however, describe both direct and indirect recombinant DNA analysis for identifying carriers unequivocally.

Prenatal diagnosis, fetal blood obtained via fetoscopy when fetal sex determination reveals a male karyotype, is available in only a few centers. Other families have elected to carry only female fetuses to term. Now, prenatal diagnosis via recombinant DNA analysis is already becoming available in selected centers. It, as well as the carrier testing, is expected to be informative for at least 50% of hemophilia families, and probably will prove feasible for nearly all cases in the near future. Based on amniotic fluid or chorionic villus cells, the DNA analysis is safer than blood studies via fetoscopy and can be done earlier in the pregnancy. Family studies well before pregnancy may simplify the procedure.

PROCEDURES FOR DIAGNOSTIC CONFIRMATION

Partial thromboplastin time (PTT, elevated in defects in intrinsic pathway)
Specific assay of clotting factor VIII coagulant activity
Laboratory study to determine presence of inhibitor
Correction of prolonged PTT by administration of factor VIII concentrate
Recombinant DNA analysis in selected centers

CONSIDERATIONS IN MANAGEMENT

Genetic
- Establish type of factor VIII deficiency (clinical and laboratory criteria)
- Establish probability of carrier status in female relatives via family history and carrier testing (factor VIII activity/factor VIII-related antigen ratio; available in selected centers); investigation of current status of carrier testing by recombinant DNA analysis
- Determination of bleeding liability in carrier females
- Risk counseling
- Investigation of current status of prenatal diagnosis by recombinant DNA analysis
- Consideration of prenatal diagnosis for fetal sex determination and/or consideration of fetoscopy to rule out factor VIII deficiency in fetal blood sample in male pregnancies at risk
- Discussion of sharing information with other family members to encourage carrier testing and risk counseling

Psychosocial, educational, familial
- Family counseling: maternal guilt and overprotection, paternal rejection (frequent pattern)
- Family education for effective home care
- Assessment of family/community resources (e.g., arrangements for source of medications, eligibility for public funds, support groups)

Medical
- Replacement therapy with factor VIII concentrate (self-infusing lyophilized; cryoprecipitate in mild cases) for symptomatic, prophylactic, and presurgical care
- Drug treatment with vasopressin analogue, DDAVP for mild and moderate cases
- Physical therapy to maintain joint function
- Orthopedic management of hemarthroses
- Control of pain: some analgesics may alter platelet function
- Increased risk of hepatitis and tetanus

Dental
- Prophylactic dental care
- Replacement therapy for dental procedures

VON WILLEBRAND'S DISEASE

ETIOLOGY AND PATHOGENESIS

Von Willebrand's disease is another clotting disorder that is associated with defective synthesis of factor VIII. Although heterogeneity is considerable in expression of this disease, the concentration of the entire factor VIII complex is generally reduced.

Three components of factor VIII can be regarded separately in terms of function and laboratory analysis: coagulant activity; immunologic properties—factor-VIII-related antigen; and ristocetin cofactor activity—an expression of platelet aggregation function. Variant forms of von Willebrand's disease, with a normal range of coagulant activity, factor-VIII-related antigen, and ristocetin cofactor, are known.

GENETIC CHARACTERISTICS AND MODE OF INHERITANCE

Von Willebrand's disease is usually inherited as an autosomal dominant disorder.

- Children of an affected parent have a 50% risk of inheriting the disease: On average, half the offspring of an affected parent will be affected.
- However, genetic heterogeneity apparently has been reported with other forms of transmission and variable clinical expression. Accurate diagnosis and pedigree analysis can help establish the pattern of transmission in specific cases.

CLINICAL NOTES

Von Willebrand's disease resembles mild hemophilia A but can be distinguished by several criteria. A prolonged bleeding time is characteristic, contrary to mild hemophilia A, in which bleeding time is normal. This reflects defective primary hemostasis—namely, formation of the platelet plug.

Response to transfusion differentiates the two diseases. The elevation of factor VIII activity level after transfusion persists for von Willebrand's disease patients, and peaks six to eight hours' postinfusion, whereas it falls rapidly in hemophilia A.

Severity of symptoms is not consistent among affected members of the same family, nor in the same patient at different times. Rarely, patients with factor VIII levels as low as 1% can simulate the hemorrhagic patterns of hemophilia A.

The disease has been detected more often among women than men, possibly because menorrhagia is a frequent problem. Titer of factor VIII in von Willebrand's patients rises during pregnancy.

Bleeding is typically cutaneous and mucosal, with gastrointestinal manifestations the most severe. Postsurgical and post-traumatic bleeding are also typical.

PROCEDURES FOR DIAGNOSTIC CONFIRMATION

Bleeding time
Partial thromboplastin time
Adhesion of platelets to glass beads
Plasma ristocetin cofactor activity assay
Specific assay of clotting factor VIII coagulant activity
Determination of factor VIII-related protein (antigen)

CONSIDERATION IN MANAGEMENT

Genetic
  Exclusion of diagnosis of mild hemophilia A via family history and laboratory findings

Evaluation of asymptomatic at-risk family members
Risk counseling

Psychosocial, educational, familial
Family education for home care, if indicated
Assessment of family/community resources (e.g., arrangements for source of cryoprecipitate)

Medical
Local control of bleeding epidsodes; replacement therapy (Cryoprecipitate must be used, *not* commerical concentrate)
Suppression of menorrhagia through hormonal therapy

HEMOPHILIA B (CHRISTMAS DISEASE; FACTOR IX DEFICIENCY)

ETIOLOGY AND PATHOGENESIS

Hemophilia B is the result of defective synthesis of factor IX (plasma thromboplastin component, or PTC). PTC is a relatively stable, vitamin-K-dependent protein which is not normally consumed during the clotting process.

GENETIC CHARACTERISTICS AND MODE OF INHERITANCE

Like classic hemophilia, hemophilia B is inherited as an X-linked recessive disorder (see risk figures for hemophilia A). The carrier state can be determined occasionally on the basis of low factor IX activity. Normal factor IX levels cannot preclude carrier status, however, and carrier testing is otherwise unreliable. Testing by recombinant DNA analysis is being developed.

CLINICAL NOTES

Patients with hemophilia B develop inhibitors to factor IX following repeated transfusions. In general, patients who do so lack detectable IX antigen levels.

Symptoms of hemophilia A and B are indistinguishable. There is a range of severity corresponding to factor IX levels, and all manifestations of hemophilia A are seen in B. They differ clinically only in the ability of hemophilia B patients to respond to transfusions of stored blood or plasma.

Factor IX has a longer life than factor VIII (approximately 18-20 hours), which allows longer intervals between infusions in replacement therapy. Carriers have greater bleeding liability than hemophilia A carriers.

PROCEDURES FOR DIAGNOSTIC CONFIRMATION

Partial thromboplastin time
Factor IX assay
Laboratory study to determine presence of inhibitor

CONSIDERATIONS IN MANAGEMENT

Genetic
Determination of carrier status, if possible, through factor IX assay
Risk counseling

- Consideration of prenatal diagnosis (see hemophilia A)
- Discussion of sharing information with other family members to encourage risk counseling

Psychosocial, educational, familial

- Family counseling: maternal guilt and overprotection, paternal rejection (frequent pattern)
- Family education for effective home care
- Assessment of family/community resources (e.g., arrangements for source of medications, eligibility for public funds)

Medical

- Medical management parallels that for hemophilia A, with adjustments for differing properties of replacement products

HEMOPHILIA C (FACTOR XI, OR PTA DEFICIENCY)

ETIOLOGY AND PATHOGENESIS

Hemophilia C is the least common form of hemophilia, and generally, the least severe. Factor XI, (plasma thromboplastin antecedent, or PTA) is a glycoprotein, deficient in this disease, that takes part in coagulation early in the intrinsic pathway.

GENETIC CHARACTERISTICS AND MODE OF INHERITANCE

PTA deficiency is seen with greatest frequency, though not exclusively, in individuals of Ashkenazi Jewish descent (eastern Europe). It is inherited as an autosomal recessive disorder. The carrier (heterozygote) is demonstrably half-deficient in factor XI and may be symptomatic.

- If both parents are carriers, there is a 25% risk for each pregnancy that the offspring will be affected (homozygous), and a 50% risk for each pregnancy that the offspring will be a carrier.
- Individuals are assigned to heterozygous or homozygous status on the basis of PTA level:
  - Heterozygote: 20% to 70% of normal
  - Homozygote: less than 20% of normal

CLINICAL NOTES

Clinical expression is variable, generally mild to moderate; but extremes on both sides of the range are seen—asymptomatic and severely affected. Severity does not necessarily correspond to PTA level.

Bleeding tends to be mucocutaneous; menorrhagia may be a problem. The disease may be discovered as an incidental finding on routine blood studies, or appear following surgery, trauma, or dental extraction.

Replacement therapy is used for severe cases. PTA has a relatively long half-life of 30 to 70 hours.

PROCEDURES FOR DIAGNOSTIC CONFIRMATION

Partial thromboplastin time
Factor XI assay

CONSIDERATIONS IN MANAGEMENT

Genetic
- Determine homozygote or heterozygote (carrier) status by assay
- Risk counseling

Psychosocial, educational, familial
- Family education re-avoidance of trauma
- Assessment of family/community resources for source of replacement products in severe cases

Medical
- Control of acute bleeding episodes; PTA replacement, if indicated
- Suppression of menorrhagia through hormonal therapy

SUGGESTED READING

Antonarakis SE, Waber PG, Kittur, SD, et al: Hemophilia A. Detection of molecular defects and carriers by DNA analysis. *N Engl J Med* 1985;313:842.

Firshein SI, Hoyer LW, Lazarchick J, et al: Prenatal diagnosis of classic hemophilia. *N Engl J Med* 1979;300:937.

Hoyer LH, Tuddenham EGD: Excessive bleeding caused by defective coagulation, in Lichtman MA (ed): *Hematology for Practitioners*. Boston, Little, Brown, 1978, pp 297-324.

Inherited Blood Clotting Disorders, Report to WHO Scientific Group, World Health Organization Technical Report Series, No. 504, 1972.

Klein HG, Aledort LM, Bouma BN, et al: A co-operative study for the detection of the carrier state of classic hemophilia. *N Engl J Med* 1977;296:959.

McVerry BA, Voke J, Vicary FR, et al: Ultrasonography in the management of haemophilia. *Lancet* 1977;1:872.

Miller CH, Graham JB, Goldin LR, et al: Genetics of classic von Willebrand's disease. I. Phenotypic variation within families. *Blood* 1979;54:117.

Nachman RL: Von Willebrand's disease and the molecular pathology of hemostasis *N Engl J Med* 1977;296:1059.

Oberle I, Camerino G, Heilig R, et al: Genetic screening for hemophilia A (classic hemophilia) with a polymorphic DNA probe. *N Engl J Med* 1985;312:682

Ratnoff OD, Bennett B: The genetics of hereditary disorders of blood coagulation. *Science* 1973;179:1291.

Ratnoff OD: Antihemophilic factor (factor VIII), *Ann Intern Med* 1978;88:403.

Ratnoff OD: Hemorrhagic disorders: Coagulation defects, in Beeson PB (ed): *Cecil's Textbook of Medicine*. Philadelphia, WB Saunders, 1979.

Chapter 17

Sickle Cell Anemia

Sickle cell anemia, known for its increased incidence in populations of African ancestry, is characterized by episodes of vaso-occlusive crises and chronic anemia, due to a tendency of RBCs to become deformed under conditions of decreased oxygen tension.

Incidence is estimated at one in 625 births in the American black population and prevalence at one in 1,875 American blacks. Sickle cell anemia has also been noted to occur with somewhat increased incidence in populations of other historically malaria-prone areas. This distribution of the sickle cell gene is thought to be the result of a selective advantage conferred on carriers of some hemoglobinopathies, including sickle-cell disease, by an increased resistance to malaria.

The carrier state, or sickle cell trait, is found in approximately one in 10 American blacks.

Carrier testing is routinely available and prenatal diagnosis can be performed on specimens submitted to selected laboratories.

ETIOLOGY AND PATHOGENESIS

The basic defect in sickle cell anemia involves the change of a single amino acid in the hemoglobin molecule, altering its configuration, which causes the cells to sickle and to obstruct blood flow in small vessels. Ischemia of tissues and organs results. Increased cell fragility with increased phagocytosis and splenic sequestration of the fragile cells produce anemia. Aplastic response to infection can also cause or contribute to the anemia.

GENETIC CHARACTERISTICS AND MODE OF INHERITANCE

- Sickle cell anemia (Hgb SS) is autosomal recessive and carries a 25% risk of occurrence for each pregnancy when both parents are carriers (Hgb AS).
- If either or both parents are carriers, the risk to have a carrier child is 50% for each pregnancy.
- When both parents are carriers, unaffected children have a 67% risk of being carriers.
- When sickle-cell disease or carrier status has been diagnosed in a relative, especially a first-degree relative, carrier testing is suggested for other family members to identify carrier couples at risk for having affected offspring.

Although hemoglobin SS is the most common form of the disease, other mutant hemoglobins can be inherited and interact with hemoglobin S to produce clinical disease—for example, hemoglobin C or one of the thalassemias. Carriers of those genes can be detected (see Chapter 18).

Prenatal diagnosis for Hgb SS is available via amniocentesis (at about 16 weeks' gestation) and direct, site-specific recombinant DNA analysis, at selected medical centers. For prenatal diagnosis of a sickle cell gene in combination with another hemoglobinopathy, family studies of DNA polymorphisms by restriction enzyme analysis may be prerequisite. Evaluation for the latter should be made prior to, or very early in the pregnancy.

Chorionic-villus sampling, now becoming available, offers the potential for prenatal diagnosis in the first trimester.

CLINICAL NOTES

Clinical manifestations of sickle cell anemia may include: ischemic organ damage, musculoskeletal infarcts, delay in growth, delay in development of secondary sex characteristics, increased susceptibility to osteomyelitis, opsonization defect leading to pneumococcal sepsis, pulmonary and other infarcts, cardiac high-output failure, CVAs, seizures, ocular abnormalities (proliferative retinopathy, retinal detachment), skin ulcers, and gallstones. Many of these manifestations are associated with severe pain, but the range and severity of symptoms is highly variable.

Carriers do not exhibit clinical manifestations of disease except possibly with extreme hypoxia.

PROCEDURES FOR DIAGNOSTIC CONFIRMATION

Preliminary screening with test tube solubility test for case finding only

Hemoglobin electrophoresis: Cellulose acetate and citrate agar for definitive diagnosis

Tests to quantitate percentage of Hgb F, A_2 and S for diagnostic refinement

Newer techniques, involving isoelectric focusing, have come into use in selected laboratories

CONSIDERATIONS IN MANAGEMENT

- Genetic
  - Screening for trait carriers in at-risk population and among relatives of affected individuals
  - Consideration of further diagnostic testing for individuals with positive results and their mates, to rule out carrier status for sickle-cell disease or another mutant hemoglobin in both partners
  - Counseling for at-risk couples
  - Consideration of prenatal diagnosis when both parents are carriers for a mutant hemoglobin
- Psychosocial, educational, familial
  - Counseling and support for problems encountered with severe and chronic disease, e.g., pain, disability, depression, disruption of schooling
- Medical
  - No effective antisickling agent available
  - Surveillance to detect growth and development delay; damage to tissues and organs; endocrine, hepatic, renal, neurologic, and ocular impairments

Management of vaso-occlusive crises which may include control of pain, hypertonicity, acidosis, hypoxia

Control of anemic crises

Research efforts for effective treatment include studies on marrow transplantation and on mechanisms for reactivation of fetal hemoglobin production

SUGGESTED READING

Chang JC, Kan YW: A sensitive new prenatal test for sickle cell anemia. *N Engl J Med* 1982;307:30.

Dean J, Schecter A: Sickle cell anemia: molecular and cellular basis of therapeutic approaches. *N Engl J Med* 1978;299:752.

Desforges J, Milner P, Wethers D, et al: Sickle-cell disease: Tell the facts, call the fables. *Patient Care* 1978;June 15.

Goossens M, Dumez Y, Kaplan L, et al: Prenatal diagnosis of sickle cell anemia in the first trimester of pregnancy. *N Engl J Med* 1983;309:831.

Kramer MS, et al: Growth and development in children with sickle cell train. *N Engl J Med* 1978;299:686.

Motulsky AG: Frequency of sickling disorders in U.S. blacks. *N Engl J Med* 1973;288:31.

Orkin SH, Little PFR, Kazazian HH, Jr, et al: Improved detection of the sickle mutation by DNA analysis. *N Engl J Med* 1982;307:32.

Schechter AN, Bunn FH: What determines severity in sickle cell disease? *N Engl J Med* 1982;306:295.

Chapter 18

The Thalassemias

A group of complex inherited hematologic disorders, the thalassemias, involve defects in normal hemoglobin production.

They are more prevalent among populations with origins in malaria-prone areas such as the Mediterranean countries, the Middle East, parts of Africa, India, and the Orient. This distribution is thought to be due to an increased resistance to malaria enjoyed by carriers of the milder forms of thalassemia. The best known type of thalassemia is beta thalassemia, which has an increased incidence in populations of Greek and Italian descent. In the United States, it is estimated that 1/2,500 to 1/800 individuals of such ancestry have beta thalassemia major, also known as Cooley's anemia. About 4% in these ethnic groups are carriers (thalassemia minor). Among U.S. blacks, the incidence of thalassemia genes is thought to be higher than in the general population, and may influence expression of the sickle cell gene.

Carrier testing is routinely available and prenatal diagnosis can be performed on specimens submitted to selected laboratories.

ETIOLOGY AND PATHOGENESIS

The main adult hemoglobin, hemoglobin A, contains four globin polypeptide chains—two alpha and two beta..Fetal hemoglobin contains two alpha and two gamma globin chains. A deletion or mutation in the DNA coding for any of these chains can result in deficient or absent globin chain synthesis. In contrast to sickle cell anemia, abnormal polypeptide chains are usually not synthesized. The form of the disorder produced will be determined by the specific globin chain involved. The following are examples of thalassemia:

Beta Thalassemia may be expressed in a number of ways: In *beta thalassemia major* (Cooley's anemia), the homozygous condition, both genes coding for the beta chain are abnormal. Either no beta chains are detectable (beta thalassemia^0), or beta-chain synthesis occurs at a greatly reduced rate (beta thalassemia$^+$). Alpha-chain synthesis is normal. However, having few or no beta chains to pair with, alpha chains form unstable globins which aggregate or precipitate in the red cells. Cell-membrane injury, decreased red cell survival time, hemolysis, and anemia follow.

In *beta thalassemia minor* (carrier-heterozygous condition) the mutation or deletion is present only at one of the two genes coding for the beta chain. There may be mild anemia.

The beta thalassemia gene can be associated with other beta hemoglobin variants such as hemoglobins (Hgbs) S and C. *Beta thalassemia/Hgb S* will present as sickle-cell disease (see Chapter 18); *thalassemia/Hgb C* will present as Hgb C disease.

Alpha Thalassemia, seen mainly in Oriental populations, has several variants caused by deletion(s) on genes coding for alpha-chain synthesis. There are four such genes in most populations; classification is based on the number of deletions present:

One deletion: silent carrier state; reduced globin synthesis
Two deletions: alpha thalassemia trait; morphologic changes in red cells
Three deletions: hemoglobin H disease, unstable beta chains produce mild hemolytic anemia
Four deletions: Hydrops fetalis, lethal in utero or stillborn

Other thalassemias include:
Hemoglobin Lepore—fusion of two genes (also presents as Cooley's anemia)
HPFH (hereditary persistence of fetal hemoglobin), which in some cases compensates for defective beta-chain synthesis.

GENETIC CHARACTERISTICS AND MODE OF INHERITANCE

- Beta thalassemia major is autosomal recessive and carries a 25% risk of occurrence for each pregnancy when both parents are carriers (beta thalassemia minor).
- If either or both parents are carriers, the risk for having a carrier child is 50% for each pregnancy.
- When both parents are carriers, unaffected children have a 67% risk of being carriers.
- If one parent is a beta thalassemia carrier and the other a carrier of the gene for either Hgb S or Hgb C, the risk of occurrence for thal/Hgb S or thal/Hgb C, respectively, is 25%.
- The inheritance of alpha thalassemia is complicated by the fact that four genes are involved. When both parents are alpha-thalassemia-trait carriers, with deletions at two loci on the same chromosome, they have a 50% risk for offspring with hydrops fetalis.

CLINICAL NOTES

Beta thalassemia major produces severe anemia with onset at age 2 or 3 months. It is most severe in beta thalassemia^0, in which no beta globin is synthesized at all. The disorder is characterized by stunted growth; frontal and parietal skull bossing; increased susceptibility to infection (a common cause of death); overactivity of bone marrow, which causes marrow expansion, bone deformities and pathologic fractures; delayed or absent sexual maturation; hepatomegaly; splenomegaly; and a markedly shortened life span—death usually occurs by the second or third decade.

Increased tissue deposits of iron (hemochromatosis), due both to hemolysis and repeated transfusion therapy, result in organ damage. Cardiac failure is another common cause of death.

Beta thalassemia minor generally produces few or no clinical problems. Mild hypochromic anemia may be present which can worsen during illness or pregnancy. This anemia does not respond to iron therapy; therefore, if continued, the therapy may produce an iron overload. It is thus prudent to consider carrier testing for thalassemia in patients with mild anemia who are unresponsive to iron supplementation.

The clinical picture of *other thalassemias* will depend on the nature of the hemoglobin deficiency and the action of compensatory mechanisms.

PROCEDURES FOR DIAGNOSTIC CONFIRMATION

CBC with red cell indices

Peripheral blood smear for morphology

Quantitative hemoglobin electrophoresis (for increased A_2 and F hemoglobin).

Iron: total and iron-binding capacity

Beta thalassemia major will show marked reduction or absence of Hb A, severe anemia (hemoglobin range of 6 to 8 gm/100 ml), gross anisocytosis and poikilocytosis, hypochromia, target cell formation, and basophilic stippling of RBCs.

Beta thalassemia minor will show low mean corpuscular volume (MCV), low mean corpuscular hemoglobin (MCH), reduction in beta-chain synthesis, microcytosis, mild hypochromia, some basophilic stippling, and resistance to osmotic fragility.

CONSIDERATIONS IN MANAGEMENT

Genetic
- Carrier testing in at-risk populations
- Risk counseling
- Consideration of prenatal diagnosis via amniocentesis or chorionic villus sampling and restriction enzyme analysis at selected medical centers. Applicability of this test in a given case may be subject to findings in family studies of DNA polymorphisms. Evaluation is best made prior to or very early in the pregnancy.
- Prenatal diagnosis via fetoscopy, using fetal blood samples, is available at selected centers. The procedure carries a higher risk than amniocentesis or chorionic villus sampling, but may be applicable for cases without informative family polymorphisms.
- Chorionic villus sampling, currently coming into use, offers potential for prenatal diagnosis in the first trimester.

Psychosocial, familial, educational
- Counseling and support for problems encountered in severe chronic disease, e.g., pain, disability, depression, disruption of schooling

Medical
- Transfusions
- Therapy to minimize iron storage
- Control of intercurrent infections
- Research efforts for effective treatment include studies on marrow transplantation and on mechanisms for reactivation of fetal hemoglobin production

SUGGESTED READING

Bank A, Mears JG, Ramirez F: Disorders of human hemoglobin. *Science* 1980;207:486.

Boehm CD, Antonarakis SE, Phillips JA III, et al: Prenatal diagnosis using DNA polymorphisms. *N Engl J Med* 1983;308:1054.

Kazazian HH, Jr, Phillips J, Boehm CD, et al: Prenatal diagnosis of beta-thalassemia by amniocentesis: Linkage analysis, using multiple polymorphic restriction endonuclease sites. *Blood* 1980;56:926.

Mentzer WC: Thalassemia, in Rudolph, AM (ed): *Pediatrics*, 16th ed. New York, Appleton-Century-Crofts, 1977, p 1155.

Rowley PT: The diagnosis of beta-thalassemia trait: A Review. *Am J Hematol* 1976;1:129.

Kan YW: The thalassemias, in Stanbury JB, Wyngaarden JB, Fredrickson DS, et al (eds): *The Metabolic Basis of Inherited Disease*, 5th ed. New York, McGraw-Hill Book Co., 1983.

Chapter 19

Predisposition to Early Coronary Artery Disease

The genetic basis of the well-established familial aggregation of coronary artery disease involves several and probably many heterogeneous predisposing conditions, among which the *hyperlipidemias*, here discussed, are prominent. In addition, many mendelian genetic syndromes are associated with cardiovascular disease (see Table 19-1).

Three major autosomal dominant disorders of lipid metabolism have been delineated. They present as familial hypercholesterolemia (FHC, formerly considered synonymous with hyperlipoproteinemia, type IIa), familial hypertriglyceridemia, and familial combined hyperlipidemia. Different genetic mechanisms are responsible for rarer disorders of lipid metabolism and are also implicated in elevated cholesterol levels not diagnosed as FHC. Certain other genetic factors in atherosclerosis are partially or completely independent of lipid abnormalities. Some of these influence other cardiovascular risk factors, such as high blood pressure, obesity, and diabetes. However, even those risk factors which might appear to be exclusively environmental, such as cigarette smoking or personality type, may involve a genetically determined vascular or metabolic sensitivity to the environmental agent. The anatomic pattern of the coronary arterial vasculature is probably also affected by inherited factors.

A much clearer picture promises to emerge as a result of current research on the identification of the genes underlying lipoprotein function and other factors in atherogenesis.

ETIOLOGY AND PATHOGENESIS

Familial Hypercholesterolemia. The precise etiology of familial hypercholesterolemia has been elucidated: a dominant mutation in the gene coding for cell-surface receptors for low-density lipoproteins (LDLs) results in deficient binding of plasma LDL in heterozygotes and absent binding in the rare homozygote.

LDL is approximately 25% protein and 75% lipid, nearly two-thirds of which is cholesterol. About 70% of the total plasma cholesterol is normally carried in the LDL fraction. LDL functions to transport cholesterol to peripheral tissues, where it interacts with a specific LDL receptor on the cell surface, is internalized by endocytosis and degraded in secondary lysosomes. The cholesterol ester in

TABLE 19-1
Common Genetic Disorders Associated With Cardiovascular Disease

DISORDER	CARDIOVASCULAR FINDINGS	OTHER CHARACTERISTICS	MODE OF INHERITANCE
Cutis laxa	Pulmonary hypertension, peripheral pulmonary artery stenosis	Cutaneous laxity, deep voice, reduced numbers of elastin fibers	AR
Cystic fibrosis	Cor pulmonale	Viscid mucus secretions, pulmonary disease, pancreatic deficiencies, high sweat electrolytes	AR
Friedreich's ataxia	Myocardiopathy	Ataxia, nystagmus, dysarthria, muscle weakness and atrophy, kyphoscoliosis	AR
Glycogenosis	Myocardiopathy	Glycogen-storage diseases; hepatomegaly, low blood glucose, no response to glucagon	AR
Homocystinuria	Coronary and other vascular thromboses	Skeletal abnormalities, ectopia lentis, mental retardation, urinary excretion of homocystine, elevated plasma concentrations of homocystine and methionine	AR
Mucolipidosis III	Aortic valve disease	Early-onset joint stiffness, short stature, coarseness of facial features, mild mental retardation	AR
Hurler's syndrome	Coronary artery disease (CAD), aortic and mitral valve insufficiency	Coarse facial features, depressed nasal bridge, corneal clouding, hepatosplenomegaly, joint stiffness, kyphosis, fine body hair, mucopolysacchariduria, mental retardation	AR
Scheie's, Morquio's, Maroteaux-Lamy syndromes	Aortic valve disease, CAD	Mucopolysaccharidoses, corneal clouding, no serious mental retardation, other features vary	AR
Pseudoxanthoma elasticum	Coronary insufficiency, mitral insufficiency, hypertension	Angioid streaks; yellowish nodular or reticular thickening of skin, especially around neck and in the axilla, due to degeneration of elastic fibers	AR

TABLE 19-1 *Continued*

DISORDER	CARDIOVASCULAR FINDINGS	OTHER CHARACTERISTICS	MODE OF INHERITANCE
Sickle-cell disease	Myocardiopathy, MI, vaso-occlusive crises	Chronic anemia, ischemic organ damage, increased susceptibility to infection	AR
Thalassemia major	Myocardiopathy	Severe anemia, abnormal red cell morphology, pallor, jaundice, hepatosplenomegaly	AR
Ehlers-Danlos syndrome	Rupture of large blood vessels	Joint laxity, skin friability, cutaneous hyperextensibility	AD
Marfan's syndrome	Great artery aneurysms: aortic insufficiency, MI	Tall thin habitus, arachnodactyly, ectopia lentis	AD
Myotonic dystrophy	Myocardiopathy	Early-adult-onset wasting and weakness of facial, neck, and distal limb muscles; ptosis; EMG changes; cataracts	AD
Neurofibromatosis	Pulmonic stenosis, pheochromocytoma with liver hypertension; coarctation of aorta	Neurofibromas, café-au-lait spots, axillary freckling, intracranial tumors, bony deformities, variable expression	AD
Tuberous sclerosis	Myocardial rhabdomyoma, aortic aneurysm	Ash-leaf spots, adenoma sebaceum (angiofibromas), seizures, mental retardation, phakomas	AD
Hunter's syndrome	CAD, valve disease	In male: past infancy, with coarsening facial features, joint stiffness, growth failure, excessive fine body hair, hepatosplenomegaly, hearing loss, mucopolysacchariduria, mental retardation	XR
Duchenne's muscular dystrophy	Myocardiopathy	In male: progressive muscular weakness; onset in first decade; pseudohypertrophy of calf muscles, wasting of others; elevated CK; death, usually in second decade	XR

(Continued)

DISORDER	CARDIOVASCULAR FINDINGS	OTHER CHARACTERISTICS	MODE OF INHERITANCE
Incontinentia pigmenti	Pulmonary hypertension; patent ductus arteriosus	Mostly females: bandlike pattern of hyperpigmentation; patchy alopecia; dental, ocular, and skeletal anomalies Usually lethal in males	XD

Key: AR = autosomal recessive; AD = autosomal dominant; XR = X-linked recessive; XD = X-linked dominant

LDL is hydrolyzed, and the resultant free cholesterol is utilized by the cell for membrane formation and other purposes.

While cellular requirements for cholesterol could be met by endogenous cholesterol biosynthesis in the cells, the primary source is exogenous cholesterol delivered in the form of LDL, which turns off endogenous synthesis at the HMG-CoA-reductase step. Heterozygotes with familial hypercholesterolemia maintain intracellular cholesterol homeostasis only at a greatly elevated plasma LDL level, which results from decreased clearance from the plasma.

The high LDL leads to increased LDL flux into arterial walls, by pathways independent of the LDL receptor, and this increased LDL accumulation acts as an atherogenic agent. (The condition does not, of course, account for all cases of elevated cholesterol, which may be due to a polygenic mechanism or other genetic or environmental predisposing factors.)

Familial Hypertriglyceridemia. By contrast, the pathogenesis of familial hypertriglyceridemia is poorly understood. Overproduction and/or underutilization of triglycerides occurs, and there appears to be heterogeneity of mechanisms and mutations.

Familial Combined Hyperlipidemia. Similarly, the mechanism of familial combined hyperlipidemia is not known. The disorder is associated with a variety of plasma lipoprotein phenotypes within a single family or even in one individual at different times. Approximately one third of patients have elevated cholesterol, one third elevated triglycerides, and one third have both. Some researchers also find consistent elevations in apolipoprotein B levels.

Broad-Beta Disease. Among the rare autosomal recessive disorders of lipid metabolism, there is one which predisposes to early coronary heart disease, peripheral vascular disease and frequent tuberous xanthomas. This disorder, called broad-beta disease, is due to a deficiency of apoprotein E3 (or functionally similar E4) caused by the presence of two defective apoprotein E2 alleles, plus some additional genetic and/or environmental factors. The molecular defect in apoprotein E2 has been identified as a single arginine to cysteine amino acid substitution, which greatly affects receptor binding.

GENETIC CHARACTERISTICS AND MODE OF INHERITANCE

Familial hypercholesterolemia is one of the most common clinically important genetic disorders, and is a clear-cut autosomal dominant condition with a 50% risk for first-degree relatives. Rare homozygotes for this mutation born only to parents who are both heterozygotes have extremely severe and early coronary artery disease. The penetrance (expression of elevated lipids) of FHC is very high in childhood and complete by adulthood.

Familial hypertriglyceridemia is also inherited as an autosomal dominant disorder.

Familial combined hyperlipidemia probably has autosomal dominant inheritance, but there is some dispute and much uncertainty.

Elevated cholesterol is more often due to a polygenic mechanism than to these dominant disorders. Such a mechanism was inferred by the distribution of cholesterol levels in relatives of patients with hypercholesterolemia who could not be classified as having the monogenic disorders.

Contrary to FHC, penetrance of the other disorders is low before adulthood.

CLINICAL NOTES

Familial hypercholesterolemia occurs in about one in every 200 persons. The disorder should be suspected in individuals with plasma cholesterol values greater than 350 mg/dl but may occur with lesser elevations of cholesterol and LDL-cholesterol.

Distinctive findings on physical examination are tendon xanthomas, particularly in the Achilles tendons and the extensor tendons of the hands, sometimes complicated by arthritis or tenosynovitis in the ankles and knees. However, absence of such xanthomas does not rule out the diagnosis. Xanthelasmas and arcus corneae occur but are not specific. Premature development of coronary heart disease is common. Males have a 50% risk of heart disease by age 50 and women about 10 years later. The hyperlipidemia is usually detectable in childhood and has been diagnosed from cord blood samples in a high percentage of affected newborns studied.

Rarely, an individual can inherit the abnormal FHC gene from each parent and thus be homozygous for the abnormality. These individuals have plasma cholesterol in the 600-1,000 mg/dl range, develop xanthomas before age 10 years (sometimes they are even present at birth), suffer coronary disease and myocardial infarction (MI) in the first or second decade of life, and develop aortic stenosis secondary to deposition of cholesterol in the valve leaflets. Prenatal diagnosis has been reported for this form of the disorder.

Unlike those with other forms of hyperlipidemia, patients with hypercholesterolemia do not have a higher incidence of obesity, high blood pressure, hyperuricemia, or diabetes mellitus.

Diet therapy for heterozygotes with FHC produces a 10% to 15% reduction in plasma cholesterol, although homozygotes respond poorly to dietary management. Further reduction can be achieved with drug therapy.

Bile acid sequestrants, such as cholestyramine or colestipol increase catabolism of LDL, and nicotinic acid causes decreased production of LDL from very low density lipoproteins (VLDL) due to increased VLDL synthesis. These agents together can reduce plasma cholesterol in heterozygotes by roughly 40% below levels reached on diet alone. In addition, nicotinic acid (and possibly the bile acid sequestrants) increases high density lipoprotein (HDL) levels, which may confer additional protection against ischemic heart disease in these patients, as an elevated HDL does in those without hyperlipidemias.

Ileal bypass surgery has reduced high cholesterol levels by 15% to 30%, but is still a strictly experimental approach. Portocaval shunts, plasma exchange, and plasmaphoresis with selective removal of LDL have also been used in treatment of homozygotes.

In *familial hypertriglyceridemia* triglycerides become elevated in adolescence or young adulthood. Diabetes mellitus, obesity, hyperinsulinemia due to insulin resistance, high blood pressure, and hyperuricemia are frequently associated with this disorder. The risk for MI is enhanced only modestly, if at all. Pancreatitis can occur. Xanthomas are very unusual, unless accompanying very high triglyceride concentrations caused by high chylomicron levels. Tendon xanthomas are not found.

Familial combined hyperlipidemia is associated with considerable enhancement of risk for MI, but can be diagnosed only by finding significant elevation of both cholesterol and triglycerides in the index patient or a family member. Xanthoma formation is unusual, although yellow discoloration of palmar creases may be seen.

The dietary approach for both familial hypertriglyceridemia and combined hyperlipidemia is similar to that for hypercholesterolemia but with less severe restrictions of cholesterol and saturated fat and more attention to restricting calories to achieve desirable weight. The drug of choice for hypertriglyceridemia is clofibrate or gemfibrozil. Nicotinic acid is less well tolerated but equally effective in lowering triglyceride levels and has the additional benefit of controlling hypercholesterolemia, if present. Alternatively, concomitant hypercholesterolemia may be controlled by addition of bile acid sequestrants to clofibrate or gemfibrozil.

PROCEDURES FOR DIAGNOSTIC CONFIRMATION

Determination of age-adjusted and sex-specific cholesterol and triglyceride levels after a 12-hour fast may be diagnostic (representative values have been published, but must be viewed in terms of "norms" established in each individual laboratory). Note that plasma lipids may be depressed for up to three months after MI; test results in survivors will be informative after that interval. A more detailed profile including division of cholesterol into HDL, LDL and VLDL subfractions can be extremely helpful both in initial diagnosis and in assessment of therapeutic response. Lipoprotein electrophoresis patterns are of limited value.

Determination of LDL receptor activity might be informative. However, this is still primarily limited to research applications.

Family studies might serve to clarify a differential diagnosis or elucidate the mode of inheritance, and may additionally identify asymptomatic relatives in whom early intervention is indicated. Measurement of apoprotein levels, especially Apo AI and Apo B, is now becoming available in certain areas and may be very useful.

Direct identification of gene carriers through recombinant DNA technology lies in the forseeable future.

CONSIDERATIONS IN MANAGEMENT

- Genetic
  - Cholesterol (total and subfractions) and triglyceride determinations for relatives at risk. (Note that in hypertriglyceridemia and in combined hyperlipidemia the diagnosis often cannot be made until adulthood)
  - Family history analysis to investigate mode of inheritance
  - Risk counseling—with attention to associated risk factors for coronary artery disease: obesity, diabetes mellitus, high blood pressure, stress, and, especially, cigarette smoking
  - Consideration of prenatal diagnosis for pregnancies at risk for the severe (homozygous) form of hypercholesterolemia
- Psychosocial, educational, familial
  - Family counseling to reinforce risk information and importance of testing for other family members
  - Education to promote increased compliance with management strategies
- Medical
  - Familial hypercholesterolemia:
    - Heterozygotes:
      - Dietary restriction of cholesterol and saturated fats; enhanced intake of fish high in omega-3 fatty acids
      - Drug therapy with a bile sequestrant, such as cholestyramine or colestipol, or with nicotinic acid
      - Drug therapy with agents to increase catabolism of LDL-cholesterol or inhibit biosynthesis of cholesterol (several agents under study)
      - Ileal bypass surgery (strictly experimental, with many adverse effects)
    - Homozygotes:
      - Dietary measures are essential but always insufficient, as is currently avail able drug therapy. Portocaval shunts, ileal bypass, plasma exchange and plasmaphoresis therapy have been utilized with the latter two showing the greatest clinical utility
      - Polygenic hypercholesterolemia:
      - Treat as FHC heterozygotes
      - Hypertriglyceridemia:
      - Dietary management with restriction of calories (obesity is common) and cholesterol, and reduction of saturated, compared with unsaturated, fats
      - Increase of fish intake (high in omega-3 polyunsaturated fatty acids)
      - Drug therapy with clofibrate, gemfibrozil or nicotinic acid (Note: many additional drugs are under study)
      - Careful control of diabetes mellitus, if present
      - Discontinuation of oral contraceptives or estrogens
      - Consideration of avoidance of alcohol
      - Drug therapy to control hypercholesterolemia, if present

Degree of control indicated may vary depending on presence and severity of personal and/or family history of atherosclerosis

Familial combined hyperlipidemia:

Treatment has not yet been standardized but probably best includes empiric use of as many of the above measures as appear both necessary and sufficient.

SUGGESTED READING

Brown MS, Goldstein JL: Familial hypercholesterolemia, a genetic defect in the low-density lipoprotein receptor. *N Engl J Med* 1976;294:1386.

Dujovne CA, Krehbiel P, Decoursey S, et al: Probucol with colestipol in the treatment of hypercholesterolemia. *Ann Intern Med* 1984;100:477.

Lipid Research Clinics Program: The lipid research clinics coronary primary prevention trial results. I. Reduction in incidence of coronary heart disease. II. The relationship of reduction in incidence of coronary heart disease to cholesterol lowering. *JAMA* 1984;251:351,365.

Motulsky AG: Current concepts in genetics: The genetic hypderlipidemias. *N Engl J Med* 1976;294:823.

Rao DC, Elston RC, Kuller LH, et al (eds): *Genetic Epidemiology of Coronary Heart Disease: Past, Present, and Future*. New York, Alan R. Liss, Inc, 1984.

Chapter 20

Hypertension

An estimated 35 million Americans require treatment to manage hypertension, and an additional 25 million have high blood pressure at a borderline level. Estimates of the prevalence of hypertension in children and adolescents range from 1% to 11%. The increased prevalence of hypertension in black Americans has been well documented, apparently with no sex predilection. However, among whites under 50, hypertension is more frequent in men than in women; this reverses after age 50. Blood pressure levels are thought to be determined by an interaction among strong genetic influences and various environmental, physiologic, and psychosocial factors.

This chapter will outline some of the genetic aspects of the condition.

ETIOLOGY AND PATHOGENESIS

About 90% to 95% of hypertension is classified as primary or essential, i.e., having no identifiable cause. Controversy still exists as to whether a primary, heritable, defect of a kidney is responsible for essential hypertension in some individuals.

On the whole, hypertension in the black population appears to differ from that in whites. The incidence among blacks is higher, as are the morbidity and mortality. It starts earlier in life, and the frequency of strokes, cardiac enlargement, and renal damage is also greater than in whites. On the other hand, coronary artery disease and frequency of myocardial infarcts appears to be higher in whites with hypertension. Black patients seem to respond better to treatment with diuretics than with beta-blockers, which may be more effective in white patients. A postulated reason for the difference is that blacks in hot, dry climates may have evolved a genetic mechanism to conserve sodium, and this mechanism can lead to a problem with sodium excretion in a temperate climate and modern society, with its increased ingestion of sodium.

The remaining 5% to 10% of hypertension is secondary to a host of conditions, both genetic and acquired.

GENETIC CHARACTERISTICS AND MODE OF INHERITANCE

Essential hypertension seems to be inherited as a multifactorial trait. No accurate figures are available for recurrence risk counseling. However, first-degree relatives should be aware of their increased risk so that periodic evaluation, preventive measures, and early treatment, when necessary, can be instituted.

Hypertension associated with genetic disorders is summarized in Table 20-1. Recurrence risks depend on those for the underlying disorder, as determined by its mode of transmission.

CLINICAL NOTES

Since hypertension tends to be asymptomatic, it often goes untreated. Untreated hypertension is the single most important contributor to the occurrence of stroke, and contributes to heart disease and kidney failure. Maternal hypertension increases maternal and fetal risks during pregnancy. Thus it becomes important to take note of all possible predictors for the condition. These include a family history of hypertension and obesity; high normal levels of blood pressure (especially in young individuals); spikes of hypertension in youth or young adulthood; obesity; rapid heart rate; possible glucose intolerance; and possible hyperuricemia. Any or all of these factors may have a genetic basis.

Among the predictors for genetic disorders, a family history of early stroke may be an indication for especially careful surveillance for hypertension, and may be a clue to the presence of an underlying genetic disorder, such as homocystinuria (autosomal recessive) or hyperlipoproteinemia (autosomal dominant).

The presence of neoplasms, particularly those interfering with physical or hormonal renal function, may indicate a genetic basis for the resultant hypertension (e.g., pheochromocytomas in neurofibromatosis or the multiple endocrine adenomatoses, both autosomal dominant).

Adult polycystic kidney disease is often complicated by portal hypertension and berry aneurysms. Some individuals also exhibit hepatic cysts. Onset is usually in the fourth decade and is consistent within families. Inheritance is autosomal dominant.

Childhood or infantile polycystic kidney disease is a condition associated with early hypertension, and may be congenital or have onset in childhood. Renal cysts and hepatic fibrosis both exist with varying severity. Inheritance is autosomal recessive.

Congenital anomalies of the urinary tract are very common and can lead to obstructions resulting in hypertension. While some may be familial, most occur sporadically.

The adrenogenital syndromes are a group of disorders due to inborn errors in steroid metabolism, characterized by ambiguous genitalia. The 11-hydroxylase and 17-hydroxylase deficiencies (non-salt-losing forms) are frequently accompanied by hypertension. Inheritance of all the adrenogenital syndromes is autosomal recessive.

PROCEDURES FOR DIAGNOSTIC CONFIRMATION

Aside from sphygmomanometry, additional screening procedures may help to identify underlying genetic disorders. Among them are:

Chest x-ray, electrocardiogram to assess organ involvement
Urinalysis, serum creatinine concentrations for renal function

TABLE 20-1
Genetic or Familial Disorders Associated with Hypertension

DISORDER	OTHER CHARACTERISTICS	MODE OF INHERITANCE
Coarctation of aorta	Absent or greatly reduced femoral pulses	MF; may be associated with a syndrome, e.g., Turner's syndrome
Polycystic kidney disease		
Adult	Bilateral renal cysts, progressive renal dysfunction, portal hypertension, hepatic cysts	AD
Childhood or infantile	Renal cysts, hepatic fibrosis, congenital or childhood onset	AR
Alport's syndrome	Hematuria, renal dysfunction, nerve deafness, occasional ocular abnormalities	AD, XR
Familial amyloidosis	Amyloid deposition in tissues, hepatosplenomegaly, nephrotic syndrome	AD, AR
Pheochromocytoma, isolated	Tachycardia, headache, diaphoresis, hypertensive crises, increased urinary vanillylmandelic acid	AD in some families
Neurofibromatosis	Neurofibromas, café-au-lait spots, pheochromocytomas, renal artery stenosis	AD
von Hippel-Lindau disease	Retinal and cerebellar hemangiomas, pheochromocytomas	AD
Familial hyperaldosteronism	Dexamethasone suppressible	AD, rare
Acute intermittent porphyria	Drug sensitivity, acute neuropathic attacks, abdominal pain, dark urine, psychiatric manifestations, cutaneous lesions not seen	AD
Neuroblastoma	Diverse features <10% have hypertension	AD in some families
Adrenogenital hyperplasias	Virilization of male and female genitalia, incomplete virilization of male genitalia, electrolyte abnormalities, hypertension in some types	AR
Multiple endocrine adenomatoses	Pheochromocytoma, medullary thyroid carcinoma, parathyroid adenoma, mucosal neuroma, Cushing's syndrome	AD
Wilms' tumor	Nephroblastoma, highest incidence at 2 to 4 years of age	⅓ of cases AD with reduced penetrance
Homocystinuria	Vascular disease, early stroke, dislocated ocular lenses, malar flush, mental retardation, seizures	AR
Familial hypertriglyceridemia	Diabetes, obesity, hyperuricemia, xanthomas unusual	AD

Key: AD = autosomal dominant; AR = autosomal recessive; XR = X-linked recessive; MF = multifactorial

Serum sodium and potassium concentrations for hyperaldosteronism
Serum triglycerides for hyperlipidemia
Retinal funduscopy for accelerated or malignant hypertension
Intravenous pyelography for renal abnormalities
Renal arteriography for renal artery disease
Renal CT scan, renal sonography, e.g., for polycystic kidneys
Urinary and plasma steroid analysis, e.g., for adrenogenital syndromes
Urinary vanillylmandelic acid analysis (VMA), e.g., for pheochromocytoma, neuroblastoma

CONSIDERATIONS IN MANAGEMENT

Genetic
- Establish underlying diagnosis
- Pedigree analysis
- Risk counseling, depending on diagnosis, and considering associated lifestyle risks, e.g., diet, weight, exercise levels, smoking, stress
- Consideration of prenatal diagnosis for pregnancies at risk for some specific syndromes

Psychosocial, educational, familial
- Encouragement of sharing information with relatives regarding importance of testing and institution of preventive measures
- Education to promote increased compliance with management

Medical
- Management of the hypertension as indicated by current practice
- Therapy and/or management for any underlying genetic defect that may have been identified

SUGGESTED READING

Giovanelli G, New MI, Gorini S, (eds): *Hypertension in Children and Adolescents*. New York, Raven Press, 1981.

Nora JJ, Nora AH: *Genetics and Counseling in Cardiovascular Diseases*. Springfield, Ill, Charles C. Thomas, 1978.

Sing CF, Skolnick M, (eds): *Genetic Analysis of Common Diseases: Applications to Predictive Factors in Coronary Disease*. (Proceedings workshop.) Snowbird, Utah. New York, Alan R. Liss, 1979.

Chapter 21

Congenital Heart Defects

A congenital heart defect (CHD) is found in about 8/1,000 live-born infants. Approximately 50% of children dying from a congenital malformation die of a CHD, and 15% of all infant deaths are attributed to CHDs. A CHD can occur in isolation, but 20% to 45% of infants who have one have been reported to have other abnormalities, and 10% exhibit a syndrome complex. The majority of CHDs are thought to have occurred under some degree of genetic control.

ETIOLOGY AND PATHOGENESIS

The embryonic period between the second and eighth week of gestation is especially sensitive to genetic and environmental influence on cardiac development. Structural maldevelopment may be the outcome of various insults at this stage. Postnatal persistence of fetal structures can also lead to a CHD. Etiology involves a continuum of components ranging from primarily genetic to predominantly environmental and is not always clear in a given case. About 2% of CHDs are thought to be due to environmental factors, including rubella, maternal insulin-dependent diabetes, maternal lupus, and exposure to thalidomide and diphenylhydantoin. See Table 21-2 for a listing of known or suspected cardiac teratogens. About 5% are due to recognized chromosomal abnormalities; about 3% to single gene disorders. The remaining 90% of CHDs have no identifiable cause and are generally considered to be of multifactorial origin, with multiple gene-environment interactions.

GENETIC CHARACTERISTICS AND MODE OF INHERITANCE

For purposes of risk assessment, precise diagnosis is necessary, because when CHDs are part of other syndromes, risk of recurrence will be based on that of the syndrome and may be as high as 50%.

When environmental insults cause a CHD, recurrence risks are very low, provided the insult can be avoided in future pregnancies.

Chromosome disorders commonly associated with CHDs are listed in Table 21-3. Table 21-4 provides a summary of single gene disorders involving CHDs, and those syndromes of unknown etiology associated with CHDs are listed in Table 21-5.

Isolated CHDs, not known to be part of a syndrome or environmentally caused, are counseled as being multifactorial, unless the family history indicates a specific pattern of transmission. The estimated general-population risk and re currence risks based on empiric data are summarized in Table 21-1.

TABLE 21-1
Population Incidence* and Empiric Recurrence Risks for Isolated CHDs

AFFECTED RELATIVE	RECURRENCE RISK TO OFFSPRING (%)†
One affected child	2-5
Affected parent (father)	3
Affected parent (mother)	3‡
Two affected children	10-15
More than two affected first-degree relatives	⩾50
Second- or third-degree relatives—isolated case	Risk has not been reported to be increased over the general population risk

*General population incidence: 1%
†Recurrence may be of the same CHD or of a different one.
‡Most reports list recurrence risks for offspring of mothers with CHD at about 3%. Some recent investigations have suggested that they may be higher.

TABLE 21-2
Known or Suspected Cardiac Teratogens

AGENT	HEART DEFECT	COMMENTS
Alcohol abuse	VSD, PDA, ASD	Fetal alcohol syndrome in offspring of chronic alcoholic, spectrum of defects
Estrogens/progestins	VSD, TGA, tetralogy of Fallot	Debatable risk: low if it exists
Lithium	Ebstein's anomaly, tricuspid atresia, ASD	Inadequate data exist to confirm risk
Maternal diabetes	TGA, VSD, coarctation of aorta	Low risk, higher risk for cardiomegaly and cardiomyopathy
Maternal lupus	Fetal heart block	Unknown risk, not uncommon
Maternal phenylketonuria	Tetralogy of Fallot, VSD, ASD	Degree of risk depends on maternal levels of phenylalanine
Rubella	Peripheral pulmonary artery stenosis, PDA, VSD, ASD	Fetal rubella syndrome, defects vary with month of gestational infection
Trimethadione	TGA, tetralogy of Fallot, hypoplastic left side of heart	Known teratogen
Hydantoin	Pulmonary stenosis, aortic stenosis, coarctation of aorta, PDA	Fetal hydantoin syndrome, low risk of heart defects
Thalidomide	ASD, VSD, tetralogy of Fallot	Known teratogen

Key: TGA = transposition of great arteries; VSD = ventricular septal defect; PDA = patent ductus arteriosus; ASD = atrial septal defect

Source: Nora JJ, Nora AH: *Genetics and Counseling in Cardiovascular Disease.* Springfield, Ill., Charles C. Thomas, 1978, p 150.

CLINICAL NOTES

Some general clues which raise suspicion for a CHD in an infant or child include cyanosis, dyspnea, feeding difficulties, excessive perspiring, failure to gain weight, recurrent infections, and squatting after exercise.

Genetic forms of CHDs can occur in isolation, but the presence of other associated anomalies raises suspicion for a syndrome complex. Commonly associated anomalies include renal and urinary tract malformations, limb anomalies, facial clefts, and diaphragmatic hernia.

While evaluation for CHD is generally done by a cardiologist, valuable genetic diagnostic information based on careful prenatal history, family history analysis, and evaluation for extracardiac malformations can be provided by the primary physician. By the same token, family-history analysis may reveal a certain mode of inheritance for defects of unknown etiology, or malformations that are not always transmitted in the same way.

Some common heart defects are summarized below. See Tables 21-3, 4, and 5 for further detail.

Ventricular septal defects are the most common isolated CHDs, but may also be found as part of a syndrome, e.g., chromosomal abnormalities, particularly trisomy 18.

Patent ductus arteriosus is typically associated with fetal rubella syndrome, prematurity, and hypoxia. It is also found in chromosomal syndromes and single-gene disorders such as Treacher Collins syndrome.

Autosomal dominant inheritance has been reported in some families with isolated *atrial septal defects*. These defects are also found in single gene syndromes such as Holt-Oram (heart, limb, and digital and radial defects) or Ellis-van Creveld (dwarfism, cleft lip, and polydactyly).

About one third of *endocardial cushion defects* are associated with Down's syndrome. The other two thirds usually exhibit multifactorial inheritance, although they may be associated with the rare syndromes involving asplenia or polysplenia.

TABLE 21-3
Chromosomal Abnormalities Associated With Congenital Heart Defects

DISORDER	HEART DEFECT
Down's syndrome	Endocardial cushion defect, VSD
Trisomy 18	VSD, PDA, PS
Trisomy 13	VSD, PDA, dextrocardia
Turner's syndrome (45,X)	Coarctation of aorta, hypoplastic left side of heart, aortic stenosis, ASD
Cri-du-chat syndrome (5p-)	VSD, PDA, ASD
4p- syndrome	VSD, PDA, ASD

Key: VSD = ventricular septal defect; PDA = patent ductus arteriosus; PS = pulmonic stenosis; ASD = atrial septal defect

TABLE 21-4
Mendelian Syndromes Associated With Congenital Heart Defects

SYNDROME	HEART DEFECT	OTHER CHARACTERISTICS	MODE OF INHERITANCE
Marfan's	Aortic or mitral valve abnormalities, aortic aneurysm	Skeletal abnormalities, lens dislocation, variable expression	AD
Ehlers-Danlos	Aortic or mitral valve abnormalities, rupture of large and intermediate vessels	Thin skin, easy bruising, bowel perforation	AD, AR
Holt-Oram	ASD, VSD	Upper limb defects, thumb anomaly	AD
Noonan's	PS, others	Turnerlike phenotype, normal chromosomes, short stature, +/− mental retardation, webbed neck	AD
Osteogenesis imperfecta	Aortic regurgitation	Skeletal anomalies	AD, AR
Phakomatoses	Rhabdomyoma, aortic aneurysm	Neurocutaneous tumors Neurofibromatosis Tuberous sclerosis von Hippel-Lindau	 AD AD AD
Ellis-van Creveld	ASD, VSD, single atrium	Dwarfism, cleft lip, polydactyly	AR
Mucopolysaccharidoses	Cardiomyopathy, coronary and valvular deposits of acid mucopolysaccharides	Coarse facies, mental retardation, seizures, skeletal anomalies, hearing loss, hepatosplenomegaly	AR, XR
Treacher Collins	VSD, PDA, ASD	Micrognathia, downslating palpebral fissures, hypoplastic zygomatic arches, dysplastic ears, lower lid coloboma	AD
Pompe's disease	Glycogen-storage disease of heart	Muscle weakness	AR
Friedreich's ataxia	Cardiomyopathy	Truncal and appendicular ataxia, areflexia, muscle weakness	AR, AD
Kartagener's	Dextrocardia	Situs inversus, thick nasal secretions, infertility	?AR
Duchenne's muscular dystrophy	Cardiomyopathy	Muscle weakness	XR
Jervell-Lange-Nielsen	Prolonged QT, sudden death, syncopal attacks	Sensorineural deafness	AR

TABLE 21-4 *Continued*

SYNDROME	HEART DEFECT	OTHER CHARACTERISTICS	MODE OF INHERITANCE
LEOPARD	Pulmonic stenosis, prolonged PR interval	Ocular hypertelorism, genital abnormalities, growth retardation, deafness, variable expression	AD
QT (Romano-Ward)	Conduction defects, prolonged QT, sudden death, syncopal attacks		AD
Idiopathic hypertrophic subaortic stenosis	Obstructive myocardial disease, left ventricular hypertrophy		AD

Key: ASD = atrial septal defect; VSD = ventricular septal defect; PDA = patent ductus arteriosus; PS = pulmonic stenosis; AR = autosomal recessive; AD = autosomal dominant; XR = X-linked recessive

Aortic stenosis may be transmitted as an isolated disorder, or associated with a syndrome, e.g., supravalvular aortic stenosis syndrome, which is characterized by mental retardation, unusual facies, and hypercalcemia.

Idiopathic hypertrophic subaortic stenosis usually occurs sporadically, although autosomal dominant inheritance has been documented. Virtually all individuals have obstructive myocardial disease.

Coarctation of the aorta is three times more common in males than in females. However, it is the most common heart defect in Turner's syndrome, which should be considered when coarctation of the aorta is combined with short stature in a female patient.

Pulmonic stenosis occurs in isolation and may be associated with syndromes such as Noonan's, fetal rubella, or multiple lentigines. Noonan's syndrome resembles Turner's syndrome but is distinguished by normal chromosomes, occurrence in both sexes, possibility of reproduction in females, dominant inheritance, and its characteristic heart defect.

As a higher incidence of *endocardial fibroelastosis* has been observed in colder climates, a viral or infectious etiology has been postulated. Most cases are sporadic, although rare families may show a single-gene inheritance pattern.

Tetralogy of Fallot can occur in isolation, but can be associated with other syndromes, such as Goldenhar's and Klippel-Feil anomaly.

Fetal heart block should be ruled out when a pregnant woman has lupus erythematosus. The risk is unknown, but the defect is not uncommon in such pregnancies.

Hypoplastic left heart is invariably fatal. It is thought to be a common cause of neonatal death in Turner's syndrome.

A prolonged Q-T wave, with syncopal attacks and sudden death in a child with deafness may be a sign of Jervell-Lange-Nielsen syndrome. Since this is a

TABLE 21-5
Syndromes of Unknown Etiology With Associated Congenital Heart Defects

SYNDROME	HEART DEFECT	OTHER CHARACTERISTICS	MODE OF INHERITANCE
Pierre Robin anomaly	Variable, ASD, PDA, VSD	Cleft palate, micrognathia, glossoptosis	Etiologically heterogeneous, can be associated with Stickler's syndrome (AD), Camptomelic syndrome (AR)
Asplenia and polysplenia	Dextrocardia and other complex defects	Situs inversus, visceral anomalies	Most cases sporadic, some appear AR
VATER association	Variable	Renal, vertebral, anal, esophageal, and radial-limb defects	Sporadic
Klippel-Feil anomaly	Septal defects, tetralogy of Fallot	Fusion of cervical vertebrae	Usually sporadic, subtypes may be AD, AR
DiGeorge's syndrome	Anomalies of great vessels and others, VSD	Absent thymus and parathyroid, ear and nose abnormalities	Sporadic
Goldenhar's syndrome	Variable, tetralogy of Fallot, VSD, ASD	Eye, ear, and facial anomalies	Sporadic
Williams' syndrome	Aortic or pulmonic stenosis	Elfin facies with full lips, stellate iris pattern; hypercalcemia; mental retardation	Sporadic

Key: VSD = ventricular septal defect; PDA = patent ductus arteriosis; ASD = atrial septal defect; AD = autosomal dominant; AR = autosomal recessive; Sporadic = unknown etiology, single occurrence in a family, thought to be nongenetic

recessive condition, cardiac evaluation of siblings with hearing loss may be indicated.

Mitral valve prolapse (MVP), which has an estimated incidence of 5% to 10% in the general population, is more common in women than in men. Prevalence is estimated at 1% to 2% in children. Although very common, it is not always diagnosed. MVP appears to follow an autosomal dominant inheritance pattern, variably expressed for sex and age. Evaluation of other family members may lead to diagnosis and potential prevention of complications such as endocarditis and progressive mitral regurgitation. MVP may also be a component of Marfan's syndrome, osteogenesis imperfecta, fragile X syndrome, and Ehlers-Danlos syndrome, although diagnosis depends on the finding of other typical features. MVP has been reported in association with Duchenne's and Becker's X-linked muscular dystrophies and myotonic dystrophy.

Pregnancy in a woman with a CHD poses several risks. Some investigators have reported that recurrence risks for affected offspring seem to be higher in females than in males with CHDs. Women with nonbiosynthetic valve prostheses using anticoagulant therapy, especially Warfarin, may pose a teratogenic risk to the fetus. In addition, pregnancy itself may exacerbate a cardiac condition. The size of the infant may be decreased in mothers who are cyanotic, and fetal mortality is reported to be increased in mothers with compromised cardiac function.

Prenatal diagnosis may be accomplished using a variety of techniques, dependent on the type of CHD. For CHDs associated with chromosomal syndromes, fetal chromosome analysis is an option. Fetal cardiac ultrasonography and echocardiography are available at specialized centers, and may be useful for the detection of some lesions.

PROCEDURES FOR DIAGNOSTIC CONFIRMATION

- Hemogloblin, hematocrit
- Chromosome analysis (where indicated, e.g., multiple congenital abnormalities)
- Roentgenography, cardiac fluoroscopy
- Electrocardiography
- Referral for cardiologic evaluation for the following procedures, as indicated:
  - Echocardiography
  - Catheterization
  - Angiocardiography, aortography
  - Balloon septostomy (Rashkind's procedure)

CONSIDERATIONS IN MANAGEMENT

- Genetic
  - Establish diagnosis
  - Pedigree analysis
  - Recurrence risk assessment, risk counseling
  - Consideration of prenatal diagnosis for pregnancies at risk, if available (cytogenetic analysis, fetal ultrasonography and/or echocardiography)
  - Discussion of sharing information with other family members to encourage evaluation, e.g., MVP
- Psychosocial, educational, familial
  - Family counseling and support in regard to limitations of the affected individual
  - Assessment of family/community resources, such as availability of special equipment, support groups, special recreational programs
  - Psychological preparation of child and family for surgery
- Medical
  - Surgical repair
  - Drug therapy
  - Prophylactic antibiotic administration before dental work, oral surgery, or genitourinary procedures
  - Periodic evaluation in light of increased risk for developing scoliosis
  - Management of associated defects

SUGGESTED READING

Elias S, Yanagi RM: Cardiovascular defects, in Schulman JD, Simpson JL (eds): *Genetic Diseases in Pregnancy: Maternal Effects and Fetal Outcome,* New York, Academic Press, 1981.

Elkayam U, Gleicher N, (eds): *Cardiac Problems in Pregnancy: Diagnosis and Management of Maternal and Fetal Disease,* New York, Alan R. Liss, 1982.

Morgan BC, (ed): Symposium on pediatric cardiology. *The Pediatrics Clinics of North America,* vol 25, no 4. Philadelphia, WB Saunders, 1978.

Nora JJ, Nora AH: *Genetics and Counseling in Cardiovascular Disease.* Springfield, Charles C. Thomas, 1978.

Rosenquist GC, Bergsma D, (eds): *Morphogenesis and Malformation of the Cardiovascular System,* vol 14, no 7. (Birth Defects Original Articles Series.) New York, Alan R. Liss, 1977.

Chapter 22

Short Stature

Environmental and genetic factors influence the skeleton from the prenatal period through skeletal maturation in the teen years. Since most initial visits for short stature involve patients with normal variations, diagnosis begins with documentation of significantly retarded growth and/or disproportionate stature. The use of growth curves with attention to ethnic and family background, as well as anthropomorphic measurements, will help identify the patient with truly abnormal short stature. Precise diagnosis of both the newborn with retarded linear growth, and the child or adolescent with short stature, is essential for management and for assessment of recurrence risks.

ETIOLOGY AND PATHOGENESIS

Abnormally short stature may result from a primary abnormality of skeletal organization, or of growth and development, or from extraosseous disturbance. Some forms of short stature present at birth; others have later onset. Generally, abnormally short individuals with normal proportions have extraosseous defects—nutritional, teratogenic, endocrine, and cytogenetic. Those with disproportionate short stature usually have skeletal dysplasias of known (e.g., metabolic, single gene) or, more frequently, unknown pathogenesis. Disorders with similar clinical presentations may have different pathologic origins. Evaluative approaches to a newborn with retarded linear growth or a child with short stature are outlined in Figures 22-1 and 2.

GENETIC CHARACTERISTICS AND MODE OF INHERITANCE

Examples of genetic forms and sporadic forms of short stature are listed in Table 22-1.

Short people often marry other short people. Such selective mating can complicate the genetic picture for couples when:

- Both members have the same genetic condition (with increased risk of recurrence and risk for increased severity in offspring)
- Each parent has a different hereditary condition (where offspring may have no increased risk, or may be at risk for either or both disorders)
- One member has a genetic condition and the other is on the short end of the normal height distribution (where familial short stature may affect expression of any genetic disorder that offspring may inherit)

FIGURE 22-1 Evaluative Approaches to Short Stature

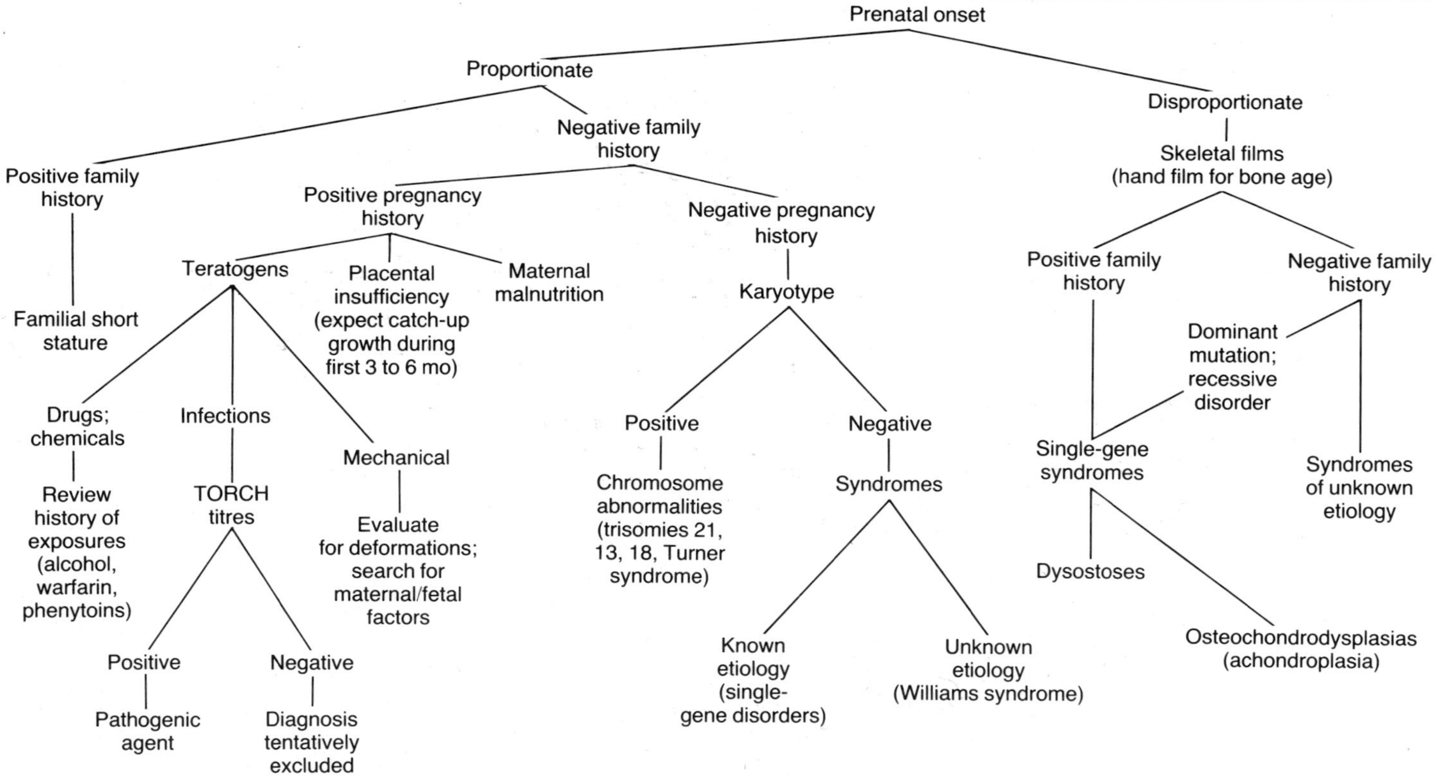

FIGURE 22-2 Evaluative Approaches to Short Stature

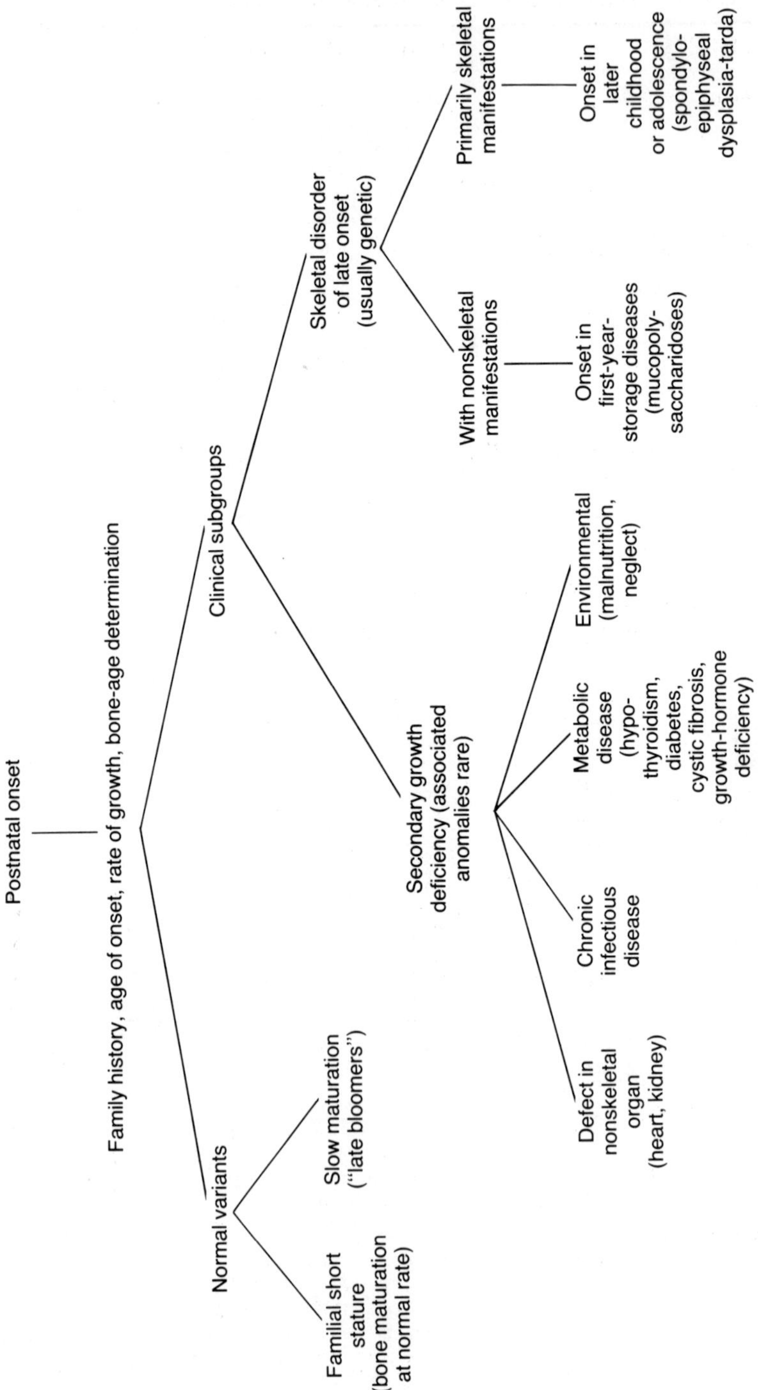

TABLE 22-1
Selected Genetic Disorders That Include Short Stature

DISORDER	OTHER CHARACTERISTICS	INHERITANCE*
Achondroplasia	Large head, frontal bossing, depressed nasal bridge, short extremities/rhizomelia, short trident hands, lumbar lordosis, oral malocclusion, orthopedic and neurologic complications	AD
Thanatophoric dysplasia	Large head, frontal bossing, flat nasal bridge, redundant skin on extremities, markedly narrow thorax, micromelic bowed limbs, hypotonia, severely deficient ossification, polyhydramnios, cloverleaf skull, incompatible with life	Heterogeneous, sporadic AR, with cloverleaf skull
Camptomelic dysplasia	Large head, short anteriorly bowed lower limbs, small chest, protuberant abdomen, small mouth and chin; infant death; genetic males may have female genitalia	?AR, heterogeneous
Acrodysostosis	Midfacial hypoplasia, hypoplastic nose, prognathism, stubby digits, acromelic shortening, mental retardation, growth failure, arthritis	Sporadic, AD
Diastrophic dysplasia	Pinnal calcification (cauliflower ear, swelling) hitchhiker-thumbs, severe clubfoot, mesomelia, progressive scoliosis/lordosis, cleft palate, infant death (in 25%)	AR
Hypochondroplasia	Normal facies, prominent forehead, macrocephaly, lumbar lordosis (30%), short broad hands and feet, short limbs, shortened long bones	AD
Dyschondrosteosis	Madelung's deformity of the wrist, mesomelia, early preadolescent epiphyseal fusion, short hands and feet, short bowed radius	AD
Spondyloepiphyseal dysplasia	Short trunk, malar hypoplasia, myopia, retinal detachment, cleft palate, lumbar lordosis, kyphoscoliosis, epiphyseal mineralization, muscle weakness in legs	2 forms (congenita and tarda), both AD
Chondrodysplasia punctata, Conradi-Hünermann type	Stippled epiphyses, congenital cataracts, (25%), hypertelorism, prominent forehead, flat face, flat nose and anteverted nares, sparse coarse hair, scaly atrophic skin, joint contractures	AD

TABLE 22-1 *Continued*

DISORDER	OTHER CHARACTERISTICS	INHERITANCE*
Chondrodysplasia punctata, recessive type	Stippled epiphyses, severe rhizomelic shortening, congenital cataracts (75%), saddle nose, cleft palate, failure to thrive, microcephaly, eczematoid dermatitis	AR
Cleidocranial dysplasia	Hypo/aplastic clavicles, increased range of shoulder motion, frontal and parietal bossing, broad depressed nasal bridge, supernumerary teeth, delayed ossification, hypertelorism, cleft palate	AD, variable expression
Hypophosphatasia infantile—severe, childhood onset, adult-most mild	Deficient alkaline phosphatase and phosphoethanolamine, gradual craniosynostosis, rachitic chest, osteoporosis	AR
Osteogenesis imperfecta type I and IV	Mild bone fragility, dentinogenesis imperfecta; type I: blue sclerae and occasional deafness	AD
type II and III	Severe bone fragility (type II: lethal, with crumpled bones), severe bowing of long bones; prenatal diagnosis available	AR
Hadju-Cheney syndrome (arthrodentoosteodyspla-sia)	Short, clubbed distal digits with osteolysis, occipital prominence, thick coarse hair, early loss of teeth, small mouth, midface hypoplasia, large nose, progressive kyphosis, 6th-nerve palsy, multiple fractures, basilar compression	AD
Oculomandibulofacial syndrome	Proportionate dwarfism, small face, beaked pinched nose, congenital cataracts, microphthalmia, small double chin, sparse hair, prominent scalp veins, hypodontia (occasionally natal teeth), hypogenitalism	Sporadic
Russell-Silver syndrome	Infants very small for gestational age, body assymetry, small triangular face with relatively large skull, micrognathia, prominent eyes, frontal bossing, café-au-lait spots, precocious sexual development (30%), mental retardation (30%)	Sporadic

(*Continued*)

DISORDER	OTHER CHARACTERISTICS	INHERITANCE*
Williams' syndrome	Wide depressed nasal bridge, anteverted nares, flat midface, thick lips, wide mouth, full cheeks, stellate iris pattern, heart murmur, hypoplastic nails, mental retardation	Sporadic
Aarskog's syndrome	Hypertelorism, broad forehead, widow's peak, prominent metopic ridge, ptosis of eyelid(s) (50%), short upturned nose, low-set cup-shaped ears, short, broad hands and feet, retarded bone age, shawl scrotum, cryptorchidism, mild mental retardation frequent, normal birth size; females less severely affected than males	?XD
Turner's syndrome	Female with 45,X chromosome constitution, webbed neck, shield chest, heart and renal anomalies, hearing loss, cubitus valgus, hypoplastic nails, peripheral lymphedema in infancy, sexual infantilism, primary amenorrhea	Chromosomal
Noonan's syndrome	Male or female with phenotype similar to Turner's syndrome, but no chromosomal abnormality; variable fertility	Heterogeneous (most sporadic)
Rubenstein-Taybi syndrome	Beaked nose, microcephaly, broad thumbs and toes, low-set ears, grimacing smile, mental retardation, hypotonicity	Sporadic
Storage diseases	Variable features include: failure to thrive or regression, coarse features, hepatomegaly, splenomegaly, neurologic deterioration, progressive psychomotor retardation; prenatal diagnosis available for some (see Chapter 33)	Variable, most AR

Key: AD = autosomal dominant (50% risk to each offspring of an affected individual); AR = autosomal recessive (25% risk to each offspring of unaffected carrier parents); XD = sex-linked dominant (50% risk to each offspring of an affected female, 100% risk of an affected male/daughters, no risk to sons); Sporadic = a single occurrence in a family is usual

Examination of seemingly unaffected parents of affected children may disclose subclinical signs of dysplasia, particularly in dominant disorders with variable expression (e.g., osteogenesis imperfecta, type I).

CLINICAL NOTES

Precise body measurements, such as upper- to lower-limb-segment ratios, hand measurements, and foot measurements are diagnostically useful, particularly in individuals with skeletal disproportion.

Not all children with chromosomal or other genetic causes for short stature are strikingly abnormal in appearance.

Teratogenic exposures, such as alcohol, may be discovered only through direct, extensive, and repeated interviews.

PROCEDURES FOR DIAGNOSTIC CONFIRMATION

Review of previous x-ray films (radiographic characteristics of many dysplasias change with time)

Complete series of skeletal films including the hand for bone age

Analysis of associated clinical findings and/or complications

Family-history analysis, including careful evaluation of parents when an autosomal dominant condition with variable expression is suspected

Pregnancy history

Endocrine evaluation

Blood chemistry

Cytogenetic studies

Immune-function studies

Biochemical studies (e.g., suspected storage disease)

CONSIDERATIONS IN MANAGEMENT

- Genetic
  - Family-history analysis
  - Pregnancy-history analysis
  - Establishment of diagnosis and assessment of recurrence risk
  - Consideration of prenatal diagnosis when available for pregnancies at risk (e.g., storage diseases, some skeletal dysplasias, chromosome disorders; see Chapter 3)
- Psychosocial, educational, familial
  - Family counseling
  - Assessment of community resources, e.g., physical therapy, special education
  - Referral to support groups
- Medical
  - Dependent on type and severity of clinical manifestations
  - For skeletal dysplasias, particular emphasis should be placed upon avoiding risk factors that may interfere with growth or damage developing bones and joints

SUGGESTED READING

Beighton P: *Inherited Disorders of the Skeleton*. London, Churchill Livingstone, 1978.

Maroteaux P: *Bone Diseases of Children*, Philadelphia, Lippincott, 1979.

Rimoin DL, Hall J, Maroteaux P: *International Nomenclature of Constitutional Diseases of Bone with Bibliography*, vol 15, no 10. (Birth Defects Original Article Series.) White Plains, N.Y., March-of-Dimes Birth Defects Foundation, 1979.

Spranger JW, Langer LO, Jr, Wiedmann JR: *Bone Dysplasias: An Atlas of Constitutional Disorders of Skeletal Development*, Stuttgart, Gustav Fischer-Verlag, 1974.

Chapter 23

Congenital Limb Defect

Limb defects became newsworthy when thalidomide was exposed as the powerful teratogen responsible for a sudden jump in the incidence of phocomelia. Actually, limb abnormalities, encompassing a profusion of possible defects and a multiplicity of causes, are not uncommon, and often constitute an obvious handicap. Many isolated defects, such as polydactyly or clubfoot, can usually be readily repaired and have little permanent clinical significance; others are more serious or may be part of a syndrome with broader implications.

Selected isolated defects, and the most common syndromes in which limb defects are a prominent feature, are presented here. (Conditions in which limb length is reduced or disproportionate, but gross limb structure is normal, are discussed in Chapter 22).

ETIOLOGY AND PATHOGENESIS

Structural limb anomalies include reduction defects, dysplasias, supernumerary growth, and other abnormal developments. They may be malformations, deformations, or disruptions (see Chapter 44), and may result from a variety of causes, which are not always identified. Among the known ones are teratogens, single-gene defects, multifactorial conditions, and chromosome abnormalities. During embryogenesis, the limb differentiates sequentially from the proximal to the distal end. Thus, the position of a limb defect often pinpoints the gestational period during which the insult, genetic or nongenetic, exerted its influence. This can be especially helpful when the obstetric history is reviewed for evidence of teratogenesis (e.g., infection, hyperthermia, maternal medication).

Since limb-bud development is induced by the mesonephros, aberrant renal development is often coupled with limb dysplasia. While many are primary malformations, limb deformities can also arise from secondary destruction (e.g., amniotic-band syndrome) and from mechanical restriction of fetal movement (e.g., clubfoot).

GENETIC CHARACTERISTICS AND MODE OF INHERITANCE

Whether or not an isolated limb defect is considered genetic may depend on a specific diagnosis, or ruling out nongenetic causes. Pedigree analysis may reveal a pattern of inheritance with attendant recurrence risks within a given family.

When a defect is part of a syndrome, recurrence risks depend, of course, on the genetic nature of the underlying defect. Characteristic examples of genetically

TABLE 23-1
Differential Diagnosis of Limb Defects: Selected Disorders

DISORDER	LIMB DEFECTS	OTHER CHARACTERISTICS	INHERITANCE
Holt-Oram syndrome	Absent thumb, fingerlike thumb	Heart defects (ASD, VSD)	AD, variable expression
Moebius syndrome	Hypodactyly to amelia	VI-VII cranial-nerve palsies → masklike facies, Poland's anomaly, mental retardation	Sporadic, AD
Ectrodactyly	Partial to complete absence of central digit(s), lobster-claw deformity	–	Sporadic, may be AD, variable expression
Ectrodactyly, ectodermal dysplasia clefting syndrome	Partial-to-complete ectrodactyly	Absent lacrimal puncta, cleft lip/palate, sparse hair	AD, variable expression incomplete penetrance
Robert's syndrome	Tetraphocomelia, ectrodactyly	High neonatal mortality, bilateral cleft lip/palate, ocular defects, mental retardation, marked growth deficiency	AR, sporadic
de Lange's syndrome	Proximal thumbs, limb hypoplasia, ectrodactyly, elbow contractures	Growth deficiency, mental retardation, hirsutism, synophrys, urogenital anomalies	Most are sporadic
Poland's anomaly	Syndactyly, brachydactyly, symphalangism	Unilateral hypoplasia of anterior chest wall, hypoplasia/absent nipple	Sporadic
Amniotic-band syndrome	Amputations, deformations	Facial deformations (clefts, constrictions), oligohydramnios	Sporadic
Acrocephalosyndactyly (Apert's) syndrome	Cutaneous/osseous syndactyly	Craniosynostosis with turribrachycephaly, mental retardation, cleft palate	AD
Saethre-Chotzen syndrome	Partial cutaneous syndactyly, brachydactyly	Craniosynostosis, strabismus/ptosis, dysplastic ears, beaked nose	AD, variable expression
Fanconi's pancytopenia	Hypoplastic/absent thumb, supernumerary/bifid thumb, hypoplastic radius	Skin hyperpigmentation urogenital anomalies, childhood-onset anemia, chromosome breakage	Sporadic, AR, AD

(Continued)

DISORDER	LIMB DEFECTS	OTHER CHARACTERISTICS	INHERITANCE
Thrombocytopenia-aplastic radius syndrome	Radial aplasia (thumbs are present), upper limb hypoplasia	Thrombocytopenia, cardiac anomalies	AR
Orofaciodigital syndrome, Type I	Syndactyly/ brachydactyly	Midline facial clefts, mental retardation (in 50%), failure to thrive	XD, lethal in male conceptions
Trisomy 13	Postaxial polydactyly	Microcephaly, microphthalmia/eye anomalies, cleft lip/palate, renal anomalies, heart defects	Chromosomal
Meckel's syndrome	Polydactyly (usually postaxial)	Microcephaly (encephalocele common), cleft palate, polycystic kidneys, stillbirth, neonatal death	AR
Polysyndactyly	Preaxial polysyndactyly	–	AD, variable expression
Isolated polydactyly	Postaxial polydactyly	–	AD
Diastrophic dysplasia	Hitch-hiker thumbs, severe clubfoot	Pinnal calcifications (cauliflower ear), very short stature, scoliosis, cleft palate, infant death	AR, variable expression
Multiple exostoses	Bony exostoses (cartilage-capped)	Osteoma	AD
Tricho-rhino-phalangeal syndrome, Type I	Clinobrachydactyly	Sparse, brittle scalp hair, bulbous nose, tented alae, kyphoscoliosis, small stature	AD, AR
Isolated arthrogryposis	Distal joint contractures	Skin dimples	Sporadic, AD recurrence risk 5%
Arthrogryposis association	Joint contractures	Skin dimples, CNS dysfunction, organ malformations	Inheritance varies

Key: AD = autosomal dominant (50% risk to each offspring of an affected individual); AR = autosomal recessive (25% risk to each offspring of unaffected carrier parents); XD = sex-linked dominant (50% risk of an affected female for each conception; Sporadic = a single occurrence in a family is usual

Syndactyly. *Isolated syndactyly* is usually dominantly inherited, and is most common between the third and fourth digits, or second and third toes. A diagnos tic component of Apert's syndrome (acrocephalosyndactyly), syndactyly may also be a feature of Poland's anomaly, amniotic-band deformation, and other disorders.

Clubfoot is often caused by uterine constraint and oligohydramnios, but genetic factors are also important. Overall recurrence risk to siblings is about 3%. Clubfoot may be associated with neural-tube defects as well as with a variety of other conditions (see Chapter 47).

Radial Defects can be isolated, or may be a diagnostic clue to an underlying syndrome. Most isolated cases are sporadic. Commonly associated anomalies involve the cardiovascular, genitourinary, and hematologic systems. Radial aplasia occurs in Holt-Oram syndrome, Fanconi's anemia, TAR (thrombocytopenia with absent radius) syndrome, fetal thalidomide syndrome, and the VATER association.

Limb-reduction Defects run the gamut, from minor finger defects to complete amelia, including deformities like ectrodactyly, as well as simple absence of part or all of an extremity. Most are thought to be primary defects of development, often of genetic origin. A high proportion of isolated bilateral defects are inherited as single-gene traits; unilateral or asymmetric ones are more likely to be nongenetic; for example, amniotic-band syndrome. Limb-reduction defects are characteristic of many syndromes—e.g., Robert's , Holt-Oram, and Fanconi's pancytopenia.

CLINICAL NOTES

Since the kidneys and limbs are developmentally related, consider renal sonogram and/or IVP for individuals with limb defects.

When transverse limb-reduction defects, in which the limb has a smooth distal end, affect a single limb only, that end should be carefully scrutinized to identify small nubbins or radiographic evidence of finger rays that rule out amputation defects.

Hereditary conditions that include limb defects often have highly variable expression. Thus, it is especially important to search for subtle symptoms in seemingly unaffected relatives (e.g., mild cutaneous syndactyly, shortened third digit, displaced thumb, and hypoplastic nails).

PROCEDURES FOR DIAGNOSTIC CONFIRMATION

Physical examination of affected individuals and close relatives
X-ray studies
Careful prenatal history
Family-history analysis (with possible scrutiny of family photographs)
Renal sonogram/IVP
Cytogenetic studies (when a chromosome disorder is suspected)

CONSIDERATIONS IN MANAGEMENT

Genetic
- Establish diagnosis
- Assess recurrence risks
- Risk counseling
- Consider reproductive options, including prenatal diagnosis for at-risk pregnancies—sonography, fetal radiography, fetal chromosome analysis, fetoscopy, as indicated

Psychosocial, educational, familial
- Family counseling—acceptance of disfigurement, disability, and/or limited lifestyle
- Assessment of community resources for physical and rehabilitation therapy, prosthetics, special education
- Referral to support groups

Medical
- Dependent on type and severity of clinical manifestations

SUGGESTED READING

Beighton P: *Inherited Disorders of the Skeleton,* London, Churchill Livingstone, 1978.

Goodman R, Gorlin R: *The Malformed Infant and Child, an Illustrated Guide,* New York, Oxford University Press, 1983.

Rimoin DL, Hall J, Maroteaux P: *International Nomenclature of Constitutional Diseases of Bone with Bibliography,* vol 15, no. 10. (Birth Defects Original Article Series.) White Plains, N.Y., March-of-Dimes Birth Defects Foundation, 1979.

Chapter 24

Cystic Fibrosis

Cystic fibrosis (CF) is the most common genetic disease in the Caucasian population. Incidence among American whites is 1/2,000 to 1/1,600; the carrier rate is assumed to be one in 20. Incidence in American blacks is only 1/17,000. CF is characterized by the triad of chronic pulmonary disease, pancreatic insufficiency, and elevated sweat electrolytes. The devasting effects of CF on the whole family are exacerbated by the fact that, in spite of extensive research and many promising starts, no reliable carrier or prenatal diagnostic test is yet available. However, experimental work is once again raising hopes for definitive carrier testing and prenatal diagnosis. Meanwhile, new treatment modalities are steadily improving the prognosis for affected individuals.

ETIOLOGY AND PATHOGENESIS

The basic biochemical defect is not known but there is now evidence that the gene for CF is located in the middle of the long arm of chromosome 7. Dysfunction of the exocrine glands is manifested by mucous-secretion, mucociliary-transport abnormalities, and electrolyte changes in sweat. Mucus is thick. Serum contains a factor which inhibits normal ciliary activity, eventually destroying bronchial and pulmonary cilia. Pancreatic-duct obstruction eventually results in fibrosis, which may destroy the islets of Langerhans. Pancreatic trypsin, lipase, and amylase are deficient or absent, and reabsorption of sodium and chloride in sweat and salivary glands is reduced.

GENETIC CHARACTERISTICS AND MODE OF INHERITANCE

Cystic fibrosis is inherited as an autosomal recessive disease. In the absence of carrier testing, the first case in a family is usually completely unexpected. Risk estimates for another case in the family must be based on family history and population carrier frequency (one in 20). Some examples of risks for affected offspring are listed in Table 24-1.

CLINICAL NOTES

Respiratory system: Most morbidity and virtually all mortality result from respiratory failure caused by chronic bronchopulmonary involvement. Generalized bronchial obstruction may lead to pulmonary hypertension and cor pulmonale. *Staphylococcus aureus* and *Pseudomonas aeruginosa* are frequently present in the nasopharynx and sputum. Nasal polyps are common. The chest diameter may be increased and fingers and toes may be clubbed. At all ages, prognosis is poorer for females than for males, but for both, survival to adulthood is becoming more common with improved management.

Gastrointestinal system: Meconium ileus in the neonate is virtually pathognomonic of CF and occurs in 10% to 16% of patients. Pancreatic insufficiency is present in 80% to 85% of patients. CF is also a common cause of rectal prolapse in children and young adults.

Other Systems: Heat prostration and hyponatremia may result from heavy salt loss. Females may experience delayed puberty, decreased fertility due to increased mucus content and viscosity in cervical glands, and associated inflammatory changes in the cervix. More than 95% of males are sterile due to absence or atresia of the vas deferens, an absent epididymal body, and atresia of efferent ducts. Diabetes mellitus occurs with increased frequency.

Though primarily a disease of childhood, mild cases of CF may not become apparent until adulthood. However, in carriers the single CF gene has no known clinical effect.

PROCEDURES FOR DIAGNOSTIC CONFIRMATION

Sweat test—quantitative pilocarpine iontophoresis—although accurate in children, may produce equivocal results in adults.

Despite numerous claims, no other laboratory test for affected individuals, nor any test for carriers or prenatal diagnosis, has been confirmed as reliable. Recent studies on amniotic fluid alkaline phosphatase levels seem promising, but

TABLE 24-1
Recurrence Risks for Cystic Fibrosis

PARENT STATUS	RISK/PREGNANCY FOR AN AFFECTED CHILD (%)
Both parents known carriers*	25.0
One parent with affected sibling, other parent a known carrier*	16.7
Both parents with affected siblings	11.1
One parent affected, other parent no known family history	2.5
One parent known carrier,* other parent no known family history	1.25
One parent with affected sibling, other parent no known family history	.83
One parent with sibling's child affected, other parent no affected relatives	.63
One parent with uncle/aunt affected, other parent no affected relatives	.42

If one or both parents are known carriers,* the risk for offspring also to be a carrier (1 gene for CF) is 50%.

The risk for a phenotypically normal child of two carriers* also to be a carrier is 67%.

*Since there is no carrier test, the only known carriers are parents of a child with CF

await confirmation. Nevertheless, prenatal diagnosis based on them is already offered in specialized centers. At the same time recombinant DNA researchers are screening DNA and protein markers to identify those suitable for carrier identification and prenatal diagnosis. Several restriction fragment linked polymorphisms on chromosome 7 are under consideration.

CONSIDERATIONS IN MANAGEMENT

Genetic

Risk counseling:

Because of high carrier frequency—one in 20 in the white population—and lack of a reliable carrier test, caution is indicated for the option of artificial insemination. If offered at all, it must be accompanied by appropriate risk information; (the offspring of a known carrier woman and a randomly chosen white man has a one in 80 risk for being affected.)

Accurate risk information for offspring of unaffected siblings should be stressed, since it is often overestimated.

Investigation of current status of carrier testing and/or prenatal diagnosis.

Psychosocial, educational, familial

Family counseling:

Counseling and support for problems encountered in chronic disease with a rigid regimen; problems associated with lack of carrier and prenatal tests; psychologic problems brought on by the diagnosis, such as depression, denial, guilt, blame, anxiety, family or marital discord, etc.

Education for postural drainage and inhalation therapy

Referral to cystic fibrosis support group, if desired

Medical

Control of respiratory infections and prophylactic maintenance with antibiotics; postural drainage and inhalation therapy

Pancreatic enzyme-replacement therapy

SUGGESTED READING

Brock DJH, Bedgood D, Barron L, et al: Prospective prenatal diagnosis of cystic fibrosis. *The Lancet* 1985;1(8439):1175.

Committee for a Study for Evaluation of Testing for Cystic Fibrosis, an NAS/NRC report. Evaluation of testing for cystic fibrosis. *J Pediatr* 1976;88:711.

di Sant'Agnese P, Davis PB: Research in cystic fibrosis. *N Engl J Med* 1976;295:481, 534, 597.

di Sant'Agnese PA, Davis PB: Cystic fibrosis in adults. *Am J Med* 1979;66:121.

Fishman SE: Psychological issues in the genetic counseling of cystic fibrosis, in Kessler S, (ed): *Genetic Counseling: Psychological Dimensions*. New York, Academic Press, 1979.

Mangos JA, Talamo RC, (eds): *Cystic Fibrosis: Projections Into the Future*, New York, Stratton Intercontinental Book Corp., 1976.

Newmark P: Testing for cystic fibrosis. *Nature* 1985;318:309.

Talamo RC, Rosenstein BJ, Berninger RW: Cystic fibrosis, in Stanbury JB, Wyngaarden JB, Fredrickson DS, et al, (eds): *The Metabolic Basis of Inherited Disease*, 5th ed. New York, McGraw-Hill, 1983.

Chapter 25

Alpha$_1$-Antitrypsin Deficiency

Alpha$_1$-antitrypsin (α_1AT) deficiency is a relatively common inborn error of metabolism, which gained attention in 1963, when the deficiency was related to the development of emphysema. Its association with juvenile cirrhosis was established several years later. Gene-frequency estimates put the likely number of individuals with marked α_1AT deficiency at approximately 1/700 whites in the United States.

ETIOLOGY AND PATHOGENESIS

Alpha$_1$-antitrypsin is a glycoprotein which accounts for approximately 90% of the trypsin inhibitory capacity of normal serum; it has inhibitory capability for a number of other proteolytic enzymes as well. Serum levels in normal individuals respond with a twofold to threefold rise with inflammation, neoplasia, or pregnancy. This protein is synthesized by hepatocytes; liver function affects serum levels. Histologic and other evidence suggests that when there is synthesis of the biochemically abnormal ''Z''α_1AT molecule, it is retained by the hepatocyte, where it forms a cytoplasmic inclusion body.

The role of α_1AT deficiency in emphysema is assumed to follow from the reduced capacity of serum to inhibit proteolytic enzymes released in the lungs at the site of inflammation or environmental irritation. Proteolysis of lung parenchyma may alter the normal pulmonary blood flow pattern and reduce elastic recoil.

GENETIC CHARACTERISTICS AND MODE OF INHERITANCE

Synthesis of α_1AT is controlled by one autosomal gene with over 20 variant alleles, some extremely rare. The alleles are codominant, that is, each independently produces its specific α_1AT variant. The allelic system has been termed the Pi (protease inhibitor) system , and each variant is designated by a letter. The most common allele is PiM; most normal individuals are homozygous PiMM. There is some geographic variation in the frequency of other alleles.

The most marked deficiency of α_1AT is produced by the Z allele. Individuals who are homozygous PiZZ (termed genotype Z) have serum levels averaging 15% of normal or less. There are reported cases of no detectable α_1AT, which may represent a gene deletion. Heterozygous PiMZ individuals have α_1AT levels approximately 60% of normal. Alleles S and F result in less marked reduction in serum levels of α_1AT. Heterozygous SZ individuals may have relatively low serum levels, averaging 38% of normal.

- Couples who are both heterozygous for the Z allele—most commonly MZ—have a 25% risk of producing offspring with the ZZ genotype.
- Carrier typing is available, and prenatal diagnosis, using recombinant DNA technology, is possible in selected laboratories.

CLINICAL NOTES

The ZZ genotype carries 60% to 70% risk of chronic obstructive lung disease. Emphysema associated with α_1AT deficiency develops earlier and has a more rapid progression than idiopathic emphysema. Its onset, generally in the third or fourth decade of life, may occur without a preceding history of chronic cough or bronchitis.

Alpha$_1$-antitrypsin deficiency is one of the most common predisposing factors in cirrhosis of infancy and childhood. About 14% of ZZ-genotype patients have been reported to have developed fatal liver disease; cholestatic jaundice is the most common presenting symptom. Also, the homozygous PiZZ infant has been reported as small for gestational age.

The occurrence of either emphysema or liver disease is extremely variable, and ZZ genotype individuals may be asymptomatic. The reasons for such variability are not yet known. Smoking is known to be especially damaging in these patients.

Whether genotypes other than the homozygous Z are at increased risk of morbidity has been debated. This issue is especially important for MZ heterozygous individuals, who are thought to represent 2% to 3% of the white population in the United States. They may have an increased risk for emphysema if the MZ genotype is combined with smoking.

PROCEDURES FOR DIAGNOSTIC CONFIRMATION

Assay of trypsin inhibitory capacity of serum

α_1AT (Pi) typing by electrophoresis

Lung scan

Liver-function analysis

Blood-gas studies (may remain normal until late in the course of the disease)

CONSIDERATIONS IN MANAGEMENT

Genetic
- Pi typing for family members
- Risk counseling on the basis of Pi typing
- Consideration of prenatal diagnosis via restriction enzyme analysis for families at risk for ZZ-genotype offspring

Psychosocial, educational and familial
- Education about increased risk of smoking
- Counseling on possible sensitivity to air pollutants

Medical
- Symptomatic treatment of pulmonary and liver disease

Aggressive treatment of respiratory infections
Experimental use of danazol and of parenteral α_1AT replacement therapy have shown some promise for the development of effective treatment
Consideration of a transplant in end-stage liver or pulmonary disease

SUGGESTED READING

Gadek JE, Crystal RG: α_1-Antitrypsin deficiency, in Stanbury JB, Wyngaarden JB, Fredrickson DS, et al, (eds): *The Metabolic Basis of Inherited Disease,* 5th ed. New York, McGraw-Hill, 1983.

Hood M, Koep LJ, Peters RL, et al: Liver transplantation for advanced liver disease with alpha-1-antitrypsin deficiency. *N Engl J Med* 1980;302:272.

Kidd VJ, Golbus MS, Wallace RB, et al: Prenatal diagnosis of α_1-antitrypsin deficiency by direct analysis of the mutation site in the gene. *N Eng J Med* 1984;310:639.

Lieberman KJ, Gaidulis L, Garoutte B, et al: Identification and characterestics of the common alpha-1-antitrypsin phenotypes. *Chest* 1972;62:557.

Morse JO: Alpha-1-antitrypsin deficiency, Parts I and II. *N Engl J Med* 1978;299:1045, 1099.

Sharp HL: The current status of alpha-1-antitrypsin, a protease inhibitor, in gastrointestinal disease. *Gastroenterology* 1976;70:611.

Sveger T: Liver disease in alpha-1-antitrypsin deficiency detected by screening of 200,000 infants. *N Engl J Med* 1976;294:1316.

Tobin MJ, Hutchinson DES: α_1-Antitrypsin deficiency, current and future therapeutic strategies. *Drug Therapy* 1983, May.

Chapter 26

Cystic Kidney Disease

Heterogeneity is the hallmark of cystic kidney disease. Several hereditary disorders give rise to renal cysts, each generally distinctive in histopathology, age of onset, and subsequent course. Cysts can appear in any portion of the nephron, but each disorder has a characteristic predilection. Genetic forms can be cortical, medullary, or both. However, not all renal cysts are genetic in origin.

ETIOLOGY AND PATHOGENESIS

Cystic changes can occur early in fetal development, or arise in fully formed nephrons. With the exception of known nephrotoxic agents, the basic defect(s) inducing cyst formation of any type, genetic or nongenetic, is unknown.

GENETIC CHARACTERISTICS AND MODE OF INHERITANCE

The two principal categories of genetic cystic disorders are distinguishable by site and distribution of cysts. Each can be further subdivided by average age of onset and mode of inheritance:

Polycystic Kidney Disease. The kidneys are characteristically enlarged by a diffuse distribution of cysts, with bilateral involvement. Adult polycystic kidney disease is autosomal dominant with a 50% recurrence risk for offspring of affected individuals.

Infantile polycystic kidney disease is autosomal recessive. A 25% recurrence risk exists for each pregnancy of parents who have had one affected child. Prenatal diagnosis is available.

Medullary Cystic Disease. Less common than polycystic kidneys; at least two age-related subgroups can similarly be differentiated.

Adult medullary cystic disease is autosomal dominant with a 50% recurrence risk for offspring of affected individuals.

Familial juvenile nephronophthisis is autosomal recessive with a 25% recurrence risk for each pregnancy of parents who have had one affected child.

CLINICAL NOTES

Adult polycystic kidney disease can be latent for years. Age of onset varies considerably, although families tend to be consistent within a close range. Clinical problems most frequently appear early in the fourth decade. An occasional patient develops symptoms in childhood. Portal hypertension and berry aneurysms are associated complications. Cystic nephrons can retain function for some time, although larger cysts and high blood pressure indicate a poor prognosis. Peak mortality is during the fifth decade.

Infantile polycystic kidney disease is congenital or manifests in the first few years of life. Kidneys are smooth and palpable, sometimes massively enlarged. Cysts are radically arrayed throughout the cortex and medulla. This type is always associated with hepatic lesions. Few affected newborns survive infancy, but in those who do, renal insufficiency progresses more slowly after the first year. The degree of renal involvement and clinical severity appear to be consistent within families.

Adult medullary cystic disease usually manifests in the second to third decade. Tubular damage is rapidly progressive. The disease is characterized by defective sodium and water reabsorption, and a fulminant course.

Familial juvenile nephronophthisis is clinically similar to the adult form of medullary cystic disease, but symptoms appear in the first to second decade. Anemia is frequently the presenting symptom. The disease is characterized by growth retardation and bone changes, in addition to polyuria and salt wasting. In some families an association with ocular defects, such as retinitis pigmentosa, has been reported.

A number of genetic syndromes characteristically include renal cysts as a component of the symptom complex. Some of these follow:

- Ehlers-Danlos syndrome: generalized connective-tissue disorder, with several clinical subtypes
- Tuberous sclerosis: neurocutaneous disorder characterized by sclerotic nodules in the brain
- Jeune's syndrome: asphyxiating thoracic dystrophy, fatal in neonatal period
- Meckel's syndrome: posterior encephalocele, polydactyly; death occurs in perinatal period
- Polycystic kidney disease, cataract, and congenital blindness
- Zellweger's syndrome (cerebrohepatorenal syndrome): chondral calcification, characteristic craniofacial anomalies
- Orofaciodigital (OFD) syndrome: malformations of face and skull, syndactyly, and mental retardation; lethal in males
- Trisomy 13

PROCEDURES FOR DIAGNOSTIC CONFIRMATION

Intravenous pyelogram

Renal ultrasonography

Renal CT scan

Renal-function studies

Evaluation of hepatic, cardiovascular, or other possibly associated abnormalities

CONSIDERATION IN MANAGEMENT

Genetic

Determination of mode of transmission (including consideration of other genetic syndromes with renal cysts)

Risk counseling

In infantile polycystic kidney disease, consideration of prenatal diagnosis by ultrasonography (available in selected medical centers)

In adult forms, consideration of reproductive options, such as donor insemination, adoption

Discussion of sharing information with other family members to encourage evaluation and risk counseling

Psychosocial, educational, familial

Ongoing psychologic support for families with one or more members affected with progressive disease

Crisis counseling for families with infant affected with rapidly lethal form

Ascertainment and mobilization of extrafamilial sources of support

Exploration of possibilities for compatible kidney donor in appropriate cases

Medical

Control of hypertension

Treatment of infection

Monitoring sodium balance in medullary form

Hemodialysis for acute exacerbations

Management of acute and chronic renal failure

Transplant

Treatment of associated medical problems (hepatic or cardiovascular abnormalities) or congenital anomalies

SUGGESTED READING

Blyth H, Ockenden BG: Polycystic disease of kidneys and liver presenting in childhood. *J Med Genet* 1971;8:257.

Dalgaard OZ: Bilateral polycystic disease of the kidneys, in Strauss MD, Welt LG, (eds): *Diseases of the Kidney,* 2nd ed, vol 2. Boston, Little Brown, 1971.

Gardner KD, Jr, (ed): *Cystic Diseases of the Kidney.* New York, John Wiley & Sons, 1976.

Hobbins JC, Grannum P, Berkowitz RL: Ultrasound in the diagnosis of congenital anomalies. *Am J Obstet Gynecol* 1979;134:331.

Potter EL: *Pathogenesis of cystic kidneys,* vol 10, no. 4. (Birth Defects Original Article Series.) White Plains, N.Y., March-of-Dimes Birth Defects Foundation, 1974.

Chapter 27

Ambiguous Genitalia

Ambiguous genitalia in the newborn represents a true genetic emergency. Not only may such abnormalities trigger recognition of other potentially serious birth defects, such as renal malformations, but the failure to promptly resolve the question of gender may lead to a major family psychosocial crisis. Nevertheless, precipitate and poorly thought out comments or actions can have even more disastrous consequences. Prompt evaluation and careful planning by a qualified team of consultants, including open discussion with the family, are the key to successful management.

ETIOLOGY AND PATHOGENESIS

Early in fetal development, the gonads are sexually undifferentiated and other internal genital structures are biopotential. At about the eighth week of fetal life, influenced by genes on the Y chromosome, the gonads in the male differentiate into testes. Concurrently, a müllerian inhibitory factor causes regression of female internal structures. As the testes begin to produce testosterone, the external genitalia are masculinized through fusion of the labioscrotal folds to form a scrotum, enlargement of the phallus, and fusion of the labia minora to form a penile urethra. Later, usually in the seventh to eight month, the testes descend into the scrotum.

In the absence of Y-born genetic factors, the female gonad differentiates into an ovary. Likewise, the absence of müllerian inhibitory factor allows persistence and fusion of müllerian ducts producing the uterus and the Fallopian tubes; the wolffian duct structures regress, and external genital structures remain unmasculinized.

Hence, ambiguous genitalia represent either excessive masculinization of the female, or inadequate masculinization of the male, and are best thought of as "partially masculinized."

A variety of genetic and environmental factors may disturb genital morphogenesis, thereby leading to ambiguous genitalia in the newborn (Table 27-1).

GENETIC CHARACTERISTICS AND MODE OF INHERITANCE

Disorders of genital morphogenesis are etiologically heterogeneous. Many, if not most, have a genetic basis. One notable exception is prenatal exposure of the female fetus to androgens.

The following are important genetic possibilities:

Sex-chromosome abnormalities
- Mosaicisms, e.g., XO/XY, XX/XY
- Structural abnormalities: deletions and translocations

Mendelian disorders
- Virilizing adrenal hyperplasia syndromes: mostly autosomal recessive
- Disorders of androgen steroidogenesis: mostly autosomal recessive
- Androgen insensitivity (testicular feminization) syndrome: X-linked recessive
- Some malformation syndromes: inheritance depends on syndrome

Specific recurrence-risk estimates depend upon the disorder in question. For most of the chromosomal problems the recurrence risk is low. Prenatal diagnosis is available for several of these conditions.

CLINICAL NOTES

Chromosomal sex is not the sole criterion for gender identity. Other factors include gonadal sex, internal and external genital morphology, gender of rearing,

TABLE 27-1
Classification of Disorders of Genital Morphogenesis

	QUANTITATIVE ABNORMALITIES		QUALITATIVE ABNORMALITIES IN (♀) and/or (♂)
	EXCESSIVE MASCULINIZATION (♀)	INADEQUATE MASCULINIZATION(♂)	
Genetic Etiology	1. Masculinizing adrenal hyperplasia syndromes 2. Sex-chromosome syndromes	1. Disorders of androgen steroidogenesis 2. Androgen insensitivity syndromes 3. Sex-chromosome syndromes 4. Some primary defects of testicular development, hypothalamic and/or pituitary development, or external genital development	1. Some malformation syndromes: a. Smith-Lemli-Opitz syndrome b. Opitz syndrome 2. Some chromosomal syndromes: a. 18 Trisomy syndrome b. 13 Trisomy syndrome
Non-Genetic Etiology	1. Prenatal exposure to some teratogens: a. androgens b. trimethadione	1. ? prenatal exposure to some teratogens, (trimethadione) 2. Some primary defects of testicular development, hypothalamic and/or pituitary development, or external genital development	1. Some malformation syndromes: a. Cloacal extrophy syndrome b. Sirenomelia syndrome 2. ? prenatal exposure to some teratogens: trimethadione

endocrine functional status, and personal and familial psychosocial considerations. Gender should not be assigned without considering all factors.

Thus, gender of rearing may not be the same as chromosomal sex. For example, a chromosomal male with ambiguous genitalia may be brought up as a female, if genital reconstruction is not clinically feasible. In other cases, gender of rearing may have been assigned before a disorder is recognized. For instance, patients with testicular feminization are chromosomally male, but may be born and grow up as unambiguous phenotypic females. Diagnosis is often not made until the patient presents with primary amenorrhea.

Chromosomal sex is best established by chromosome analysis (buccal-smear analysis is inadequate). Because the salt-losing and hypertensive adrenal hyperplasia syndromes can produce potentially life-threatening complications, they should be excluded early in the diagnostic evaluation.

PROCEDURES FOR DIAGNOSTIC CONFIRMATION

Cytogenetic analysis
Urinary 17-ketosteroid assay
Urinary pregnanetriol assay
Assay of sex hormones and gonadotropins
Gonadotropin stimulation (for androgen production)
Panendoscopy
Vaginography
Cystourethrography

CONSIDERATION IN MANAGEMENT

- Genetic
  - Risk counseling
  - Considerations of prenatal diagnosis (in some cases)
  - Discussion of sharing risk information with other family members
- Psychosocial, educational, familial
  - Family counseling: gender-identity concerns, coping with friends and relatives
- Medical
  - Investigation and management of associated anomalies

SUGGESTED READING

Simpson JL, Jirasek JE, Speroff L, et al: *Disorders of Sexual Differentiation—Etiology and Clinical Delineation*, New York, Academic Press, 1976

Smith DW: Approach to ambiguous (partially masculinized) external genitalia, in: *Recognizable Patterns of Human Malformation*, 2nd ed. Philadelphia, WB Saunders, 1982, pp 402-406.

Chapter 28

Genetic Disorders of the Immune System

IMMUNODEFICIENCY DISEASES

Most of the well described immunodeficiency diseases are the result of a genetic abnormality somewhere in the complex system of the immune response and are inherited as single-gene disorders. While the severe disorders are very rare, milder irregularities of immune function may be quite common. Characterized chiefly by increased susceptibility to primarily nongenetic illness, such as infection, autoimmune disorders, and malignancies, immunodeficiencies may go unrecognized, especially in a family history.

ETIOLOGY AND PATHOGENESIS

Immunodeficiencies reflect the malfunction of one or both of the major components of the immune system—cell-mediated immunity and humoral immunity. Although there are known specific abnormalities, most of the basic defects are not clearly understood.

DESCRIPTION, GENETIC CHARACTERISTICS, AND MODE OF INHERITANCE

The serious genetic immunodeficiences include the following:

Congenital X-linked Agammaglobulinemia, a severe form of a group of hypogammaglobulinemias, is an antibody deficiency characterized by recurrent pyogenic infections that begin to appear in infancy, after the protective effects of maternal antibodies have worn off. Associated complications may include malabsorption, rheumatoidlike arthritis, dermatomyositis, and malignancies. Only males are affected. In most, B lymphocytes are absent or greatly reduced, as are all gammaglobulins. Adenoids, tonsils, and peripheral lymph nodes are hypoplastic. Untreated, the disorder is lethal in infancy. However, early detection may prove lifesaving, since gammaglobulin replacement therapy is reported to provide adequate protection. The risk of occurrence in the offspring of a carrier female is 50% for each son to be affected, 50% for each daughter to be a carrier. Sons of an affected male will not inherit the gene, but daughters will be carriers.

Other Forms of Hypogammaglobulinemia have complete or selective immunoglobulin deficiency, varying degrees of clinical severity, and different estimates of genetic components. Among them, *transient hypogammaglobulinemia of infancy* is included in the differential diagnosis of congenital agammaglobulinemia, and *isolated deficiency of IgA*, found in about 1/500 normal people, may predispose to sinopulmonary infection, and is associated with increased incidence of autoimmune disease in the patient and his relatives.

Severe Combined Immunodeficiency (SCID), although rare, is the most severe and most publicized immune deficiency disorder. Historically, failure to thrive, chronic diarrhea, and extensive infections beginning early in infancy have prevented survival past age 2, unless drastic measures were taken, such as isolation in an aseptic "bubble." Recent efforts to establish immunologic reactivity with bone marrow transplants hold promise for future effective management.

SCID patients are unable to fight even the most benign infections because of absent or greatly reduced function of the lymphoid tissue and the thymus. Etiologic and genetic heterogeneity are present; both X-linked and autosomal recessive inheritance have been reported. About half of the patients with an autosomal recessive SCID have an adenosine deaminase (ADA) enzyme deficiency; in others a deficiency of purine nucleoside phosphorylase has been implicated. In families with documented ADA deficiency with 25% recurrence for each pregnancy of two carrier parents, both prenatal diagnosis of affected fetuses and carrier testing of other family members are possible.

DiGeorge's Syndrome, is heterogeneous and clinically variable, although absent or incomplete thymus and parathyroid glands characterize all cases. Circulating T-cells are absent or reduced in number; neonatal hypocalcemic tetany, characteristic facies, cardiac malformations, and failure to thrive may be among associated findings. Increased susceptibility to infection usually leads to early death; however, hormone replacement and fetal thymus transplants are reported to have corrected the abnormality. Although most cases are sporadic, families with autosomal recessive inheritance have been reported.

Chronic Granulomatous Disease is generally lethal, usually X-linked recessive, and thought to have a heterogeneous etiology—involving abnormalities in the oxygen-dependent microbicidal action of phagocytes. Frequent severe bacterial and fungal infections and the effects of chronic inflammation can greatly limit life span, but families with a milder clinical course have been described. Female carriers of the X-linked form are not affected, but some have been reported to develop a disorder resembling lupus erythematosus.

Complement Deficiencies refers to a variety of inherited deficiencies in one or more complement components. Although not all are associated with adverse clinical consequences; some seem to lead to recurrent infections or an increased frequency of autoimmune disorders. The latter may be related to a linkage of complement genes to the HLA complex. C_1 inhibitor deficiency is found in *hereditary angioedema*, an autosomal dominant disorder with potentially dangerous episodes of edema in the skin, intestinal tract, and upper-respiratory tract.

TABLE 28-1
HLA Association With Selected Diseases

HLA ANTIGEN	DISEASE	RELATIVE RISK FOR ANTIGEN CARRIERS
B27	Ankylosing spondylitis*	90×
B27	Reiter's syndrome	40
DR7	Psoriasis	43
DR3 and DR4	IDDM	33
DR4 alone	IDDM	6
DR3 alone	IDDM	3
DR3	Celiac disease	17

*See discussion, Chapter 32
Key: IDDM = Insulin-dependent diabetes mellitus
Source: Adapted from Hirschorn R: Hereditary immunodeficiency disorders: Recent developments, in Kaback MM, (ed): *Genetic Issues in Pediatric and Obstetric Practice*. Chicago, Yearbook Medical, 1981.

Accurate diagnosis may be life-saving, because treatment with attenuated androgens is effective. Testing of first-degree relatives of affected individuals may lead to early diagnosis and prevention of complications.

Increased Susceptibility to Infection is a major characteristic of a number of genetic syndromes including Down's syndrome, sickle cell anemia, Wiskott-Aldrich syndrome (X-linked recessive; additional characteristics: thrombocytopenia, eczema, absent isohemagglutinins, and lymphoid malignancy), ataxia-telangiectasia (autosomal recessive; chromosome breakage syndrome), Chédiak-Higashi syndrome (autosomal recessive; imcomplete oculocutaneous albinism, giant granules in leukocytes), and severe G6PD deficiency (X-linked recessive; hemolytic anemia after ingestion of specific precipitating agents).

HLA ASSOCIATION WITH DISEASE

The antigens of the major histocompatibility gene cluster on chromosome 6 (HLA system) are involved in cellular immune recognition and response. They seem to play a part in susceptibility and resistance to disease. Specific HLA antigens apparently influence the development of a number of common disorders, most of which are autoimmune-related and may be inherited as multifactorial traits. Table 28-1 gives examples of HLA antigens with a particularly strong association to some disorders, and lists suggested relative risks of occurrence for individuals carrying the antigen, as compared with the general population. Other antigens are thought to confer increased resistance to specific disorders, such as HLA-DR2 for insulin dependent diabetes.

CLINICAL NOTES

Although failure to thrive and recurrent or chronic infection in infancy can be complications of many other conditions, suspicion of an immunodeficiency may be raised. Suspicion is reinforced by a family history of early death due to infectious disease. The history itself may alert the physician to a potential risk for such a disorder in the family. A report of several male maternal relatives having succumbed to infection at a time when immune disorders were rarely recognized, constitutes a particularly useful clue for the X-linked recessive types.

Reviews of medical records and autopsy reports may be helpful when an unexplained history of chronic infection and lymphopenia in a previous child signal the possibility of SCID.

PROCEDURES FOR DIAGNOSTIC CONFIRMATION*

- Humoral immune function
  - X-ray studies to check for absent tonsils and adenoids
  - Quantitation of immunoglobulin content of serum (radial immunodiffusion)
  - Isohemagglutinins (quantitative titers)
  - Schick test (assuming prior immunization)
  - Antibody titers to diphtheria and tetanus following immunization (CDC)
  - Specialized evaluation
    - Immunoelectrophoresis of serum (with anti-kappa, antilambda light-chain antibody and specific anti-heavy-chain antibodies, as well as anti-human immunoglobulin)
    - Antibody response to various antigens (in vivo)
    - Enumeration of B cells
      - Membrane Ig (IgM and IgD) by immunofluorescence
      - Complement component receptors (C3b and C3d, also present on other cells)
      - EBV receptors
    - Synthesis of immunoglobulin in vitro in response to mitogens
    - Intracellular immunoglobulin
    - Immunoglobulin concentration in secretory fluids (saliva, GI, for IgA abnormalities)
  - Determination of catabolic rate
- Cellular Immune Function
  - X-ray to check for balance of thymic shadow
  - Absence of delayed hypersensitivity skin-test reactions of antigens (requires previous history of exposure;a negative test is not significant in children under 1 year)
  - Karyotype analysis: absence of mitotic figures (i.e., inability to obtain karyotype) is very crude, gross measure of T-cell function
  - Sensitization with DNCB (only after all parameters indicate markedly abnormal T-cell function)
  - Specialized in vitro evaluation
    - Enumeration of E-rosette forming cells (T cells) in peripheral blood (viral infections and malignancies also cause diminution)
    - Proliferative response in vitro to polyclonal activators (mitogens such as PHA, Con A)
    - Proliferative response in vitro to specific antigens
    - Proliferative response in vitro to allogeneic lymphocytes (HLR)
    - Production of lymphokines in vitro
    - Killer function, helper and suppressor function
- Studies appropriate for confirming other disorders with immune-related features

---

*Adapted from:
Hirschhorn R: Hereditary immunodeficiency disorders: Recent Developments, in Kaback MM (ed): *Genetic Issues in Pediatric and Obstetric Practice*, Chicago, Year Book Medical Publishers Inc, 1981.

CONSIDERATION IN MANAGEMENT

- Genetic
  - Establish diagnosis
  - Family-history analysis
  - Risk counseling
  - Carrier testing of family members, if available, e.g., ADA levels
  - Consideration of prenatal diagnosis for future pregnancies in families with ADA-negative SCID (available in specialized laboratories)
  - Consideration of alternative reproductive options, as indicated, e.g., donor insemination
  - Discussion of sharing information with other relatives to encourage evaluation and counseling (particularly for the X-linked disorders)
- Psychosocial, educational, familial
  - Family counseling: support for acceptance of grave prognosis; crisis counseling, as indicated
  - Consideration of referral for psychotherapy or support for prevention of psychologic problems brought on by diagnosis, such as marital discord, depression, guilt feelings
  - Education for effective home care
  - Assessment of family/community resources—e.g., support groups, financial assistance
- Medical
  - X-linked agammaglobulinemia
    - Antibiotic therapy
    - Humoral gammaglobulin replacement
    - Fresh frozen plasma
  - SCID
    - Protective isolation
    - Transfusion with irradiated blood
    - Transplantation of HLA compatible marrow cells
    - Enzyme replacement therapy (experimental)
  - DiGeorge's syndrome
    - Administration of calcium, vitamin D
    - Fetal thymus transplant (experimental)
  - Hereditary angioedema
    - Treatment with attenuated androgens

SUGGESTED READING

Goldman JN, Goldman MB: What the clinician should know about the major histocompatibility complex. *JAMA* 1981;246:873.

Hirschhorn R: Hereditary immunodeficiency disorders: Recent developments, in Kaback MM, (ed): *Genetic Issues in Pediatric and Obstetric Practice*, Chicago, Year Book Medical Publishers, 1981.

Rosen FS, Cooper MD, Wedgwood RJP: The Primary Immunodeficiencies (in two parts). *N Engl J Med* 1984;311:235, 300.

Stanbury JN, Wyngaarden JB, Fredrickson DS, et al (eds): *The Metabolic Basis of Inherited Disease*, 5th ed. New York, McGraw-Hill, 1983, Chp 3, pp 53, 88-90.

Stiehm ER, Fulginiti VA (eds): Immunologic disorders, in *Infants and Children*. Philadelphia, WB Saunders, 1980.

Chapter 29

Hereditary Visual Disorders

Close to $1\frac{1}{2}$ million people in the United States suffer from severe visual impairment. About 500,000 of them are considered legally blind, with 124,000 diagnosed as such before age 45. Genetic causes have been estimated to account for nearly half of the vision loss in this early-onset group. As the optic apparatus is particularly sensitive to both primary genetic influence and secondary effects of mutant genes affecting other systems, a vast array of hereditary eye disorders have been identified. These include conditions limited to the eye or the ocular region, and ocular manifestations of other heritable disorders and complex syndromes.

A detailed review of ophthalmic genetics is beyond the scope of this volume. This chapter primarily suggests guidelines for consideration of referral for an ophthalmologic genetic workup and provides a summary of common genetic conditions in which ocular abnormalities represent an important aspect.

ETIOLOGY AND PATHOGENESIS

Etiologic and pathogenic factors that can disturb the structure or function of the eye run the gamut of disease-producing mechanisms. Among the genetic causes, any condition—single gene, chromosomal, or multifactorial—which disturbs prenatal or postnatal development, may lead to anatomic or functional ocular abnormalities. In addition, storage of abnormal metabolites, chemical imbalance, connective-tissue impairment, disturbed blood supply, dystrophic elements, failure of DNA repair, malignant disease, and many other factors can underlie serious visual deficits. Hearing may be subject to the same influences as vision, therefore, association of hearing loss with visual impairment is not uncommon.

GENETIC CHARACTERISTICS AND MODE OF INHERITANCE

Eye diseases, either isolated or part of a syndrome, may be inherited as autosomal dominant, autosomal recessive, X-linked, or multifactorial conditions. For X-linked recessive eye disorders, carrier identification is often possible because changes can be detected in heterozygous women.

Table 29-1 summarizes characteristic eye findings in various categories of storage diseases and Table 29-2 lists a selection of common genetic disorders in which eye conditions may be a presenting feature or a diagnostic clue.

Special emphasis is given here to some of the most important serious genetic eye conditions.

Retinitis Pigmentosa (RP) encompasses a group of disorders which are the commonest of many retinal degenerations. Prevalence has been estimated at 1/7,000 to 1/2,000, with more than 20,000 individuals in the United States affected severely enough to be categorized as legally blind. RP is characterized by pigmentary retinal disturbances, which may be accompanied by other ocular abnormalities and may be found in association with other genetic disorders.

The earliest clinical signs, which may precede morphologic retinal changes, are decreased night vision and constricted visual fields. The abnormal fields and progressive deterioration, leading to severe vision loss, distinguish RP from stationary night blindness, a group of genetic conditions with milder manifestations.

Isolated RP includes several genetic variants categorized by mode of inheritance. Each category may be genetically heterogeneous, but, generally, the severity of the disease varies according to subtype.

Autosomal recessive RP is the most common form, estimated to account for 50% or more of cases. Onset of symptoms is during the first two decades of life, with severe vision loss usually in the fifth decade.

TABLE 29-1
Prominent Ocular Findings in Common Storage Diseases

METABOLIC DISORDER	COMMON EYE FINDINGS	MODE OF INHERITANCE	COMMENTS
Mucopolysaccharidoses (Except Hunter's syndrome)	Corneal clouding, retinal degeneration	AR	Corneal findings may be helpful as diagnostic clue; enzyme analysis is available for precise diagnosis of most disorders in category
Mucolipidoses	Corneal clouding, cherry-red spot, retinal degeneration	AR	Mild to moderate corneal clouding is a feature of most of these disorders
Sphingolipidoses	Cherry-red spot, optic atrophy	Most: AR	Specific enzyme defect known for most of these disorders; used for carrier detection, diagnosis, and prenatal diagnosis; macular cherry-red spot is often first specific finding for Tay-Sachs disease (Chapter 34)

Key: AR = autosomal recessive
Source: Adapted from Maumenee IH, Jackson LG, Heritable eye diseases, in: Jackson LG, Schimke RN, (eds): *Clinical Genetics, A Sourcebook for Physicians,* New York, John Wiley & Sons, 1979.

TABLE 29-2
Possible Ocular Findings in Common Genetic Disorders Involving Other Systems

OCULAR FINDING	DISORDER	OTHER CHARACTERISTICS	MODE OF INHERITANCE
Ocular depigmentation, photophobia, nystagmus, decreased visual acuity	Albinism	Various types; most show skin and hair depigmentation	Most: AR (ocular albinism:XR)
Heterochromia, displacement of medial canthi	Waardenburg's syndrome	Congenital nerve deafness, white forelock	AD
Congenital glaucoma, photophobia, corneal dystrophy	Xeroderma pigmentosum	Photosensitivity, early skin cancers	AR
Retinitis pigmentosa	Usher's syndrome	Congenital deafness	AR
	Bardet-Biedl syndrome	Obesity, hypogonadism, mental retardation	?AR
	Refsum's disease	Peripheral neuropathy, cerebellar ataxia	AR
Hypoplasia of the iris and anterior synechia, microcornea, blue sclerae, congenital glaucoma	Rieger's syndrome	Midfacial hypoplasia, prognathism, short philtrum, hypodontia with cone-shaped teeth	AD
Aniridia	Wilm's tumor-aniridia syndrome	May be associated with an interstitial deletion on the short arm of chromosome II	? CH
Congenital cataracts	Lowe's syndrome	Mental retardation, hypotonia, renal tubular dysfunction	XR
Cataracts	Galactosemia	Blindness and mental retardation in untreated patients; early dietary therapy can prevent manifestations	AR
	Werner's syndrome	Premature aging	AR
	Myotonic dystrophy	Progressive myotonia may begin with ptosis	AD
Corneal and lenticular opacities	Fabry's disease	Angiokeratomas, whorl-like corneal epithelial dystrophy in most carrier females	XR
High myopia	Stickler's syndrome	Epiphyseal changes; Pierre Robin syndrome	AD
	Kniest's dysplasia	Disproportionate dwarfism	? AD

TABLE 29-2 *Continued*

OCULAR FINDING	DISORDER	OTHER CHARACTERISTICS	MODE OF INHERITANCE
Bilateral ectopia lentis, high myopia	Marfan's syndrome	Arachnodactyly; cardiovascular defects; tall, thin habitus	AD
Bilateral ectopia lentis	Homocystinuria	Excretion of urinary homocystine, skeletal malformations	AR
Blue sclerae	Osteogenesis imperfecta	Group of disorders with increased susceptibility to bone fractures	Most: AD (may be AR)
Kayser-Fleischer rings	Wilson's disease	Copper storage; hepatic and/or or psychiatric symptoms	AR
Exophthalmos	Crouzon's syndrome	Premature craniosynostosis, multiple anomalies	AD
Oblique palpebral fissures, Brushfield's spots	Down's syndrome	Additional chromosome 21; flattened facies, mental retardation	CH
Ocular neurofibromas	Neurofibromatosis	Café-au-lait spots, other neurofibromas	AD

Key: AR = autosomal recessive; AD = autosomal dominant; XR = X-linked recessive; CH = chromosomal

Autosomal dominant RP, which includes about 15% of cases, also becomes symptomatic during the first two decades. However, progression is slow and central vision may remain useful to the sixth or seventh decade. Penetrance is generally complete.

Since carrier testing for autosomal recessive RP is not available, pedigree analysis must be used to distinguish between it and autosomal dominant RP. Sporadic cases, which may represent a new mutation or an isolated autosomal recessive case do not have informative pedigrees and cannot be classified on the basis of clinical features. Thus, an empiric risk must be given: For an individual with sporadic RP to transmit the disorder to offspring the risk has been estimated at about 12%, when a child is diagnosed to have RP, dominant inheritance is, of course, documented for that family.

About 5% of isolated RP is X-linked recessive. The most severe form of the disease, it produces profound visual loss by the fourth decade. Carrier detection is often possible, since heterozygous women tend to show retinal changes and may even be symptomatic.

A closely related disorder, *amaurosis congenita of Leber* is distinguished mainly by onset in infancy. Ten percent to 20% of congenital blindness is estimated to be the result of the pigmentary changes and optic atrophy of this disorder. Frequent findings of consanguinity in the families of affected children have led to the assumption of autosomal recessive inheritance, but heterogeneity is likely.

The long list of genetic disorders with associated RP includes Usher's syndrome (autosomal recessive, with profound nerve deafness), Bardet-Biedl syndrome (autosomal recessive, with obesity, hypogonadism, mental retardation, digital anomalies), several of the mucopolysaccharidoses, and Refsum's disease (autosomal recessive, with peripheral neuropathy and ataxia).

Retinoblastoma, a relatively common childhood eye cancer is discussed more thoroughly in Chapter 40. It may also enter the differential diagnosis for pseudoglioma of the retina, or Norrie's disease, which may resemble the retinal tumor but is a rare X-linked recessive disorder that causes total congenital blindness in affected males, who may also be retarded and deaf.

Retinal Detachment, when genetic, is generally secondary to another condition. It is frequently associated with high myopia, and recurrence risks are restricted primarily to relatives who also have severe myopia. Genetic syndromes in which retinal detachment is a characteristic include Stickler's syndrome (autosomal dominant, with epiphyseal changes and high myopia), sickle-cell disease, and Ehler-Danlos syndrome (a group of heterogeneous connective-tissue disorders, for the most part autosomal dominant, but with all types of inheritance patterns documented).

Glaucoma in its most common form, *primary open-angle*, is characterized by insidious onset and distinctive visual-field defects. It affects about 2% of individuals over age 40 and is known to be frequently familial. The empiric risk estimates for first-degree relatives of affected persons to develop glaucoma range from 5% to 16% for siblings, and may be equally high for children. Periodic screening of at-risk individuals may lead to prevention of blindness through early diagnosis and treatment. Open-angle glaucoma may also be associated with diabetes mellitus, high myopia, and other genetic disorders, such as mucopolysaccharidosis I-S, or Scheie's syndrome.

Angle-closure glaucoma is more acute and episodic than the open-angle form and is thought to be related to anatomic orbital factors, which are inherited as polygenic or multifactorial traits. Siblings of affected individuals are at an estimated risk of 12%.

Congenital glaucoma is found in 1/10,000 newborns; such an infant may also develop other ocular anomalies in the first weeks or months. The isolated disorder is probably autosomal recessive. In the absence of an informative pedigree, a recurrence risk of 25% has to be assigned to siblings. Research studies have suggested that the risk for children of affected individuals is less than 1%. Congenital glaucoma may be associated with a long list of genetic or teratogenic syndromes, including aniridia, neurofibromatosis, Rieger's syndrome, and fetal rubella syndrome.

Presenile Cataracts include a bewildering variety of lens opacities, with or without other eye anomalies, and with a range of effects on vision that extends from none to blindness. About 27,000 individuals in the United States are blind, as a result of cataracts other than the senile form. All types of inheritance have been reported, as have nongenetic causes. Among the genetic forms of isolated cataracts, autosomal dominant inheritance, with a 50% recurrence risk for offspring, is the most frequent.

Approximately 1/250 infants is born with a cataract, which may be genetic or may be due to other prenatal influence, such as fetal rubella syndrome. Others develop cataracts later, often in association with another underlying genetic disorder, such as galactosemia or myotonic dystrophy, an autosomal dominant disorder in which cataracts can be pathognomonic.

Galactosemia, an autosomal recessive disorder, with estimated prevalence of 1/40,000, merits special mention because it can be a major cause of mental retardation and blindness, unless treatment is provided very soon after birth. Management requires dietary exclusion of galactose-rich foods, such as milk and milk products. Whereas untreated patients have high morbidity and mortality in the first months of life, treated patients will grow and develop normally. Neonatal screening for activity of the relevant enzyme galactose-1-phosphate uridyl-transferase in erythrocytes is used in most states to identify homozygous infants, so that preventive treatment can be provided. Heterozygote testing by assay for the enzyme is also available.

Corneal Dystrophies include a profusion of genetic types. Most are very rare; their inheritance tends to be mendelian, but specialized diagnosis may be needed to establish recurrence risks. A pedigree of the patient may help to define the pattern in his family.

The corneal dystrophies are characterized by haziness, decreased visual acuity, blurring, or dazzling. Age of onset varies, but may be similar within families.

Corneal clouding and opacities are also associated with a number of other conditions and may serve as diagnostic features for such disorders as mucopolysaccharidoses and mucolipidoses.

Dislocated Lenses can be inherited in an autosomal dominant pattern when associated with spherophakia. Generally, however, they are thought to represent a manifestation of a wider disorder, such as connective-tissue disorders like the autosomal dominant Marfan's syndrome, or the autosomal recessive error of amino acid metabolism, homocystinuria.

Strabismus and Nystagmus may both be features of genetic, neurologic, or neuromuscular disorders. They are inherited in the same manner as the underlying disease.

Strabismus may also be the result of a congenital ocular structural abnormality. When it occurs in a family as an isolated primary problem, it is thought to be inherited as a polygenic condition, although families with autosomal dominant inheritance have been reported. Empiric recurrence risks have been estimated at 15% for siblings of an affected child. If a parent is also affected, the recurrence risk is reported to be about 40%.

All types of mendelian inheritance have been reported for isolated nystagmus. Pedigree analysis may help to define inheritance in a given family. However, most pedigrees are compatible with X-linked inheritance.

The various forms of *albinism* are prominent among disorders with which nystagmus is associated. Nearly 7,000 legally blind individuals in the United States have visual impairment due to a form of albinism. Ocular albinism is an X-linked recessive disorder, whereas most of the oculocutaneous types are autosomal recessive. Eye manifestations generally include photophobia, nystagmus, reduced visual acuity, and reduced or absent binocular vision. The clinical finding of a pale iris, pale fundus, and nystagmus suggests albinism (see Chapter 31).

Aniridia, Coloboma of the Iris, and Heterochromia are among the numerous hereditary iris anomalies. Most of them are inherited as autosomal dominant traits with variation in clinical expression. As such, they do not have great pathologic significance. However, they are frequently associated with other ocular abnormalities, such as cataracts, macular hypoplasia, or optic nerve colobomas. They may constitute part of a syndrome, such as heterochromia in Waardenburg's syndrome—an autosomal dominant disorder that includes congenital sensorineural deafness and often a white forelock or early graying. Since the association of aniridia with Wilms' tumor may reflect a deletion in the short arm of chromosome 11, chromosome analysis for an infant with isolated aniridia may reveal the deletion and thus, a risk for Wilms' tumor.

Optic Atrophy, although rare, is an important cause of severe visual impairment in childhood. Both autosomal dominant and autosomal recessive inheritance have been described for the isolated disorder. The autosomal recessive forms are particularly rare and likely to be associated with other symptomology such as mental retardation, hypertonia, and ataxia (Behr's syndrome).

Leber's optic atrophy is a heterogeneous disorder with onset mostly in the second or third decade. It is inherited in an unusual manner resembling X-linkage but not conforming to the usual X-linked pattern. Maternal inheritance via mitochondrial genes has been suggested as a possible mode of transmission.

Anophthalmia, Microphthalmia, and Cryptophthalmia are all extremely rare, yet the prevalence of legal blindness in the United States due to malformed or missing eyes is estimated at more than 3,000. More often than not, the eye anomalies are associated with other abnormalities and may be part of a genetic syndrome, for example, chromosomal defects, or Lenz's syndrome (X-linked recessive, with cataracts, mental retardation, and digital and genitourinary malformations). The eye anomalies may also result from teratogenesis, as in fetal-rubella or fetal-toxoplasmosis syndromes. Mental retardation is a frequent association. Isolated anophthalmia and cryptophthalmia have been reported as autosomal recessive disorders.

Refractive Errors, correctible by glasses, affect about 50% of the population. Familial incidence is well recognized. Except when associated with another disorders, this type of visual impairment has a continuous distribution, possibly related to anatomic variation in the eye or orbit and is usually a multifactorial trait. However, mendelian forms of isolated high myopia, hypermetropia, or astigmatism also occur. Individual pedigrees may suggest single-gene inheritance of other refractive errors in specific families.

Abnormalities of the Ocular Region are associated with a multitude of complex syndromes. Rarely presenting symptoms, they may be an aid to making a specific diagnosis. Among them are hypertelorism, hypotelorism, short or slanted palpebral fissures, epicanthal folds, shallow or prominent orbital ridges, synophrys, and ptosis. Isolated congenital ptosis, a common autosomal dominant trait with complete penetrance, may be unilateral or bilateral within one pedigree and is generally benign. However, ptosis may also be an early sign of myotonic dystrophy.

CLINICAL NOTES

Severe visual impairment or blindness in childhood has been estimated to affect 1/2,000 in the populations of developed countries. With the growing control of problems such as retrolental fibroplasia and fetal rubella, probably well over 50% of this early vision loss is now genetic in nature. Choroidoretinal degeneration and cataracts rank high among the many genetic causes. Associated symptomology, often in the form of mental retardation, may or may not be readily apparent.

In children severe bilateral reduced acuity or visual-field loss is likely to be noted early, but a unilateral problem may go undetected for years. A number of general clues may raise suspicion of an ocular lesion and lead to a more extensive ophthalmologic evaluation and identification of a possible genetic etiology. Among them are strabismus and nystagmus, a white reflex in one or both eyes, photophobia, pain, a small or malformed eye, a pale iris and fundus, corneal clouding, or a cherry-red spot. Other eye findings may also be of diagnostic value for common genetic disorders (see Tables 29-1 and 29-2).

Although referral to a specialist is indicated for a thorough ophthalmologic examination, the diagnosis of genetic disease may rest on valuable additional information gathered by the primary physician. For instance, an extensive family history may reveal a pedigree pattern; a careful search for associated symptomology may define a complex syndrome; and laboratory evaluation may document a metabolic error or identify a gene carrier.

Specific diagnosis is essential for accurate counseling and is particularly important for the growing number of conditions that are becoming subject to successful intervention (e.g., the dietary regimen for galactosemia, safe removal of cataracts, effective management of retinoblastoma, and prenatal diagnosis of metabolic disorders).

PROCEDURES FOR DIAGNOSTIC CONFIRMATION

Referral for ophthalmologic evaluation for the following procedures, as indicated:

- Funduscopy
- Electroretinography
- Slit-lamp examination
- Fluorescein angiography
- Perimetric studies; tangent screen

Tonometry

Gonioscopy

Ocular ultrasonography

Assorted diagnostic procedures to investigate the possibility of other genetic disorders underlying the eye lesion—e.g., amino acid screen, enzyme assays, chromosome analysis

CONSIDERATIONS IN MANAGEMENT

Genetic
- Pedigree analysis
- Establish diagnosis and assess recurrence risk
- Establish carrier status of other family members, if indicated and obtainable (e.g., Hurler's syndrome)
- Risk counseling
- Consideration of reproductive options, if indicated by diagnosis
- Consideration of prenatal diagnosis for pregnancies at risk, if available (e.g., chromosomal syndrome, metabolic error)
- Discussion of sharing information with family to encourage evaluation, available carrier testing, and counseling

Psychosocial, educational, familial
- Family counseling; support for acceptance of limited lifestyle for individual with visual impairment with or without other handicaps
- Family education for dealing with handicap
- Assessment of special educational facilities, educational materials, visual aids
- Career counseling
- Assessment of family/community resources, such as special services for the blind, support groups, recreational opportunities

Medical
- Medical intervention as indicated by diagnosis and clinical manifestations
- Periodic screening of asymptomatic individuals at risk for onset of visual impairment (e.g., open-angle glaucoma)

SUGGESTED READING

Cotlier E, Maumenee IH, Berman ER, (eds): *Genetic Eye Diseases: Retinitis Pigmentosa and other Inherited Eye Disorders*. (Birth Defects Original Article Series) vol 18, no. 6. New York, Alan R. Liss, 1982.

Goldberg MF, (ed): *Genetic and Metabolic Eye Disease*, Boston, Little, Brown & Co, 1974.

Krill AE: *Hereditary Retinal and Choroidal Diseases,* vol 1. Evaluation. New York, Harper and Row, 1972.

Krill AE, Archer D: *Hereditary Retinal and Choroidal Diseases,* vol 2. Clinical Characteristics. New York, Harper and Row, 1977.

Suger A, Podof S: Ophthalmic aspects of inborn errors of metabolism, in Mausolf FE, (ed): *The Eye and Systemic Disease*. St. Louis, CV Mosby, 1980.

Chapter 30

Hereditary Hearing Loss

Approximately 1/1,000 infants in the United States is born deaf or develops profound hearing loss in early childhood. An equal number of individuals become deaf or severely hard of hearing by age 16. Including later-onset hearing loss, 10% to 15% of the adult population is estimated to be hearing impaired.

Although the many causes of hearing loss include teratogenesis, infections, trauma, drugs, noise pollution, and other environmental effects, a high proportion of hearing impairment is genetic. The degree of disability in genetic forms can range from mild to profound and often constitutes deafness— strictly speaking: hearing loss so severe as to preclude successful processing of linguistic information, with or without a hearing aid. Genetic causes are particularly implicated in congenital or early childhood deafness; they are estimated to account for about 50% of cases.

Well over 100 inherited types of serious hearing loss have been described, including both isolated auditory impairment and syndromes in which deafness plays a major part. In addition, loss of hearing is a potential factor in a host of other genetic conditions. Table 30-1 lists common genetic disorders in which hearing impairment is characteristic (an exhaustive compilation of recognized syndromes and types of hearing impairment can be found in the book *Genetic and Metabolic Deafness,* see Suggested reading).

When congenital deafness is not associated with a known syndrome, it may easily escape attention in early infancy. It may be difficult, as well, to establish a precise diagnosis for any isolated hearing loss, once it is identified, because similar phenotypic features may result from environmental causes and form defects in one or more different genes involved in the development of hearing.

This chapter will summarize common types of hereditary hearing impairments, discuss genetic considerations, and give some guidelines for referral for an audiologic genetic workup.

ETIOLOGY AND PATHOGENESIS

Detailed descriptions of known pathologic lesions in auditory disorders can be found in standard texts on the subject. In general, hearing impairment can be divided into two major categories: conductive and sensorineural. Either may be responsible for hearing deficit, but both may be present in the same individual. The lesion may be due to developmental malformation or to degenerative changes occurring later in life. The genetic defect underlying the hearing loss may be

TABLE 30-1
Hearing Impairment as a Major Aspect in Genetic Disorders

DISORDER	TYPE OF HEARING LOSS	OTHER CHARACTERISTICS	MODE OF INHERITANCE
Pigmentary Disorders			
Multiple lentigines syndrome	Congenital sensorineural deafness	Multiple lentigines, ECG abnormalities, ocular hypertelorism, pulmonary stenosis, genital abnormalities, growth retardation, variable expressivity	AD
Cutaneous albinism and deafness	Congenital sensorineural deafness	Symmetric areas of hypo-melanosis and hypermelanosis, white scalp hair with patches of pigmentation, no heterochromia	XL
Waardenberg's syndrome	Congenital sensorineural deafness	Displacement of inner earth, white forelock, heterochromia	AD
Ocular Disorders			
Usher's syndrome	Congenital sensorineural deafness	Retinitis pigmentosa resulting in blindness	AR
Stickler's syndrome	Progressive childhood-onset sensorineural hearing loss	Epiphysal changes, Pierre Robin syndrome, high myopia	AD
Refsum's disease	Progressive childhood- or later-onset sensorineural deafness	Peripheral neuropathy, cerebellar ataxia, retinitis pigmentosa, phytanic acid storage	AR
Wildervanck's syndrome	Severe congenital sensorineural deafness	Retraction of eyeball, cervical vertebral fusion; mostly affected females, ? lethal in male	XD Polygenic
Norrie's disease	Late-onset severe sensorineural deafness	Congenital bilateral pseudoglioma of the retina	XR
Skeletal Disorders			
Otopalatodigital syndrome	Conductive hearing loss	Radiographic alterations of hands and feet, cleft palate, pugilistic facies	XR
Osteogenesis imperfecta	Hearing loss, may be diagnosed as otosclerosis	Group of disorders with increased susceptibility to bone fractures, blue sclerae; wide range of clinical variability	most: AD (may be:AR)
Treacher Collins syndrome	Congenital conductive hearing loss	Downward slanting palpebral fissures, lower-eyelid coloboma, dysplastic ears, micrognathia, malocclusion	AD

TABLE 30-1 *Continued*

DISORDER	TYPE OF HEARING LOSS	OTHER CHARACTERISTICS	MODE OF INHERITANCE
Crouzon's syndrome	Conductive hearing loss	Premature craniosynostosis, frontal bossing, exophthalmos, shallow orbits, maxillary hypoplasia, mandibular prognathism	AD
Multiple synostosis syndrome	Conductive hearing loss	Hand and foot malformations, characteristic facies	AD
Klippel-Feil sequence	Conductive or sensorineural deafness	Short neck, low hairline, fused cervical vertebrae, torticollis, facial asymmetry; usually sporadic	?(may be AD)
Others			
Alport's syndrome	Progressive sensorineural hearing loss	Hematuria, proteinuria, nephritis; more common in males; nephritis and deafness can occur separately in affected individuals	?AD
Pendred's syndrome	Profound congenital sensorineural deafness	Goiter, perchlorate or thiocyanate discharge of radioiodide	AR
Jervell-Lange-Nielsen syndrome	Profound congenital sensorineural deafness	Prolonged QT interval in ECG; risk of sudden death in syncopal attack	AR
Fetal rubella syndrome	Congenital conductive and/or sensorineural deafness	Low birth weight, transient hepatosplenomegaly, purpura, bulging anterior fontanelle, congenital heart disease; unexplained deafness may be only finding	Viral teratogen

Key: AR = autosomal recessive; AD = autosomal dominant; XL = X-linked recessive; XD = X-linked dominant

expressed, as well, in other systems. Hearing deficit has been described in syndromes involving a wide array of other abnormalities, including external-ear malformations; skin, eye, or metabolic disorders; muscle and skeletal abnormalities; neurologic deficits; and renal problems.

DESCRIPTION, GENETIC CHARACTERISTICS AND MODE OF INHERITANCE

Hearing impairment may be inherited as an autosomal dominant, autosomal recessive, X-linked (rare), or multifactorial condition. Although the presence of a known syndrome may clarify etiology and inheritance, often a specific diagnosis cannot be made. In such a case, it may be possible to infer inheritance from the family history. Otherwise, genetic counseling must be based on empiric risk estimates. Risk estimates can change drastically when the birth of additional children with or without hearing loss provides further pedigree information. The distribution of the types of inheritance varies between congenital deafness and later-onset hearing loss.

Severe Congenital or Early Childhood Deafness is more likely to be sensorineural than conductive and, if genetic, is likely to be autosomal recessive. It has been estimated that 40% to 60% of cases with unknown etiology are autosomal recessive, 20% to 30% are autosomal dominant, about 2% are X-linked, and 20% to 30% are due to environmental or multifactorial causes.

When parents with normal hearing have more than one deaf child, recessive inheritance is likely. Many extremely rare forms of hearing loss have been identified in such families. Consanguinity is often a finding—both parents having inherited the same rare recessive gene from a common ancestor. When a deaf child is born to unrelated hearing parents with no informative family history of deafness, and the etiology is not identified, recurrence risks can range from minimal—e.g., when an environmental teratogen or new dominant mutation was the actual cause—to 50%, such as for an autosomal dominant disorder, not penetrant in a gene-carrying parent.

If two generations are affected, the disorder is usually dominant. When one parent is deaf, the empiric chance that the disorder will be transmitted is only about 5%, but if one deaf child is born, the recurrence risk approaches 50%, depending on the penetrance of the particular disorder.

Often, two congenitally deaf people marry. Seventy percent to 80% of such marriages produce only hearing children. Although most of these parents are likely to have a recessive disorder, affected children are born only when the same recessive gene is responsible for the deficit of both parents. In such cases, all children will be deaf. This happens in only 5% to 14% of such marriages, because several different recessive genes can cause congenital deafness. Empiric risk figures must be used until the hearing status of progeny can be considered (see below). In the remaining marriages between deaf individuals, one or both of the parents may have an autosomal dominant disorder; there will be both hearing and deaf children. Estimated recurrence risks under various circumstances are given in Table 30-2.

Mild, Partial, and Later-onset Hearing Losses include many different kinds of deficits. Compared with congenital deafness, the impairment is more likely to be autosomal dominant and more likely to be due to a conductive lesion. Such hearing loss has been classified according to age of onset, rate of progression, severity, and range of tones involved. However, making an accurate diagnosis is highly complex and not always possible when associated symptoms do not suggest a known genetic syndrome. A high proportion of sporadic cases are environmentally caused but may be difficult to distinguish from the genetic forms. When there are other cases in the family, there may be wide variability of expression. Therefore, careful examination of family members is warranted for identification of minimally affected relatives. Pedigree analysis may then reveal a pattern of transmission within a family.

In addition to the syndromes in Table 30-1, some of the most common well-defined genetic forms of hearing impairment are given special emphasis below.

Otosclerosis is one of the most common causes of hearing impairment among adults. Caused by gradual immobilization of the ossicles in the middle ear, which can also affect the inner ear, it is characterized by progressive conductive or mixed conductive and sensorineural hearing loss. Onset of the hearing deficit can be in childhood but usually occurs in the second or third decade of life. Pregnancy can exacerbate the impairment.

About 3/1,000 Caucasians and 3/10,000 Afro-Americans are clinically affected, but the histologic lesion has been estimated to be present in about 8% of whites and 1% of blacks. The isolated impairment is autosomal dominant with about 40% penetrance in families with the clinical disorder. Thus, while offspring of an affected individual are at a 50% risk to inherit the gene for otosclerosis, the risk for developing hearing loss is only about 20%. Children who inherit

TABLE 30-2
Recurrence Risks for Profound Congenital or Early Childhood Deafness of Unknown Etiology

AFFECTED INDIVIDUALS	RISK FOR EACH PREGNANCY (%)
Unaffected parents	
One child affected, parents unrelated	10
One child affected, parents related	10-25
One child affected, two or more unaffected children	5
Two children affected	25
One parent affected	
One parent affected, no family history	5
One parent and one child affected	50
One parent and one grandparent affected	50
One parent and sib(s) of parent affected, no other family history	1
Other family members affected	
Sib(s) of parent affected, parent unaffected	<1
Both parents affected; not related	
No children born yet	10
One unaffected child	5
One affected child	>50
Two or more affected children	50-100
Both parents affected; related	
No children born yet	>50
One unaffected child	10
One or more affected children; no unaffected children	75-100

the gene but have normal hearing are at risk to pass the gene to their children, who may then develop clinical symptoms.

Otosclerosis may also be diagnosed in patients with osteogenesis imperfecta (OI), a group of mostly autosomal dominant disorders (characterized by bone fragility, blue sclerae, and other anomalies). However, the actual cause of the hearing loss may be related to fractures and subsequent deformities of the bones in the ear rather than to true otosclerosis. Since OI has a wide spectrum of severity, ranging from neonatal death to unrecognized minimal expression, a careful medical and family-health history and examination of a young adult diagnosed with otosclerosis may reveal the presence of this genetic disorder in a family (see also Chapter 32).

Alport's Syndrome represents about 1% of genetic deafness. Although association of nephritis and sensorineural deafness characterizes this syndrome, either disorder can occur separately in different affected family members. Expression is variable, and ocular defects may be present. Males are generally affected more severely, often dying of uremia before age 30. The disorder is autosomal dominant with an unusual pattern of inheritance, in that affected women apparently pass on the gene to more than 50% of their offspring; and among the children of affected men, daughters are more likely to be affected than sons. In family-histo-ry analysis, relatives who have either nephritis or are deaf are considered to be affected.

Waardenburg's Syndrome is responsible for an estimated 2% of congenital deafness. It is characterized by lateral displacement of the medial canthi, heterochromia, and congenital sensorineural deafness. There may be an associated white forelock, early graying, and other anomalies, including a cleft palate, which may complicate diagnosis of the hearing loss. Inheritance is autosomal dominant with a 50% risk to pass the gene on to a child. However, since both penetrance and expression are variable, apparently unaffected individuals can pass on the disorder. Careful examination for minimal expression may identify those relatives of an affected individual who may be at risk for having affected children.

Usher's Syndrome is a genetic association of moderate to profound congenital deafness and retinitis pigmentosa, with onset in the first or second decade. This is probably one of the most severe types of inherited deafness—affected individuals become blind, as well as deaf. Additional problems, including mental retardation, have been reported in some families. The syndrome is thought to account for the 10% of retinitis pigmentosa patients who have severe hearing loss and for the 5% to 10% of congenitally deaf persons who develop retinitis pigmentosa. Inheritance of Usher's syndrome is autosomal recessive, with a 25% risk of occurrence for children of carrier parents.

Malformations or missing ears are noted in about 35/10,000 live births. Minor ear malformations are often sporadic and isolated defects. Careful examination may reveal audiologic or branchial arch abnormalities. In addition, ear malformations—including total anotia, abnormal placement of the ear, abnormal or large auricles, branchial fistulas, and preaurical pits and tags—are associated with a host of genetic and some teratogenic syndromes. These syndromes are usually characterized by other serious symptoms, such as renal malformations, cardiovascular defects, skeletal dysplasia, genital abnormalities, mental retardation, or multiple anomalies associated with chromosomal disorders. Sometimes the ear malformation may be a diagnostic clue, e.g., in 10% to 20% of cases, ultrasonic examination may reveal renal malformation, thus leading to medical management of the kidney problem. Large, protruding ears in a mentally retarded male may indicate fragile X syndrome.

Malformations of the external ear, particularly anomalies of the pinna, may also reflect anomalies of the middle or inner ear and are thought to be associated with 2% or more of congenital conductive or mixed hearing loss. A family history of ear malformations and/or hearing loss may reveal an inheritance pattern and identify family members who may benefit from an audiologic evaluation.

CLINICAL NOTES

Most commonly, congenital deafness is sensorineural, bilateral, and symmetric. Since there is usually some retention of hearing, particularly in the low-frequency tones, very early use of a hearing aid and auditory training can help to build otherwise unobtainable speech and language skills and prevent emotional problems. Such measures are ideally begun before age 6 months and are less successful after age 2.

Unfortunately, since even deaf babies make cooing noises until about 9 months of age, the diagnosis is often made only when lack of speech development becomes obvious. Therefore, it is important to consider referral for evaluation by an otolaryngologist, if any one of a number of factors raises the suspicion of deafness. Any sign of parental concern should be heeded, and a suggestion of developmental retardation or infantile autism should be viewed as a possible reflection of deafness. Any infant with a family history of deafness should be considered at high risk and should be examined for hearing loss as early as possible. Since onset of profound deafness may occur during early childhood, periodic reassessment is indicated.

Different types of congenital hearing loss may be clinically indistinguishable. Although a precise diagnosis is not necessary for the introduction of management, a careful search for associated abnormalities may reveal a genetic form of deafness, for which specific counseling is available.

Occasionally, awareness of an associated defect can prove to be lifesaving. For example, the untreated cardiovascular anomaly (prolonged QT interval in an EKG) in Jervell-Lange-Nielson syndrome can lead to sudden death. Since effective treatment is available, consideration of cardiac evaluation has been suggested for children with profound sensorineural deafness, even though the occurrence of syndrome is rare.

PROCEDURES FOR DIAGNOSTIC CONFIRMATION

Referral to otolaryngologist and audiologist:
- Behavorial screening (useful for infants under 6 months)
- Auditory evoked potential (useful for infants under 6 months)

Air- and bone-conduction threshhold studies
Speech threshhold studies
Speech discrimination assessment
Impedance audiometry, including tympanometry and acoustic reflex evaluation
Radiology of ears and skull

Assorted diagnostic procedures to investigate associated symptomology for possible other genetic disorders underlying the hearing impairment (e.g., endocrine evaluation, neurologic examination, chromosome analysis)

CONSIDERATIONS IN MANAGEMENT

Genetic
- Pedigree analysis
- Hearing test of parents and siblings to rule out unrecognized hearing loss (unimpaired hearing in parents suggests autosomal recessive hearing loss in offspring)
- Established etiology, if possible
- Risk counseling (use an interpreter, if necessary, when counselee is deaf; consider that deaf counselees may not view deafness as a serious handicap and thus may not express interest in risk counseling or early diagnosis)
- Consideration of reproductive options (e.g., donor insemination for future pregnancies of carrier couples for recessive deafness; adoption)
- Discussion of sharing information with other relatives to encourage evaluation, risk assessment, early diagnosis, and treatment

Psychosocial, educational, familial
- Family counseling; support for acceptance of limits for individual with hearing impairment with or without other handicaps; consideration of psychotherapy
- Education for early auditory training and use of hearing aids
- Assessment of special education facilities: preschool and school programs for hearing impaired; programs for teaching sign language and lip-reading; mainstream education with special classes for speech and language skills
- Career counseling
- Assessment of family/community resources, such as special telephone services for the hearing impaired, support groups, recreational facilities

Medical
- Medical or surgical intervention as indicated by diagnosis and clinical manifestations
- Periodic screening of individuals at risk for hearing loss

SUGGESTED READING

Fraser GR: *The Causes of Profound Deafness in Childhood.* Baltimore, Johns Hopkins University Press, 1976.

Konigsmark BW, Gorlin RJ: *Genetic and Metabolic Deafness.* Philadelphia, WB Saunders, 1976.

McCurdy, JA, Jr: Congenital hearing impairment. Am Fam Physician 21:101, 1980.

Chapter 31

Genetic Disorders of the Skin

Collectively, genetic disorders of the skin represent a substantial proportion of dermatologic problems. Many are single-gene disorders and, because of their high visibility, can be readily documented in the family history. Although seldom fatal, they confer a burden of cosmetic disfigurement that may have psychologic effects disproportionate to their medical severity.

This chapter will discuss the most important categories of genetic skin disease, and will tabulate selected common genetic disorders and syndromes in which skin manifestations are prominent or diagnostic (see Table 31-1).

ETIOLOGY AND PATHOGENESIS

Genetic skin disorders, as well as many genetic syndromes, have cutaneous manifestations. The specific gene action and pathogenic mechanisms vary with the underlying defect.

TABLE 31-1
Common Genetic Multisystem Disorders With Dermatologic Findings as a Presenting or Diagnostic Feature

DISORDER	DERMATOLOGIC FINDING	OTHER CHARACTERISTICS	MODE OF INHERITANCE
Eruptions, Cysts, Nevi, Growths			
Gardner's syndrome	Epidermal and sebaceous cysts, inclusion cysts, mainly on face and scalp	Colonic polyps with high risk of malignancy, osteomas, desmoid tumors	AD
Acrodermatitis enteropathica	Bullous or verrucous eruptions on extremities, pustular around orifices; alopecia	Diarrhea, general debility; malabsorption of zinc	AR
Hypercholesterolemia	Xanthoma on tendons and eyelids in third to fourth decade; earlier in homozygote	Elevated serum cholesterol, corneal changes, coronary disease; more severe in homozygotes who are not expected to survive young adulthood	AD

(Continued)

DISORDER	DERMATOLOGIC FINDING	OTHER CHARACTERISTICS	MODE OF INHERITANCE
Fabry's disease	Dark-red or black papules or macules, particularly on thighs, scrotum, and periumbilical area	Decreased α-galactosidase A activity, acroparesthesia, corneal opacities, renal-cardiovascular problems	XR
Pseudoxanthoma elasticum	Yellowish nodular or reticular thickening of skin, especially on neck and in axilla; angioid streaks	Hypertension, GI bleeding, vascular changes	AR(AD)
Basal cell nevus syndrome	Basal cell nevi and/or carcinomas, dyskeratotic pits on hands and feet; dental anomalies	Skeletal anomalies, jaw cysts	AD
Pigmentation Anomalies			
Multiple lentigines (LEOPARD) syndrome	Multiple dark lentigines, especially on neck and trunk	ECG abnormalities, sensorineural deafness, ocular hypertelorism, pulmonary stenosis, genital abnormalities, growth retardation, variable expressivity	AD
Adrenocortical unresponsiveness to ACTH	Hyperpigmentation	Lethargy, feeding problems, hypoglycemic episodes, seizures, psychomotor retardation	XR (AR)
Fanconi's anemia	Brownish pigmentation of skin	Small stature, pancytopenia, hypoplasia to aplasia of thumb, increased chromosome breakage	AR
Peutz-Jeghers syndrome	Brown-black macules on lips, buccal mucosa, and hands	Rectal bleeding, rectal prolapse, intestinal polyps (particularly in jejunum), ovarian tumors	AD
Incontinentia pigmenti	Neonatal linear inflammation and bullae, later turning into hypertrophic verrucous bands; then bandlike pattern of hyperpigmentation, patchy alopecia; dental anomalies	Ocular, CNS, and skeletal anomalies, most patients female; ?lethal in male	XD

TABLE 31-1 *Continued*

DISORDER	DERMATOLOGIC FINDING	OTHER CHARACTERISTICS	MODE OF INHERITANCE
Neurofibromatosis	Neurofibromas, nodules along peripheral nerves, café-au-lait spots, axillary freckles	Intracranial tumors, bony deformities, hypertension; variable expression; slow progression	AD
Hemochromatosis	Metallic-grey hue or bronze skin with areas of hyper-pigmentation, loss of pubic or axillary hair	General weakness, loss of libido, diabetes, secondary hypogonadism, liver cirrhosis	AR
Tuberous sclerosis	Ash-leaf spots, adenoma sebaceum	Seizures, mental retardation, phakomas	AD
Waardenburg's syndrome	Patchy, depigmented macules, white forelock	Laterally displaced inner canthi, heterochromia, sensorineural deafness	AD
Albinism	Depending on type, various degrees of hypopigmentation, general or partial, of skin, hair, eyes; some improve with age	Nystagmus, photophobia and other visual problems in most types; sensorineural deafness in one type; bleeding disorders in one type	Varied
Chédiak-Higashi syndrome	Depigmentation and grey sheen of skin and hair	Giant granules in leukocytes of peripheral blood, recurrent infections, ocular complications, neuropathy, shortened life span	AR
Photosensitivity			
Xeroderma pigmentosum	Photosensitivity leading to premalignant and malignant skin lesions, pigmentation anomalies	Photophobia, defective DNA repair after ultraviolet exposure	AR
Bloom's syndrome	Sun-sensitive telangiectasic erythema, mainly on face	Short stature, characteristic facies, increased susceptibility to malignancy, increased sister-chromatid exchange	AR
Porphyria, erythropoietic and variegate	Photosensitivity, leading to bullous lesions with ulceration and disfiguring scars; hypertrichosis; discoloration of teeth	Abnormality of phorphyrin metabolism hemolysis, splenomegaly, neurologic dysfunction	AR AD

(Continued)

DISORDER	DERMATOLOGIC FINDING	OTHER CHARACTERISTICS	MODE OF INHERITANCE
Erythropoietic protoporphyria	Photosensitivity, burning erythema, mild, chronic skin lesions	Increased erythrocyte protoporphyrin level, ? increased incidence of gallstones	AD
Hartnup disease	Dry, scaly skin, photosensitivity	Aminoaciduria, diarrhea, pellegralike intermittent psychiatric disorder; may be asymptomatic	AR
Others			
de Lange's syndrome	Hirsutism, synophrys	Physical and mental retardation, microcephaly, long philtrum, carplike mouth	?
Nail-patella syndrome	Hypoplastic nails	Absent or hypoplastic patellae, webbing of elbow, nephropathy, skeletal abnormalities	AD
Ectrodactyly-ectodermal dysplasia–clefting syndrome	Mild hyperkeratosis, light complexion, thin, light hair, missing or malformed teeth	Ectrodactyly, absent lacrimal puncta, cleft lip +/− cleft palate, photophobia, ocular anomalies	AD
Ehlers-Danlos syndrome	Easy bruising, friability and stretchability of skin	Joint laxity, hiatal hernia, bowel rupture, pneumothorax, CNS bleeding	Varied
Ataxia-telangiectasia	Telangiectasia	Chromosome instability, cerebellar ataxia, immunoglobulin deficiencies, neurologic incapacitation, pulmonary or neoplastic disease	AR
Refsum's disease	Ichthyotic skin changes	Abnormal plasma or tissue concentrations of phytanic acid; retinitis pigmentosa, peripheral neuropathy, cerebellar ataxia	AR

Key: AD = autosomal dominant; AR = autosomal recessive; XR = X-linked recessive; XD = X-linked dominant

DESCRIPTION, GENETIC CHARACTERISTICS, AND MODE OF INHERITANCE

Dermatologic conditions often vary and overlap. Therefore, precise diagnosis and/or family-history analysis may be required before the mode of inheritance is clarified and recurrence risks can be established.

The major forms of genetic skin disorders can be classified clinically and etiologically:

Scaling Keratoses. Trivial mild scaling can occur in normal newborns. However, a persistant unidentified scaling disorder in an infant, particularly on the extremities, suggests further evaluation, as it may be associated with bone abnormalities and mental retardation. Also, scaling keratoses accompany a host of inherited syndromes.

Psoriasis is the most common and most benign scaling keratosis. It has been estimated to affect 1% to 3% of the white population, and a lesser proportion of blacks. In most patients, the lesion is mild and confined and expressed only during stress. In a few persons, however, the disorder is serious, chronic, and potentially disabling.

All forms of inheritance have been reported. However, except for families with obvious autosomal dominant pedigrees, polygenic multifactorial inheritance is considered the most likely. Empiric recurrence risks are given in Table 31-2.

TABLE 31-2
Empiric Recurrence Risks for Psoriasis

AFFECTED RELATIVE(S)	RECURRENCE RISK (%)*
One first-degree relative	7.5-17
Two first-degree relatives, other than both parents	16-31
One parent in a family with an autosomal dominant inheritance pattern and reduced penetrance	Approaching 50
Both parents	50-75

*Severity may be greater or lesser than that in the index case(s)

Hereditary ichthyosis encompasses scaling keratoses ranging from mild, easily managed cosmetic problems to a neonatally lethal disease. Manifestations range from varying degrees of dry flaky skin to conditions with thick rhomboidal scales. Inheritance depends on the specific type, with the milder varieties more likely to be autosomal dominant, and the severe ones autosomal recessive. Clinically, however, some types may be difficult to distinguish. A family history may be useful, for both genetic counseling and prognosis. The following are the principal forms of isolated ichthyosis.

Ichthyosis vulgaris (autosomal dominant): With onset in infancy or later, the disorder affects approximately 1/300 individuals. The relatively mild cosmetic lesions are mostly on the extremities. Associated atopic dermatitis may lead to complicating infections.

X-linked ichthyosis (X-linked recessive): Onset is at birth or in early infancy with roughly 1/6,000 males thought to be affected. It resembles ichthyosis vulgaris but has a wider distribution, seldom predisposes to atopic dermatitis, and carries increased risk for cataracts.

Lamellar ichthyosis (autosomal recessive): One in 3,000,000 newborns has this severe form. The infant may be a ''collodion baby,'' that is, covered with a smooth, glistening layer of epidermis, which is shed in two to three weeks. Although the underlying skin may be normal, the entire body is often covered with thick scales. These may clear up spontaneously in early infancy or remain as a serious ichthyosis.

Epidermolytic hyperkeratosis (autosomal dominant): This rare, congenital generalized or localized disorder may also present as a collodion baby. The disorder is distinguished by hyperkeratosis, erythema, and erosions. Severity decreases with age.

Harlequin fetus (autosomal recessive): In this rare, lethal form of congenital ichthyosis, the infant has armadillolike scales and deep fissures. The disorder must be distinguished from lamellar ichthyosis, which has a far better prognosis.

Darier's disease (autosomal dominant) is a form of scaling keratosis. It is characterized by a greasy appearance, with lesions in seborrheic areas. Bullous lesions may also occur.

Keratosis palmaris et plantaris (tylosis) is a thickening of the palms and soles. A horny epidermis develops, usually at about six months or later. It may have only trivial effects, or become thick enough to be crippling, with involvement of fingers, toes, and nails. The inheritance is usually autosomal dominant with incomplete penetrance.

Blistering Disorders appear as bullosa and other forms, dominant or recessive, of varied origin, onset, severity, and symptoms.

Epidermolysis bullosa includes several forms, ranging from mild to lethal. Their common characteristic is the formation of blisters in response to otherwise trivial trauma. The forms differ in their severity; in the skin layer in which blisters originate; and in the nature of healing, with or without scars. The mode of inheritance depends on the subtype. Principal forms are:

Epidermolysis bullosa simplex (autosomal dominant): This nonscarring form is most common. It may be generalized, but more likely is restricted to areas exposed to trauma, such as hands and feet. Onset is in childhood, and severity may decrease at puberty.

Epidermolysis bullosa letalis (autosomal recessive): This type is also nonscarring, but severe deformations occur (already in utero). Ubiquitous bullae, renal, and cardiovascular anomalies lead to spontaneous abortion, stillbirth, and death in infancy.

Epidermolysis bullosa dystrophica (autosomal dominant and autosomal recessive forms): This serious scarring disorder may be congenital or of later onset. It can lead to severe disfigurement, including epidermal overgrowth of digits to form a mitten-like hand, stenosis of the alimentary tract, mental retardation, early death, or lifelong disability.

Other blistering disorders of genetic origin include the following four. *Hailey-Hailey disease* (benign familial pemphigus), is an autosomal dominant condition with recurrent lesions on the neck and in the groin and axillae. It is important to distinguish this entity from yeast infections, which it may resemble. *Acrodermatitis enteropathica*, with onset in childhood, is thought to be autosomal recessive. It is characterized by vesicobullous lesions, diarrhea, alopecia, and other symptoms related to faulty absorption of zinc. The condition is notable because, while it carries a very poor prognosis if not recognized, it responds remarkably well to zinc supplementation, which can lead to normal development. Among the disorders of prophyrin metabolism, *erythropoietic porphyria* (autosomal recessive) and *variegate porphyria* (autosomal dominant) have, in addition to other symptoms, disfiguring blistering ulceration and scarring when skin is exposed to sunlight.

Pigmentation Disorders. Hypopigmentation or hyperpigmentation of the skin in genetic disorders is generally found in combination with other symptoms. Some of the hereditary conditions in which skin coloration is a major feature, are discussed here.

Albinism (abnormal melanin metabolism) is one of the most widely recognized and striking genetic disorders. Of a number of types of albinism that have been identified, all but one (ocular albinism) involve generalized depigmentation.

Most forms (including those discussed below) are inherited as autosomal recessive traits. Carrier testing is available for some types of albinism. Selective mating for skin pigmentation can complicate the genetic picture.

- When both parents are unaffected, but carry the same recessive gene, the risk is 25% in each pregnancy for an affected child.
- When both parents are affected with the same type of albinism, all children will also be affected.
- When each parent is affected with a different type of albinism, offspring are not at increased risk for either type.
- When one parent has albinism and the other parent is an (unaffected) carrier of the same type of albinism, the risk is 50% in each pregnancy for an albino child.
- When one parent has albinism and the other parent does not carry the gene for that type of albinism, offspring are not at an increased risk for albinism.

Although there are clinical differences, it is not always easy to distinguish the various forms of albinism without biochemical analysis.

Tyrosinase-negative oculocutaneous albinism occurs in roughly 1/39,000 American whites and 1/28,000 American blacks. In the absence of tyrosinase activity, skin and hair stay white throughout life and the deficiency of melanin leads to an increased suseptibility to skin cancer. Eye color is translucent gray to blue, without pigment accumulation and with a red reflex. Ocular manifestations, such as photophobia and nystagmus, are severe. Carriers can be identified through hair-bulb analysis, and prenatal diagnosis has been achieved with analysis of hair bulbs obtained by fetoscope. A rare variant (Hermansky-Pudlak type) is associated with a bleeding tendency that is both progressive and potentially life-threatening.

Tyrosinase-positive oculocutaneous albinism is estimated to occur in 1/15,000 blacks and 1/40,000 whites in the United States and is particularly common in some Indian tribes. Although early clinical characteristics are very similar to the tyrosinase-negative form, partial activity of tyrosinasc permits gradual accumulation of pigment with age, so that hair becomes yellow or reddish and the skin may show freckles or pigmented nevi. Ocular symptoms are less severe and improve with age.

Chédiak-Higashi syndrome is a rare form of tyrosinase-positive albinism, in which hair color tends to be blond or light brown with a gray sheen, and the skin may also be grayish. While oculocutaneous symptoms are moderate, the disorder is usually fatal in childhood due to leukocyte abnormalities that lead to repeated infections, neuromuscular manifestations, and a leukemialike condition.

Vitiligo, generally not genetic, may be associated with autoimmune endocrine disorders, which may be autosomal dominant. Splotchy depigmented macules may also be a feature of the autosomal dominant *Waardenburg's syndrome,* which comprises mainly laterally displaced inner canthi, heterochromia, a white forelock, and sensorineural deafness (see Chapter 30).

Other abnormal pigmentation may be diagnostic of a genetic disease. For example, pigmented lines, forming after healing of neonatal inflammation and bullae, are the hallmark of *incontinentia pigmenti.* This disorder, which may also include alopecia, tooth anomalies, and other abnormalities, is found almost exclusively in females. It is thought to be X-linked dominant and lethal in the male. Recurrence risk for offspring of an affected female is 50% for daughters and negligible for sons surviving to term.

Ash-leaf-shaped depigmented spots, angiofibromas, and seizures are diagnostic of *tuberous sclerosis.* Pigmented café-au-lait spots and axillar freckling distinguish *neurofibromatosis.* These disorders are discussed in detail in Chapters 10 and 11.

Ectodermal Dysplasias. Dysplasic development of ectodermal derivatives is a frequent finding in genetic disorders of all sorts. Aberrant distribution, color, texture, or growth patterns of hair, as well as abnormal development of dentition or nails may present as isolated hereditary conditions. They may also be characteristic of metabolic disorders, or be among the clinical characteristics of assorted syndromes.

Hypohydrotic ectodermal dysplasia, is characterized by a variable degree of sweating deficiency and heat intolerance, sparse hair, missing and conical teeth, dystrophic nails, and dry, rough skin. Autosomal dominant and autosomal recessive, as well as X-linked recessive inheritance has been reported.

Anhydrotic ectodermal dysplasia is more severe, with special problems of heat regulation, as well as sparse hair, dry skin, dental abnormalities, and increased susceptibility to infection. Facial characteristics may include a short upper lip and saddle-shaped nasal bridge. Occasionally, mental retardation is associated. An X-linked recessive condition, the full-blown disorder is seen only in males, but female carriers may have missing teeth and areas on the body with absence of sweating.

Chromosomal Fragility Syndromes. Skin manifestations are consistent characteristics of the syndromes discussed below, which are all marked by DNA instability and associated with severe morbidity and a shortened lifespan. Since an increased frequency of chromosome breaks, rings, and other chromosome abnormalities can be demonstrated in cytogenetic analysis of amniotic fluid, prenatal diagnosis for these disorders is becoming a reality (see Clinical notes). They are all autosomal recessive, with a 25% recurrence risk for siblings of an affected individual. Reliable carrier identification, other than the birth of an affected child, has not yet been achieved. The chromosomal fragility syndromes mentioned below are also discussed in Chapter 5.

Xeroderma pigmentosum is characterized by photosensitivity, photophobia, abnormal skin pigmentation, and a high incidence of skin tumors after exposure to ultraviolet light. The genetic lesion is well defined as a defect in the excision of pyrimidine dimers in the repair of DNA damage.

Bloom's syndrome. Telagiectasic lesions, as well as short stature and characteristic facies, are associated with *Bloom's syndrome*. Patients with this disorder are highly sensitive to ultraviolet light and have an increased susceptibility to lymphoproliferative malignant disease.

Fanconi's anemia demonstrates cutaneous hyperpigmentation and an increased susceptibility to malignancies that accompany pancytopenia and abnormalities of the skeleton, heart, and kidneys. *Dyskeratosis congenita*, an X-linked disorder, characterized also by nail dystrophy and leukoplakia of the oral mucosa, may resemble Fanconi's anemia and must be distinguished from it.

Ataxia telangiectasia. Progressive ataxia and oculocutaneous telangiectasia, as well as immune deficiencies and susceptibility to lymphoproliferative neoplasm, are characteristic of *ataxia telangiectasia,* which is associated with a sensitivity to ionizing radiation.

Immune-related Skin Lesions include frequently seen atopic dermatitis and some uncommon manifestations with apparent family ties.

Atopic dermatitis, one of the most common skin problems, has been estimated to affect 0.7% of the population. An allergic phenomenon, its effects are strongest in childhood and often resolve by puberty. A strong familial incidence exists, but its nature is not clear. Although no two studies agree on recurrence risks, it is known that if both parents are affected, offspring are at increased risk. Affected individuals frequently have a strong family history of asthma and/or other allergies, and are themselves at risk for yet another type of allergy.

Phenylketonuria and ataxia telangiectasia are two of the other genetic disorders whose skin lesions resemble atopic dermatitis.

Scleroderma and lupus erythematosus both have skin manifestations and have been reported occasionally as familial disorders. Specific immune-system abnormalities and evidence for a clear-cut genetic predisposition are currently under investigation.

Connective-tissue Disorders. This large group of genetic conditions is discussed in Chapter 32, but merits mention here, because skin manifestations may be a primary feature. *Pseudoxanthoma elasticum*, for example, may present, usually in the second decade, with characteristic angioid streaks and skin lesions of yellowish nodular or reticular thickening, particularly around the neck and the axilla. Weakening of connective tissue threatens vision, and may cause GI bleeding and arterial rupture. Recognition of the syndrome, which has been reported as both autosomal dominant and autosomal recessive, may lead to early management, such as vigorous control of blood pressure, as well as genetic counseling.

The genetically heterogeneous *Ehlers-Danlos syndrome* may feature skin friability and cutaneous hyperextensibility. *Cutis laxa* is characterized by generalized looseness and redundancy of the skin. Both may be associated with serious visceral complications.

Congenital vascular lesions, such as hemangiomas, commonly raise concern about potential recurrence, but are usually not genetic.

CLINICAL NOTES

About 7% of newborns have been estimated to have congenital skin lesions. Consideration of a genetic condition may be important for prognosis, recurrence-risk assessment, and for potential prenatal diagnosis, which is becoming increasingly available for serious skin disorders.

Many cutaneous disorders have overlapping signs and symptoms and there are not always reliable clinical guidelines for suspicion of genetic etiology. However, a family history of a particular disorder or a careful search for anomalies associated with unexplained skin lesions may identify a genetic syndrome.

Prenatal diagnosis is becoming more available for diagnosing hereditary skin disorders in pregnancies known to be at increased risk.

Most analyses for prenatal diagnosis of genetic skin disease are currently performed only at specialized centers, where referrals or specimens may be accepted upon request. Reports of disorders with dermatologic manifestations for which diagnosis in utero has already been accomplished include several types of ichthyosis (fetoscopy); epidermolysis bullosa letalis, simplex and dystrophica (fetoscopy, sonography); xeroderma pigmentosum, ataxia-telangiectasia, Fanconi's anemia (chromosome analysis of amniocytes); congenital erythropoietic porphyria (amniotic fluid porphyrin studies); and tyrosinase-negative oculocutaneous albinism (fetoscopy). Prenatal sex determination can be offered for pregnancies at risk for incontinentia pigmenti.

PROCEDURES FOR DIAGNOSTIC CONFIRMATION

Among potentially useful procedures are:

- Wood's light examination for identification of pigmentary changes (e.g., tuberous sclerosis)
- Chromosome analysis for chromosomal breakage syndromes (e.g., Bloom's syndrome)
- Laboratory tests in specialized centers (e.g., hair-bulb test for albinism)
- Consideration of referral for dermatologic workup
- Evaluation of associated symptomatology

CONSIDERATIONS IN MANAGEMENT

Genetic
- Establish diagnosis
- Risk counseling
- Consideration of prenatal diagnosis for pregnancies at increased risk for disorders for which it is indicated and available
- Consideration of reproductive options (e.g., donor insemination for future pregnancies of carrier couples for recessive disorders, adoption), when prenatal diagnosis is not available or not desired
- Discussion of sharing information with other family members to encourage evaluation for minimal expression of carrier status, risk assessment, early diagnosis, and treatment

Psychosocial, educational, familial
- Family counseling; support for acceptance of disability and/or cosmetic disfigurement; support for acceptance of loss of a child
- Consideration of psychotherapy for patient and/or family
- Education for home care and hygiene
- Education for cosmetic management
- Assessment of available support groups

Medical
- Therapy as indicated by diagnosis

SUGGESTED READING

Blandau RJ, (ed): *Morphogenesis and Malformation of the Skin*. (Birth Defects Foundation, Original Article Series) vol 17, no 2. White Plains, N.Y., March of Dimes Birth Defects Foundation, 1981.

Esterly NB, Elias S: Antenal diagnosis of genodermatoses, *J Am Acad Dermatol* 8:655, 1983.

Holbrook K, (ed): *Prenatal Diagnosis of Inherited Skin Disease*. (Seminars in Dermatology). New York, Thieme-Stratton, 1984.

Montagna W, Pasztor L, Quevedo WC, Jr, et al: Cutaneous genetics. *J Invest Dermatol* 60:343, 1973.

Solomon LM, Esterly NB: *Neonatal Dermatology*, Philadelphia, WB Saunders, 1973.

Weston WL: *Practical Pediatric Dermatology*, Boston, Little, Brown, 1979.

Chapter 32

Heritable Disorders of Connective Tissue

Heritable disorders of connective tissue can be defined as defects in the supporting structures of the body. While supporting structures include a large variety of elements, such as skin, bone, cartilage, etc., the term connective tissue disorders is generally used to denote only those conditions in which there is malfunction of the extracellular matrix—particularly of the collagens, elastin, and the proteoglycans. Among them, more than 100 genetic diseases have been identified, with a wide array of phenotypic effects. Increasingly noted heterogeneity has led to extensive subclassification within disorders once thought to be specific, and raised awareness has led to greater ascertainment. However, connective tissue disorders are thought to be even more common than current estimates indicate, because the range of expression appears to be so great that many mild cases never come to medical attention.

This chapter will review the most common genetic disorders involving primary connective-tissue defects. Connective-tissue disease associated with other genetic disorders will be discussed briefly.

ETIOLOGY AND PATHOGENESIS

The great clinical variability of the disorders suggests a large number of biochemical alterations. A few specific basic defects have been identified, and current studies of molecular defects in the incredibly complicated development of collagens are rapidly providing further information. However, most of the disorders discussed here are not yet clearly understood. It is known that in all of them, one or more connective-tissue matrices cannot function normally. These abnormalities affect the tensile strength, elasticity, resistance to deformation, and growth patterns of relevant body structures, leading to skin friability, hyperextension, and aneurysms.

Since the essential function of most organs is not compromised, most connective-tissue disorders do not involve mental retardation, neuromuscular disease, or other systemic malfunction, except for those effects that are secondary to the structural weaknesses generated by the underlying defects—for example, intestinal rupture, aneurysms, dislocation of lenses.

DESCRIPTION, GENETIC CHARACTERISTICS, AND MODE OF INHERITANCE

The common primary heritable connective-tissue disorders are inherited as single-gene conditions. Most of the primary connective-tissue disorders discussed below are autosomal dominant with a wide range of expression.

- There is a 50% recurrence risk for each offspring of an affected individual or for a sibling of a diagnosed patient, if one of the parents is found to be affected, even minimally, on close examination.
- If the disorder can be definitely ruled out in the parents of a sporadic case, the risk for siblings is not increased over the general population risk. However, an isolated case can also be the first occurence of a recessive form.
- The few subtypes that have autosomal recessive inheritance tend to be the most severe forms and may not be compatible with survival past the neonatal period. The recurrence risk for siblings is 25%.

Marfan's Syndrome is characterized by a tall thin habitus with long extremities; arachnodactyly; kyphoscoliosis; bilateral upwardly displaced ectopia lentis, with a high risk for myopia and retinal detachment; mitral valve prolapse; aortic dilation; and aortic aneurysm leading to rupture. The differential diagnosis should include some forms of Ehler-Danlos syndrome, homocystinuria, and contractual arachnodactyly.

Marfan's syndrome is an autosomal dominant disorder with a 50% recurrence risk for offspring of affected individuals. About 15% of isolated cases are thought to be new mutations. Increased paternal age has been reported frequently in this group.

Marfan's syndrome has been estimated to have a prevalence of 1/25,000 to 1/5,000 in the United States. However, since clinical expression is extremely variable, true prevalence may be even higher than noted because many cases remain undiagnosed.

To establish the family history and to identify affected family members in whom the disorder has not yet been recognized, a careful study of relatives should be considered when a case of Marfan's syndrome has been diagnosed. Since sudden death at an early age from a ruptured aortic aneurysm may be the first indication of the disorder (average age of 32 at death has been reported), investigation of autopsy records of family members who died early may be informative.

When Marfan's syndrome is identified early, surgical and medical intervention, as well as an appropriate life-style, may improve the quality of life and extend the life span (see Management, Medical).

Restriction enzyme markers, used experimentally, are expected to facilitate linkage studies which will provide prenatal and postnatal diagnosis for informative families.

Osteogenesis Imperfecta (OI) encompasses a group of disorders that are characterized by some degree of increased bone fragility (and other multisystem anomalies), with clinical expression that ranges from neonatally lethal, to cases so mild as to escape detection. The incidence of the group of disorders as a whole is estimated at 1/25,000. Most are thought to be due to abnormalities in collagen synthesis. Extensive genetic heterogeneity is likely and may underlie the variability of expression, making classification difficult.

It has been suggested that the presence of wormian bones in the calvaria distinguishes all forms of OI from other types of osteopenia. Within that limit several classification systems have been proposed, but none have proved wholly adequate, since the great amount of reported intrafamilial variability suggests that the various subtypes are not necessarily distinct entities.

With this in mind, the most common clinical subtypes can be described. It is estimated that all but about 10% of cases are autosomal dominant, which may be subject to documentation in specific families by careful examination of apparently unaffected family members who may have blue sclerae after infancy (they are common in healthy babies), mild hearing loss, or a history of one or more bone fractures.

Type I OI is characterized by postnatal onset of bone fragility, blue sclerae, and variable degrees of dentinogenesis imperfecta and hearing loss. Transmission is thought to be autosomal dominant, even though an occasional severe, neonatal case has been reported in families with this type.

Type IV OI is usually relatively mild, autosomal dominant, with postneonatal onset and variability of expression. Dentinogenesis imperfecta is present, but sclerae are white. A more severe course with dwarfing due to multiple fractures has also been noted.

Types II and III OI are congenital and severe, and may lead to neonatal death, usually from intracranial hemorrhage or respiratory distress. Multiple bone fractures occur prenatally, producing a newborn with "crumpled" or bowed bones, limb shortening, and limb deformities. Survival past infancy results in short stature, and further deformities, such as kyphoscoliosis, pectus excavatum, and deafness. Blue sclerae may or may not be present. Inheritance is thought to be autosomal recessive, but the occurrence of such cases in kindreds with dominant pedigrees suggests that autosomal dominant cases are possible. Sporadic cases may also represent a new dominant mutation.

Prenatal diagnosis by ultrasound and/or biochemical analysis is becoming increasingly available for cases of OI with prenatal onset. Restriction enzyme linkage studies hold promise for prenatal diagnosis in informative families with dominant OI.

Ehlers-Danlos Syndrome constitutes a highly heterogeneous group of connective-tissue disorders that share similar facial features and a constellation of symptoms, principally joint hypermobility, skin fragility and hyperextensibility, and bruising. At least 10 subtypes have been described with varying proportions of these manifestations and with one or more of a number of other features, including prematurity due to early rupture of the fetal membranes, hypotonia, blue sclerae, papyraceous scars, cutaneous pseudotumors, kyphoscoliosis, heart defects, hernias, spontaneous perforations of the GI tract, arterial rupture, short stature, fragility of ocular elements, extensive periodontitis, joint dislocation, and arthritic changes. Clinical manifestations range from extremely mild hypermobility, through obvious but essentially nonalarming conditions, such as the talents of a circus contortionist, to severe, often lethal, illness.

In the more severe forms, avoidance of trauma and use of protective padding are recommended. Using special surgical and obstetric strategies may reduce the risks of these procedures.

Prenatal diagnosis is possible for those subtypes in which the basic defect is known. In some of the rarer subtypes of Ehlers-Danlos syndrome a specific abnormality in collagen biosynthesis has been identified and ongoing biochemical research promises to clarify more.

Overall incidence has been estimated at 1/150,000, but it is thought that actual incidence is much higher, since mild cases are rarely ascertained. The more common forms include:

Types I(gravis), II(mitis), III(benign hypermobolity). The basic defects of these three subtypes have not been elucidated. Together they account for 85% of diagnosed cases, with a descending order of severity. All are autosomal dominant conditions with a 50% recurrence risk for offspring of affected individuals. Careful examination of relatives of a diagnosed patient may reveal minimal signs in mildly affected family members and clarify the family history.

Type IV (ecchymotic) of Ehlers-Danlos syndrome carries the greatest clinical risks. This is the form in which intestinal or arterial rupture of tissues, too fragile for surgical repair, may lead to catastrophic death in childhood or early adulthood. Both autosomal dominant and autosomal recessive inheritance have been reported.

Type V, an X-linked recessive, accounts for about 10% of cases and is characterized mainly by skin hyperelasticity.

Cutis Laxa is a rare disorder, characterized by excessive loose skin, which does not spring back on release like the extended skin in Ehlers-Danlos syndrome. It is thought to be an elastin defect, and various modes of inheritance have been reported. The autosomal dominant form is restricted to sagging skin, whereas the severe autosomal recessive form includes the risk for hernias, emphysema, and diverticuli. An X-linked form has been described, as well.

Pseudoxanthoma Elasticum has an estimated incidence of 1/40,000. At least four subtypes have been identified, all of which have skin manifestations, such as a yellow, raised rash over flexure sites, a macular rash with stretchable skin, or angioid streaks. Eye manifestations and intestinal hemorrhage may also occur. Both autosomal dominant and autosomal recessive inheritance have been noted.

Stickler's Syndrome (arthroophthalmopathy) is characterized by congenital high myopia; radiologically demonstrable epiphyseal changes with bony swellings of the wrist, even in neonates; a high incidence of cleft palate and micrognathia (Pierre Robin anomaly); and a marfanoid habitus. It is compatible with a normal life span, but involves impaired vision resulting from the myopia, late onset sensorineural hearing loss, and increased risk for arthritis. The basic defect is unknown, but is thought to be a connective-tissue abnormality. Since variable expression is the rule, it has been suggested that this syndrome may be a common cause of severe myopia and Pierre Robin anomaly, and may be more prevalent than suspected. Inheritance is autosomal dominant with a 50% recurrence risk for offspring of affected individuals.

Secondary Defects of Connective Tissue may also occur as a result of other underlying genetic disorders.

Prominent among these are the metabolic errors, such as mucopolysaccharidoses, mucolipidoses, homocystinuria, and alcaptonuria (see Chapter 33).

Arthropathies, particularly *arthritis,* are also frequent secondary features of other genetic disorders. Notable among these are *hemochromatosis* (autosomal recessive iron-overload disorder, with bronzing of the skin, weakness, hepatomegaly, loss of libido, and diabetes, found mostly in men, and subject to effective treatment by phlebotomy); *sickle-cell disease*; *familial Mediterranean fever* (autosomal recessive, acute attacks of fever, peritonitis, pleuritis, arthritis); *Stickler*'s *syndrome* (see above); and *Larsen*'s *syndrome* (flat facies, prominent forehead, dislocations of multiple joints, genetic heterogeneity).

A number of isolated *immune-related arthropathies* also aggregate in some families. The nature and extent of genetic components in these conditions are not clearly understood, but an association with histocompatibility genes (HLA system) suggests that the inheritance of a susceptibility factor may be responsible for familial incidence. Table 32-1 gives examples of such disorders, along with the associated HLA type and the relative risk for family members who share the HLA antigen with an affected individual.

The most striking HLA association has been shown with *ankylosing spondylitis*, a progressive spondyloarthropathy, often presenting as low back pain with arthritic changes. Incidence in England has been estimated as 1/2,000 and is thought to be similar in this country. The disorder is most common in young men, and the empiric risk for siblings is 4%. However, since the presence of an HLA-B27 antigen was found in 95% of affected individuals, as compared with 7% in the general population, it has been recognized that if the affected individual is HLA-B27 positive, the risk for siblings with HLA-B27 is greater than that for other siblings. In general, among individuals with the HLA-B27 antigen, as many as 20% exhibit radiographic changes suggesting spondylitis. Having the antigen has been estimated to put them at a roughly 100-fold risk for developing the disorder, compared with those without it. Estimates indicate that first-degree relatives of affected individuals have a 9% risk to develop the disorder if they have the HLA-B27 antigen, but that their risk is less than 1% if they do not (except in rare families, in which autosomal dominant transmission is evident).

Rheumatoid arthritis also has an HLA association (Dw4, Cw3) and, although striking familial patterns are not reported, any familial incidence may be related to the HLA-antigen-connected increased susceptibility. By contrast, *osteoarthritis*, which often aggregates in families, has no known HLA association. Familial aggregation may simply be by chance, since the disorder is so common in the general population (90% of all individuals have been reported to show radiographic evidence by age 40).

CLINICAL NOTES

The lethal neonatal forms in several of the categories of connective-tissue disease constitute a small proportion of total cases, but a large part of the distress generated by this group of disorders. In addition, some of the subtypes carry the risk of unpredictable sudden death after years of apparently good health (for example, from cardiovascular complications in Marfan's syndrome or GI or vascular rupture in Ehlers-Danlos, type IV). Still, in the aggregate, the connective-tissue disorders represent a group of relatively benign conditions with a normal or near normal life span.

Quality of life may be more impaired than life span (e.g., severe myopia and threat of retinal detachment in Stickler's syndrome; abnormal bleeding and

TABLE 32-1
HLA in Association With Common Immune-Related Arthropathies

DISORDER	OTHER CHARACTERISTICS	HLA ANTIGEN	RELATIVE RISK WITH ANTIGEN (%)
Ankylosing spondylitis	Low-back pain, stiffness, spinal ankylosis leading to immobility, hip involvement, iridocyclitis; male preponderance, onset in second or third decade	B27	87.8
Reiter's syndrome	Arthritis, urethritis, conjunctivitis; common among young males	B27	35.9
Psoriatic arthritis	Rheumatoid arthritis and psoriasis	B13	4.8
Rheumatoid arthritis	Chronic inflammation of synovial membranes, spontaneous remission and exacerbations; presence of rheumatoid factor	Dw4 Cw3	3.0 2.7
Juvenile rheumatoid arthritis	Similar to adult form, with additional features, including fever, rash, leukocytosis, uveitis	B27	4.7
Systemic lupus erythematosus	Chronic inflammatory disease of skin, joints, kidney, nervous system, etc., suggesting abnormal immune response; 60% concordance in monozygotic twins	DRw2 DRw3	?

wound healing in Ehlers-Danlos syndrome; frequent fractures in osteogenesis imperfecta), but connective-tissue strength and associated complications may improve with age. The range of severity in most of the connective-tissue disorders is so broad, that many individuals may be so minimally affected that the condition escapes attention. Thus, careful examination of apparently unaffected relatives for minor symptoms is suggested whenever a case is diagnosed. Heterogeneity within disorders and crossover features between them (e.g., hypermobility in both Marfan's syndrome and Ehlers-Danlos) add further emphasis to the need for detailed examination and differential diagnosis.

Although mildly affected persons may have an excellent prognosis, some may be subject to a sudden catastrophe, as mentioned above, or progression to more serious problems—such as increased incidence of fractures resulting in short stature and/or deformity in osteogenesis imperfecta. Thus, such patients may benefit from medical monitoring to prevent severe manifestations.

Aside from ascertainment of individuals in need of care, examination of relatives will help to assess the pattern of tranmission in the family and identify those family members who are at risk for having affected offspring. This becomes particularly important in view of the fact that wide intra-familial variability has been reported in many of these disorders. Mild expression in a parent does not guarantee the same in offspring.

Clinical suspicion of connective-tissue disease may be raised in the absence of a diagnosed individual when minor symptoms are noted in one or more relatives in a family.

In any event, accurate early diagnosis can lead to available treatment, an appropriate life-style, and potential prenatal diagnosis (in disorders with a known biochemical lesion or a specific ultrasonic feature).

PROCEDURES FOR DIAGNOSTIC CONFIRMATION

Diagnosis is generally based on clinical characteristics with some potentially useful procedures for specific disorders:

- Marfan's syndrome
  - Urinalysis to rule out homocystinuria
  - Serial Monitoring of aortic root by echocardiography
  - Ocular examinations with careful check for lens dislocation
- Osteogenesis imperfecta
  - Radiographic examination for typical bony alterations
- Ehlers-Danlos syndrome
  - Echocardiography
  - Biochemical assays for subtypes with known enzyme defects:
    - Type V—lysyloxidase deficiency
    - Type VI—lysylhydroxylase deficiency
    - Type VII—procollagen peptidase deficiency
    - Type IX—defective fibronectin
  - Skin biopsy for ultra-structural and microscopic studies
- Cutis laxa
  - Elastic fiber stains
  - Type IV: Lysyloxidase assay, serum copper assay

Pseudoxanthoma elasticum
- Radiographic examination
- Microscopic examination of skin/mucous membranes (von Koss staining)
- Ophthalmoscopy
- Abdominal arteriography

Immune-related arthropathies
- HLA determination
- Autoantibody assays

CONSIDERATIONS IN MANAGEMENT

Genetic
- Establish diagnosis
- Examination of apparently unaffected family members for minimal expression of the disorder—echocardiography is especially useful for relatives of known Marfan's syndrome patients to pick up asymptomatic mitral valve prolapse
- Family-history analysis
- Risk counseling
- Consideration of prenatal diagnosis for pregnancies at risk, if indicated and available
- Discussion of sharing information with other relatives to encourage evaluation, early diagnosis, treatment, risk assessment

Psychosocial, educational, familial
- Family counseling
- Education to encourage medical monitoring and adherence to appropriate life-style for individuals at risk for complications
- Assessment of community resources and support groups

Medical
- Management as indicated by symptoms with some specific considerations depending on the disorder
  - Marfan's syndrome
    - Surgical repair of aortic lesions
    - Avoidance of strenuous exercise
    - Prophylactic antibiotics for dental procedures to prevent endocarditis
    - Consideration of hormonal induction of premature puberty, especially in girls, to reduce height and lessen kyphoscoliosis
    - Attention to ophthalmologic problems
  - Ehlers-Danlos syndrome
    - Environmental protection from trauma, especially over bony prominences
    - Exercise to strengthen muscles
    - Avoidance of surgery, if possible, or special surgical and obstetric procedures to avoid ruptures and lacerations and promote healing in type IV
    - Ascorbic acid supplementation in some types of the disorder

SUGGESTED READING

Beighton P: Connective tissue disorders, in *Inherited Disorders of the Skeleton*. New York, Churchill Livingstone, 1978.

Bluestone R, Pearson CM: Ankylosing spondylitis and Reiter's syndrome: Their interrelationship and association with HLA-B27. *Adv Intern Med* 1977;22:1.

McKusick VA: *Heritable Disorders of Connective Tissue*. St. Louis, CV Mosby, 1972.

Pinnell SR, Murad S: Disorders of collagen, in Stanbury JB, Wyngaarden JB, Fredrickson DS, et al, (eds): *The Metabolic Basis of Inherited Disease*, New York, McGraw-Hill, 1983.

Prockop DJ, Kiririkko KI: Heritable diseases of collagen. *N Engl J Med* 1984;311:376.

Chapter 33

Classification of Inborn Errors of Metabolism

The inborn errors of metabolism are a large group of hereditary disorders in which specific gene mutations cause abnormal or missing proteins that lead to altered function. Although individually rare, in the aggregate these disorders represent a significant proportion of genetic disease and may account for some unrecognized causes of death in children.

Some of the more common metabolic disorders are discussed in separate chapters, for example, phenylketonuria and Tay-Sachs. This chapter will provide a brief overview of the genetic implications of the inborn errors of metabolism and will review major clinical parameters, which raise suspicion for metabolic diseases.

ETIOLOGY AND PATHOGENESIS

The biochemical basis of approximately 250 of the errors of metabolism is known; 170 of them are enzyme defects. Overall, the metabolic abnormalities can arise through a variety of mechanisms: absent or unstable enzymes, absent or defective receptors, altered tertiary structure of proteins, or altered transport activity, all of which may lead to subsequent clinical manifestations.

GENETIC CHARACTERISTICS AND MODE OF INHERITANCE

Inborn errors of metabolism are usually autosomal recessive, with a 25% recurrence risk for offspring of carrier parents. Some are inherited as X-linked recessive conditions, including adrenoleukodystrophy, agammaglobulinemia, Fabry's disease, granulomatous disease, Hunter's syndrome, Lesch-Nyhan syndrome, and Menkes' (kinky hair) syndrome; female carriers transmit the disorders to 50% of their male offspring. A few are inherited as autosomal dominant traits, including some of the porphyrias, hyperlipidemias, and hereditary angioedema, with a 50% recurrence risk in offspring of an affected individual.

CLINICAL NOTES

Consideration of biochemical genetics has become particularly prudent in view of the rapid advances in the development of diagnostic tools and methods of intervention (e.g., carrier detection, prenatal diagnosis, treatments) for metabolic disorders. Although these disorders can be difficult to diagnose, some general clues are discussed herein that may indicate a need for metabolic screening. A brief description of clinical features and diagnostic tests for selected disorders is provided in Table 33-1.

Onset of metabolic disorders is most common in childhood, although adult onset forms can exist. Clues to an amino acid disorder in an infant can include an unusual odor of the urine or body, feeding difficulties, growth failure, seizures, severe vomiting and ketoacidosis, hyperammonemia, dislocated lenses, and progressive mental and/or motor retardation. Developmental delay or mental or psychomotor regression (with loss of previously acquired skills) in the first few years of life may indicate a storage disease. Organomegaly (particularly of the liver and spleen), arthropathy, and other dysmorphic features are suggestive of some lipidoses, mucopolysaccharidoses, or mucolipidoses.

Urinary tract stones in a child may suggest disorders like the hyperuricemias or cystinuria. Also indicative of a possible metabolic disorder are hypopigmentation, osteoporosis, excessive candidal dermatosis, or a family history of siblings dying in infancy, often of sepsis or pulmonary or ventricular hemorrhages.

Adult-onset disorders often present with progressive neurologic dysfunction, behavioral abnormalities, and dementia. Some exhibit progressive muscle weakness and cardiomyopathy.

Prenatal diagnosis and carrier testing for some of these disorders is routine; for others, testing is available only in specialized laboratories. Some of these procedures are still highly experimental, and their accuracy has not been firmly established. In addition, prenatal diagnosis using restriction enzyme analysis (e.g., for PKU and β-thalassemia), may be applicable only for families meeting specific criteria. Fetal sex determination may be an option for prenatal diagnosis in certain X-linked disorders when there is no biochemical test. In view of developing technology, it is always worthwhile to investigate the current availability of tests for a specific disorder.

PROCEDURES FOR DIAGNOSTIC CONFIRMATION

In general, metabolic diseases can be detected through biochemical tests that can identify the accumulated or missing metabolites in the blood or urine, or can monitor specific enzyme activities. These tests are being applied increasingly to gene carrier detection and prenatal diagnosis. Selected inborn errors of metabolism—e.g., PKU and galactosemia—are often diagnosed via state-mandated neonatal screening programs.

For specific tests for each disorder, see the table. There are also some nonspecific tests:

- Plasma and urine screening for amino acids (chromatography) and mucopolysaccharides
- Bacterial inhibition assays
- Ferric chloride test—for a large number of amino acid disorders 2,4-DNP for α-ketoacids
- Blood ammonia for urea cycle disorders

TABLE 33-1
Selected Inborn Errors of Metabolism

DISORDER	ENZYME DEFICIENCY OR OTHER DEFECT	DIAGNOSTIC TEST	CLINICAL CHARACTERISTICS
Amino Acid Disorders			
Phenylketonuria and other forms of hyperphenylalaninemia (see Chapter 37 for further detail)	Phenylalanine hydroxylase (in classic PKU)	↑Serum and urine phenylalanine	Mental retardation if untreated; maternal PKU can result in mental retardation of non-PKU offspring
Tyrosinemia (tyrosinosis)	p-Hydroxyphenyl pyruvic acid oxidase	↑Serum and urine tyrosin, ↑plasma methionine, ↑serum bilirubin, ↑serum hepatic enzymes	Vomiting, acidosis, diarrhea, failure to thrive, rickets, hepatic cirrhosis, Fanconi's syndrome, urine odor of rotten cabbage
Alkaptonuria (homogentisic aciduria)	Homogentisic acid oxidase	Urine homogentisic acid	Dark brown or black urine, reducing substance; dark discoloration of connective tissues; progressive arthropathy
Albinism (tyrosinase negative) (see Chapter 31 for further detail)	Tyrosinase	Hair-bulb analysis for tyrosinase	Lack of pigment in skin, hair, and/or eyes
Isovaleric acidemia	Isovaleric CoA dehydrogenase	Ferric chloride test; ↑urine isovalerylglycine	Sweaty-feet odor, ketoacidosis, seizures, mental retardation, coma, death
Maple syrup urine disease (branched chain ketoaciduria)	Branched chain 2-ketoacid decarboxylase	Plasma and urine assays for ↑amino acids, organic acids; DNP test—yellow precipitate; ferric chloride test—gray with greenish tinge	Maple-syrup odor of body or urine, ketoacidosis, developmental retardation, flaccidity, coma, death
Propionic acidemia (ketotic hyperglycinemia)	Propionyl CoA carboxylase	↑Propionic acid in urine or blood, enzyme assay	Vomiting, ketosis, thrombocytopenia, neutropenia, osteoporosis, mental retardation, death

Methylmalonic acidemia	Methylmalonyl CoA mutase apoenzyme	↑Serum and urine methylmalonate, enzyme assay	Vomiting, ketosis, thrombocytopenia, neutropenia, osteoporosis, mental retardation, death
Homocystinuria	Cystathionine synthase	Skin biopsy for enzyme; serum methionine high; homocystine and methionine in urine (+ cyanide-nitroprusside test)	Dislocated lenses, thromboembolism, bony abnormalities, mental retardation
Transport Disorders			
Familial hypophosphatemic rickets (vitamin D-resistant; X-linked dominant)	Phosphate transport	Plasma parathyroid hormone normal, ↓serum phosphate	Calciopenic rickets, bony abnormalities, short stature; muscle weakness, tetany, and convulsions
Cystinuria	Cystine and dibasic amino acid transport	Cystine in urine (+ cyanide-nitroprusside test)	Urinary tract calculi
Hartnup disease	Neutral amino acid transport	Urine for neutral amino acids	Pellagralike skin rash, ataxia, coma, some have psychiatric disturbance
Cystinosis	Unknown	Urine for excess phosphorus, bicarbonate, glucose, amino and organic acids; skin biopsy for cystine in lysosomes	Renal dysfunction, Fanconi's syndrome retinopathy, hypothyroidism, short stature, rickets
Cystic fibrosis	Unknown	Sweat test	Chronic pulmonary disease, thick mucus, pancreatic insufficiency
Blood Disorders			
Hereditary spherocytosis (autosomal dominant)	Spectrin in some families; unknown in others	RBC morphology and indices, osmotic fragility test, autohemolysis test	Weakened red cell membrane, spherocyte formation, sequestration and hemolysis
Hereditary elliptocytosis (autosomal dominant)	Unknown	Peripheral blood morphology	Weakened red cell membrane, elliptocyte formation; can lead to poikilocytosis, sequestration and hemolysis

TABLE 33-1 *Continued*

DISORDER	ENZYME DEFICIENCY OR OTHER DEFECT	DIAGNOSTIC TEST	CLINICAL CHARACTERISTICS
Glucose-6-phosphate dehydrogenase (G6PD, Chapter 50)	G6PD	Enzyme assay	Hemolytic anemia subsequent to ingestion of precipitating agents
Sickle cell anemia (Chapter 17)	β-globin	RBC physiochemistry, hemoglobin electrophoresis	Chronic anemia, vaso-occlusive crises
Thalassemia (Chapter 18)	α or β-globin	CBC and indices hemoglobin electrophoresis	Severe anemia, increased infection, stunted growth
Carbohydrate-Metabolism Disorders			
Glycogen storage disease type I (von Gierke's disease)	Glucose-6-phosphatase	Glucose tolerance, glucagon challenge, open liver biopsy for enzyme assay	Hypoglycemia, glycogen accumulates in liver and kidney, short stature, good prognosis
Glycogen storage disease type IIa (Pompe's disease)	α-glucosidase (acid maltase)	Blood sugar normal, enzyme assay, muscle biopsy	Muscle weakness, cardiomegaly, heart failure, fatal
Galactosemia	Galactose-1-phosphate uridyl transferase	Enzyme assay (filter paper blood screen)	Vomiting, hypoglycemia, cataracts, mental retardation, liver and kidney dysfunction
Galactokinase deficiency	Galactokinase	Enzyme assay (whole blood-RBCs)	Cataracts; GI dysfunction
Metal-Metabolism Disorders			
Wilson's disease (Chapter 35)	Copper metabolism	Hepatic and renal function studies, $\downarrow$plasma ceruloplasmin, $\uparrow$copper excretion after standard dose of D-penicillamine	Adolescent or later onset, Kayser-Fleischer rings in eyes, liver cirrhosis, psychosis

Menkes' disease (kinky- or steely-hair syndrome—X-linked recessive)	Copper transport	↓Serum copper, ↓serum ceruloplasmin, microscopic exam of hair, liver biopsy for ↓copper, skin and duodenal mucosal biopsy for ↑copper	Sparse hair with steely texture, pili torti, arterial and brain degeneration, hypopigmentation, bone changes
Hemochromatosis (idiopathic)	Unknown	HLA typing for linkage to iron-loading gene, plasma iron and ferritin concentrations, transferrin saturation	Abnormal absorption of dietary iron, iron overload, liver, heart, and pancreatic damage (diabetes, cirrhosis, hyperpigmentation)
Lysosomal Storage Diseases			
Mucopolysaccharidoses (MPS)			
Hurler's syndrome	α-Iduronidase	Mucopolysaccharide (MPS) screen, enzyme assay, skin biopsy	Coarse facies, corneal clouding, cardiac defects, joint contractures, bone abnormalities, mental retardation, death in first decade
Hunter's syndrome (X-linked recessive)	Iduronate sulfatase	MPS screen, enzyme assay, normal radioactive sulfate uptake	Similar to Hurler's syndrome, milder; no corneal clouding, death usually before 15 years
Sanfilipo's syndrome	Heparan-N-sulfatase; glucosaminidase	MPS screen and enzyme assay, radioactive sulfate uptake studies	Diarrhea, mental retardation, joint stiffness; hirsutism, death by age 20
Morquio's syndrome	Galactosamine 6-sulfate sulfatase; β-galactosidase	MPS screen and enzyme assay, urine for ↑keratin sulfate	Dwarfism, coarse facies, mild corneal clouding, deafness, normal IQ, cardiovascular problems
Maroteaux-Lamy syndrome	Galactosamine-4-sulfatase (arylsulfatase B)	MPS and leukocyte inclusions, enzyme assay, radioactive sulfate uptake studies	Joint stiffness, coarse facies, corneal clouding, cardiac valvular defects

TABLE 33-1 *Continued*

DISORDER	ENZYME DEFICIENCY OR OTHER DEFECT	DIAGNOSTIC TEST	CLINICAL CHARACTERISTICS
Mucolipidoses			
I-cell disease	Unknown	MPS screen negative, serum assay for ↑arylsulfatase A and N-acetyl-beta-hexosaminidase, ↓acid hydrolases	Hirsutism, joint contractures, corneal clouding, glaucoma, developmental delay; death by 6 years
Pseudo-Hurler's polydystrophy	Unknown	Same as above; distinguished on clinical grounds	Milder joint stiffness, growth delay, cardiac defects, slight or no mental retardation
Sphingolipidoses			
GM_1 gangliosidosis	B-galactosidase	Enzyme assay, negative berry spot test, foamy histiocytes	Poor growth, brain dysfunction, bony abnormalities, seizures; death by 2 years
Gm_2 gangliosidoses: Tay-Sachs disease	Hexosaminidase A	Enzyme assay, no foamy cells in bone marrow	Onset 3 to 6 months; excessive startle reflex, slow growth, motor weakness, cherry-red macular spots; death by 4 years, blindness, deafness, seizures
Sandhoff disease's	Hexosaminidase A and B	Enzyme assay, moderately foamy cells in bone marrow	Similar to Tay-Sachs; onset before 6 months
Niemann-Pick disease Type A	Sphingomyelinase	Enzyme assay, foamy histiocytes in bone marrow	Failure to thrive, mental retardation, cherry-red macular spots
Glucosylceramine: Gaucher's disease, type I "adult form"	Glucocerebrosidase	Gaucher's cells in bone marrow, ↑serum acid phosphatase enzyme assay	Chronic nonneuropathic form, episodic fever and bone pain, hypersplenism; onset in late childhood
Gaucher's disease, type II	Glucocerebrosidase	Gaucher's cells in bone marrow, ↑serum acid phosphatase, enzyme assay	Acute neuropathic form; early onset; seizures, recurrent pneumonia; death in first year

Galactosylceramide Krabbe's disease (globoid cell leukodystropy)	Galactosylcerebrosidase	Enzyme assay	Onset 3 to 6 months; vomiting, seizures, retardation; early death
Sulfatide Metachromatic leukodystrophy	Arylsulfatase A	Enzyme assay, nerve-conduction time, peripheral nerve biopsy	Demyelination, hypotonia, neurologic deterioration, cherry-red macular spot, seizures
Neutral glycosphingolipids Fabry's disease (X-linked recessive)	α-galactosidase A	Proteinuria, enzyme assay, foamy macrophages in bone marrow	Angiokeratoma, episodic pain, renal damage, cardiac and vascular disease (skin lesions and corneal opacity in some female carriers)
Purine- and pyrimidine-metabolism			
Primary gout: Idiopathic (multifactorial)	Increased synthesis, reduced renal clearance of uric acid	Urinary uric acid	Gouty arthritis, tophi, renal disease
Lesch-Nyhan syndrome (X-linked recessive)	Hypoxanthineguanine, phosphoribosyltransferase (HPRT)	Enzyme assay	Delayed growth and development, choreoathetosis, self-mutilation, gouty arthritis, renal stones
Xeroderma pigmentosum	DNA repair and replication enzymes	Enzyme assay, chromosome analysis	Ultraviolet sensitivity, multiple skin carcinomas, variable mental retardation
Immune-System Disorders (See Chapter 28)			
Hereditary angioedema (autosomal dominant)	Inhibitor of $C\bar{1}$, a protein complement component ($C\bar{1}$ INH)	C4 immunoassay; radial immunodiffusion for $C\bar{1}$ INH protein	Recurrent edema, GI disturbance, sudden death from upper-respiratory obstruction
Agammaglobulinemia (X-linked recessive)	Absence of all serum immunoglobulins	Serum immunoglobulins, Schick test, isohemagglutinin levels, antigenic stimulation	Recurrent pyogenic infections, failure to synthesize antibodies, arthritis

TABLE 33-1 *Continued*

DISORDER	ENZYME DEFICIENCY OR OTHER DEFECT	DIAGNOSTIC TEST	CLINICAL CHARACTERISTICS
Severe combined immunodeficiency	Adenosine deaminase	Enzyme assay	Recurrent infections, diarrhea, moniliasis; death before 2 years
Chronic granulomatous disease (most families X-linked recessive, some appear autosomal recessive)	Enzymes of NADPH oxidase complex	Enzyme assay, neutrophil-function studies	Chronic infections, inflammatory masses, mild anemia
Kartagener's syndrome (immotile cilia)	Unknown	Semen analysis for immotile sperm, nasal or bronchial biopsy for ciliary defects	Ciliary immotility, bronchiectasis, abnormal sperm (male infertility), situs inversus
Amyloidoses			
Familial Mediterranean fever—autosomal recessive (all other forms, autosomal dominant)	Unknown	Biopsy for amyloid deposition	Recurrent febrile attacks, serosal inflammation, amyloid nephropathy (other forms neuropathic, nephropathic, or cardiopathic with systemic deposition of amyloid fibrils)

Key: ↑ = increased levels of activity; ↓ = decreased levels of activity

Test for reducing sugars

Berry spot test and acid albumin turbidity tests for mucopolysaccharidoses

Austin fluff test for metachromatic lipids

Thin-layer chromatography for oligosaccharides, sialyloligosaccharides

Peripheral blood smear for vacuolization in oligosaccharidoses and mucolipidoses, and for granules in ceroid-lipofuscinoses

Bone marrow biopsy for vacuolization, granules, and foam cells

Bone radiograms

Spinal-fluid protein for metachromatic leukodystrophy, Krabbe's disease

Skin biopsy for storage diseases

CONSIDERATIONS IN MANAGEMENT

Genetic
- Establish diagnosis
- Carrier testing, if available and indicated
- Pedigree analysis, recurrence-risk assessment
- Risk counseling
- Consideration of prenatal diagnosis for pregnancies at risk

Psychosocial, educational, familial
- Family counseling and support
- Education to promote increased compliance with special forms of therapy, such as, protein-restricted diet
- Assessment of community resources and support groups

Medical
- Dependent on diagnosis and severity; may involve:
  - Dietary or vitamin therapy
  - Drug therapy
  - Avoidance of toxic factors
  - Surgery

SUGGESTED READING

Ampola MG: *Metabolic Diseases in Pediatric Practice,* Boston, Little, Brown, 1982.

Kolodny EH, Cable WJL: Inborn errors of metabolism. *Ann Neurol* 1982;11(3):221.

Stanbury JB, Wyngaarden JB, Fredrickson DS, et al: *The Metabolic Basis of Inherited Disease,* 5th ed. New York, McGraw-Hill, 1983.

Chapter 34

Tay-Sachs Disease

A progressive lethal disorder of the nervous system, Tay-Sachs disease has an incidence of 1/3,600 conceptions among Jews of Ashkenazi descent (mostly Eastern and central European). Screening for carriers in this group coupled with the availability of prenatal diagnosis and therapeutic termination has substantially reduced the incidence at birth. A much larger number of potentially-at-risk individuals have been reassured when screening showed that they were not carriers.

ETIOLOGY AND PATHOGENESIS

Tay-Sachs disease is an inborn error of sphingolipid metabolism. The basic defect is an inability to synthesize biologically active hexosaminidase A (Hex A), a lysosomal enzyme required to catalyze one step in the catabolism of gangliosides (complex molecules that are constituents of cell membranes). The enzyme deficiency causes a block in the pathway, resulting in the accumulation of GM2 ganglioside in the brain, the nervous system, and, to a lesser extent, in the liver, spleen, and heart. This accumulation disrupts cell function and leads to eventual cell death.

GENETIC CHARACTERISTICS AND MODE OF INHERITANCE

Tay-Sachs disease is an autosomal recessive disorder.

- If both parents are carriers (heterozygotes), the risk of occurrence is 25% with each pregnancy. Prenatal diagnosis can identify affected pregnancies.
- If either or both parents are carriers, the risk to have a carrier child is 50% for each pregnancy.
- When both parents are carriers, unaffected children have a 67% risk of being carriers.

Carriers of the gene for Tay-Sachs disease are clinically asymptomatic individuals, in whom carrier status can be determined biochemically. The carrier test is recommended for relatives of known carriers and for all Ashkenazi Jewish couples. Carrier frequency is estimated at 1/30 for Ashkenazi Jews. The gene occurs with a much lower frequency (1/300) in most other populations, including non-Ashkenazi Jews.

CLINICAL NOTES

Age of onset is from birth to 10 months , usually by 6 months . The child generally appears normal at birth and seems to develop normally until the disease becomes manifest. The initial symptoms may include hypotonia, an exaggerated startle response, and loss of the ability to hold the head up or to sit. A cherry-red spot on the macula is visible on funduscopy. The disease is relentless, with other signs and symptoms appearing as it progresses—psychomotor retardation, convulsions, optic atrophy, blindness, and macrocephaly. Tay-Sachs is usually fatal by 4 years of age.

PROCEDURES FOR DIAGNOSTIC CONFIRMATION

Assay for Hex A activity in serum, leukocytes, or tissue samples:
- Undetectable Hex A activity = homozygote (affected)
- Approximately half-normal Hex A activity = heterozygote (carrier)
- Because pregnancy may induce a false-positive result in serum samples, only leukocyte assays are reliable in this situation

CONSIDERATIONS IN MANAGEMENT

Genetic
- Carrier testing and possible mass screening of high-risk populations
- Risk counseling and education of high-risk populations
- Prenatal diagnosis is available by assay for Hex A activity in cultured fetal cells obtained at about 16 weeks' gestation via amniocentesis. Chorionic villus sampling is expected to provide cells in the first trimester.

Psychosocial, educational, familial
- Psychologic support for family
- Exploration of facilities for custodial care

Medical
- No satisfactory treatment beyond supportive therapy is available

SUGGESTED READING

Kaback MM, O'Brien JS: Tay-Sachs: Prototype for prevention of genetic disease. *Hosp Pract* 1973;8:107.

Kaback MM, ed: *Tay-Sachs Disease: Screening and Prevention*. New York, Alan R. Liss, 1977.

O'Brien JS: The gangliosidoses, in Stanbury JB, Wyngaarden, JB, Fredrickson DS, et al, (eds): *The Metabolic Basis of Inherited Disease*, 5th ed. New York, McGraw-Hill, 1983.

Chapter 35

Wilson's Disease

Wilson's disease, (hepatolenticular degeneration), is an excellent example of a diagnosis that is seldom made unless the physician looks for it in the evaluation of patients with various neurologic or psychiatric problems or liver disease. Incidence is currently estimated at about 1/30,000 in the United States. Treatment can reverse or prevent the clinical signs of the disease. Without treatment it is invariably fatal.

ETIOLOGY AND PATHOGENESIS

Cirrhosis of the liver and degeneration of the brain, especially the basal ganglia, occurs as a result of copper deposition in these organs and in greenish-brown Kayser-Fleischer rings at the limbus of the cornea of the eye. The copper-binding protein serum ceruloplasmin is deficient, and various tissue proteins are overloaded with loosely-bound copper. However, the primary lesion in copper metabolism is still not known.

GENETIC CHARACTERISTICS AND MODE OF INHERITANCE

Wilson's disease is inherited as a classic autosomal recessive disorder.

- It carries a 25% occurrence risk when both parents are carriers.
- Unaffected siblings of affected individuals have a 67% risk of being carriers.
- Children of affected individuals are obligate carriers.
- Siblings of known carriers have a 50% risk of being carriers.

In the absence of reliable carrier detection, risk estimates for affected individuals or potential carriers to have affected offspring must be based on the estimated population carrier frequency (1/90), unless the other parent also has a personal or family history of the disorder or is a relative who might have inherited the gene from the same source. Examples of risks for affected offspring are given in Table 35-1.

CLINICAL NOTES

Psychiatric and behavioral changes, although common, vary in kind and degree. Diagnoses may include personality disturbances, hysteria, and schizophrenia before Wilson's disease is identified. In one series of patients, 60% had significant psychologic manifestations as the first clinical indication; usually these

TABLE 35-1
Recurrence Risks for Wilson's Disease

PARENT STATUS	RISK/PREGNANCY
Both parents affected	All affected
One parent affected; other with affected sibling	1/3
Both parents carriers	1/4
Both parents with affected siblings	1/9
One parent affected; other, no family history	1/180
One parent carrier; other, no family history	1/360
One parent with affected sibling; other, no family history	1/540
One parent with carrier sibling; other, no family history	1/720

are recognized as Wilson's disease in only about 20% of patients before the neurologic and liver-disease signs emerge. Intellectual capacities are maintained intact, though observers may be misled by childish personality changes, drooling, difficulties with verbal communication, and eventually, a masklike expressionless face.

A case history, described by Cartwright (see Suggested reading), will demonstrate potential diagnostic pitfalls. A 24-year-old woman became nervous and developed a tremor. Her physician made a diagnosis of "nervous exhaustion" and prescribed tranquilizers. She became depressed and attempted suicide. A psychiatrist treated her with imipramine and then chlorpromazine, but she became psychotic. Another psychiatrist diagnosed acute schizophrenic reaction and ordered 11 electro-convulsive treatments, without benefit. An internist was then called to see her because of abnormal liver-function-test results and attributed the abnormalities to chlorpromazine toxicity. The saga continued with a neurologist. His diagnosis was also chlorpromazine toxicity. By this time she had increased salivation, a masklike expression, severe difficulty swallowing, tremor, and dystonia; she was almost completely unable to care for herself. Yet another psychiatrist was called, who diagnosed "conversion reaction" and prescribed doxepin and diazepam. At last she was admitted to a psychiatric ward because of dehydration; her diagnosis was "hysterical neurosis/conversion type", though "catatonic schizophrenia" could not be ruled out. Amazingly, an alert on-call physician examined her carefully and recognized the distinctive Kayser-Fleischer rings in her eyes. He diagnosed Wilson's disease. It was 32 months after the onset of clinical symptoms.

The patient was told that her disorder was treatable and D-penicillamine was started. Her depression and psychiatric signs disappeared. Her neurologic signs improved slowly; but dystonia and spasticity persist, and she walks with some difficulty. Nevertheless, she returned to college, graduated, remarried, and seemed well-adjusted.

The neurologic signs vary. One form of lenticular degeneration is dystonic, with spasticity, rigidity, dysarthria, and an unrelenting downhill course with febrile episodes. This form occurs predominantly in young adults. Another type of neurologic involvement is pseudosclerosis, with flapping tremors and less spasticity. In undiagnosed patients with cirrhosis of the liver, the neurologic manifestations may be thought, incorrectly, to be due to hepatic encephalopathy—thus further delaying the proper diagnosis. In children the disease may begin with chronic active hepatitis, cirrhosis, or an acute hemolytic anemia. The Kayser-Fleischer ring does not appear before age 7 years.

PROCEDURES FOR DIAGNOSTIC CONFIRMATION

Tests of hepatic and renal function

Slit-lamp examination for Kayser-Fleischer rings

Low plasma ceruloplasmin levels

Excessive excretion of copper after a standard dose of D-penicillamine

CONSIDERATION IN MANAGEMENT

Genetic
- Risk counseling
- Suggest seeking out siblings for testing. The disease is reversible and can be diagnosed during the asymptomatic phase
- Discussion of sharing information with other family members to encourage testing, even when risks are quite low, because of treatable nature of the disorder

Psychosocial, educational, familial
- Family education for effective use of medication
- Supportive services, as needed

Medical
- Decoppering treatment with D-penicillamine or other agent.
- It is essential to sustain treatment even though symptoms and signs may worsen during the first six to eight weeks' before gradual improvement of neurologic function is noted.

SUGGESTED READING

Cartwright GC: Current concepts—diagnosis of treatable Wilson's disease. *N Engl J Med* 1978;298:1347.

Source: Foundation for the Study of Wilson's Disease, Inc., New York, (212) 892-5119.

Francone CA: My battle against Wilson's disease. *Am J Nurs* 1976;76:247.

Sass-Kortak A, Bearn AG: Diseases of copper metabolism, in Stanbury JB, Wyngaarden JB, Frederickson DS, (eds): *The Metabolic Basis of Inherited Disease,* 4th ed. New York, McGraw-Hill, 1978.

Scheinberg IH, Sternlieb I: *Wilson's Disease*. Philadelphia, WB Saunders, 1983.

Chapter 36

Diabetes Mellitus

Diabetes mellitus is the term for an etiologically heterogeneous group of disorders characterized by abnormalities in glucose homeostasis. Together, they are among the most common chronic disorders, estimated to affect 5% or more of the adult population of the western world. Manifestations can range from asymptomatic glucose intolerance, to an acute medical emergency, to chronic complications. As characterized in Table 36-1, primary diabetes mellitus is currently classified into two broad categories that aggregate separately in families, and are thus considered separate disorders.

TABLE 36-1
Differentiating IDDM and NIDDM Diabetes

	IDDM	NIDDM
Other nomenclature	Juvenile-onset DM	Maturity-onset DM
Clinical characteristics	Thin, ketosis-prone; insulin required for survival	Frequently obese; ketosis-resistant; often treatable by diet or drugs
Age of onset	Predominantly childhood and early adulthood	Predominantly after 40
Family studies	Increased prevalence of juvenile or Type I diabetes	Increased prevalence of maturity or Type II diabetes
Twin studies	$<$ 50% concordance in monozygotic twins	Nearly 100% concordance in monozygotic twins
Insulin response to glucose load	Flat	Variable
Associated with other autoimmune endocrine diseases and antibodies	Yes	No
Islet-cell antibodies and pancreatic-cell mediated immunity	Yes	No
HLA associations	Yes	Perhaps in some subtypes
Possible etiologies	Viral infections, autoimmunity, β-cell toxins	Insulin resistance, premature aging of the β-cells

Key: IDDM = insulin-dependent diabetes mellitus, Type I; NIDDM = noninsulin-dependent diabetes mellitus, Type II

Insulin-dependent diabetes mellitus (IDDM or Type I) accounts for nearly 10% of recognized cases of diabetes. The classic patient is a thin, ketosis-prone juvenile who requires insulin to survive. Non-insulin-dependent diabetes mellitus (NIDDM or Type II) comprises more than 90% of cases. The typical patient is the overweight, adult-onset diabetic, often managed by diet or oral hypoglycemics, who may, however, also be getting insulin, but who will not go into ketoacidosis on insulin withdrawal. It is important to emphasize that either type can occur at any age. A subgroup of NIDDM, maturity-onset diabetes of the young (MODY), is an example of an NIDDM presentation, often with a young age of onset. Diabetes or impaired glucose tolerance may also be associated with genetic syndromes or other genetic disorders. See Table 36-2 for a list of genetic conditions predisposing to diabetes or glucose intolerance (for further details see relevant chapters).

TABLE 36-2
Genetic Syndromes Associated With Diabetes or Glucose Intolerance

DISORDER	MODE OF INHERITANCE
Alstrom's syndrome	AR
Ataxia-telangiectasia	AR
Cockayne's syndrome	AR
Cystic fibrosis	AR
Friedreich's ataxia	AR (AD)
Glucose-6-phosphate dehydrogenase deficiency	XR
Type 1 glycogen storage disease	AR
Hemochromatosis	AR
Hereditary relapsing pancreatitis	AD
Familial hypertriglyceridemia	AD
Isolated growth hormone deficiency	Varied
Laurence-Moon-Biedl syndrome	AR
Lipoatrophic diabetes	AD (AR)
Muscular dystrophy	XR (AD) (AR)
Myotonic dystrophy	AD
Ocular hypertension induced by dexamethasone	AD
Optic atrophy	AR
Pineal hyperplasia and diabetes	AR
Acute intermittent porphyria	AD
Pheochromocytoma	AD
Prader-Willi syndrome	?CH
Retinitis pigmentosa, neuropathy, ataxia, and diabetes	AD
Schmidt's syndrome	?AR (AD)
Werner's syndrome	AR
Turner's syndrome	CH
Klinefelter's syndrome	CH
Down's syndrome-trisomy 21	CH

Key: AR = autosomal recessive; AD = autosomal dominant; CH = chromosomal; XR = X-linked recessive

The major categories of diabetes are discussed in this chapter, as are the teratogenic risks for offspring of diabetic mothers.

ETIOLOGY AND PATHOGENESIS

In both types of diabetes, the chronic hyperglycemia, which occurs to a greater or lesser degree even in treated patients, is thought by many to be responsible for the long-term complications of diabetic microvascular (retinopathy, nephropathy) and macrovascular (atherosclerosis) disease. The proximate cause of the hyperglycemia itself is an absolute or relative insulin deficiency of heterogeneous etiology in both IDDM and NIDDM.

IDDM insulin deficiency appears to result from destruction of the insulin-producing beta cells of the pancreatic islets, which leads to absolute deficiency and dependence on exogenous insulin for survival. Different etiologic factors are thought to be capable of destroying the beta cells. Studies have suggested that the process is autoimmune in nature, perhaps triggered by a viral infection interacting with a genetic susceptibility. Other environmental agents, such as beta-cell toxins have also been implicated.

The autoimmune pathogenesis is supported by the finding of antibodies and cell-mediated immunity to the pancreas in IDDM patients, and by the high frequency of other autoimmune endocrine disorders in such patients—for example, Graves' disease, chronic thyroiditis (Hashimoto's), idiopathic hypothyroidism, Addison's disease, and pernicious anemia-atrophic gastritis.

The genetic susceptibility is demonstrated by the strong association of IDDM with certain HLA alleles. The HLA region on chromosome 6 contains the loci of several genes coding for different alleles for leukocyte antigens and complement components. Some of these antigens, particularly DR3 and DR4 are found with increased frequency in IDDM patients as compared with controls (90% of IDDM patients have one or both). Other HLA alleles, for example, DR2, found with decreased frequency in IDDM, may contribute a resistance to the disorder. Two siblings, both affected with IDDM (concordant), are more likely to be HLA identical (share both HLA haplotypes) than pairs of siblings with only one affected (discordant). By the same token, HLA identical siblings of affected individuals are reported to be at much greater risk for IDDM than HLA nonidentical siblings. Individuals with more than one of the alleles associated with IDDM, i.e., DR3 and DR4, seem to be at particularly high risk for the disorder. It is still not clear whether the susceptibility is due to the HLA genes, or to genes located nearby and inherited with them, or whether there are contributions from genes on other chromosomes. HLA typing is still considered primarily a research tool.

NIDDM continues to have an ill-defined pathogenesis. Most likely the disorder represents a number of diseases with different etiologies. Insulin resistance, perhaps due to a decrease in insulin receptors, plays an important role in many patients; relative insulin deficiency is a critical factor in others. Interaction of the two with each other, with other genes, or with environmental factors may be responsible for the development of disease, as well. Diverse defects in the molecular machinery mediating the action of insulin on target cells could thus lead to a common result: inadequate insulin action and chronic hyperglycemia with its associated complications.

It has been suggested that a premature aging process may be responsible for both the relative insulin deficiency and some of the complications, such as cataracts and premature atherosclerosis. Recently an association of NIDDM with a DNA polymorphism in a flanking region of the human insulin gene has been described as a possible genetic marker, but this has not been confirmed in subsequent studies. Association with HLA antigens generally has not been noted, but may be present in some subtypes.

GENETIC CHARACTERISTICS AND MODE OF INHERITANCE

Familial aggregation of diabetes mellitus has long been noted, but clear definition of genetic parameters has remained elusive. Although many underlying disorders with different modes of inheritance may be involved, it is possible to make some general statements for each of the major types of diabetes, but genetic counseling for them still depends, for the most part, on empiric recurrence risks (Table 36-3).

TABLE 36-3
Recurrence Risks for Relatives of Diabetes Patients*

DISORDER IN PROBAND	RISK TO SIBS (%)	RISK TO OFFSPRING (%)
IDDM (Type I)†	5-10	2-5
NIDDM (Type II)†		
Risks for overt disease	5-10	5-10
Risks for abnormal glucose tolerance	15-25	15-25
MODY (autosomal dominant mature-onset diabetes of the young)	50	50
Diabetes, secondary to other genetic condition	Susceptibility inherited along with underlying condition	

*Recurrence risks cited are those for the same form of the disorder as diagnosed in the proband. Risk for other forms is not increased over population risk.
†Empiric risks

IDDM, the disease, is not inherited; rather, one inherits susceptibility to the disease. Since environmental factors are suspected to contribute to the etiology and more than one gene may well be involved, it is not surprising that a clear-cut inheritance pattern has not been established. That genes play only a partial role is underscored by the 50% or smaller concordance rate reported in monozygotic twins, who share all their genes.

Empiric risks for IDDM to relatives of individuals with IDDM have been estimated by the degree of relationship, the number of affected persons, and the age of the relative, as well as the age of onset in the proband. Such detailed risk figures are complicated and still considered tentative, but may become more clearly defined when HLA typing becomes more available for routine use. The general recurrence risks now quoted are somewhat reassuring to many families.

- The risk to the siblings of an IDDM patient is about 5% to 10%.
- The risk to the offspring of an IDDM patient to have IDDM is about 2% to 5% (see below the discussion of other risks for offspring of diabetic mothers).
- There is no increased risk for NIDDM in IDDM families.

NIDDM, although likely to represent many different modes of inheritance, seems, in the aggregate, to carry a greater genetic component than IDDM, since monozygotic twins appear to be concordant nearly 100% of the time. Nevertheless, environmental factors are important, as reflected in the rapid changes in frequency seen in migrant populations and in the major effect of obesity on the occurrence and clinical course.

Empiric recurrence risks are used for relatives of patients with NIDDM (except in families with MODY).

- The overall risk for first-degree relatives is about 5% to 10% for clinical diabetes and 15% to 25% for impaired glucose tolerance.
- Since most cases of NIDDM are adult onset, the risk for recurrence increases with increasing age of the relative in question.
- There seems to be no increased risk in this population for IDDM.

MODY, a generally milder form of diabetes, is transmitted as an autosomal dominant condition in some families. It frequently presents in adolescence or young adulthood, with a 50% risk for recurrence in siblings and offspring of an affected individual. Such families are generally identified by pedigree analysis, which reveals the inheritance pattern.

CLINICAL NOTES

Although it is usually clear, the distinction between IDDM and NIDDM can occasionally be difficult to note. Most cases of IDDM occur in childhood, adolescence, and early adulthood with relatively acute onset; polyuria, polydipsia, weight loss, severe hyperglycemia, and ketoacidosis. Dependence on exogenous insulin appears rapidly, and the risk is high for chronic complications, such as nephropathy, neuropathy, retinopathy, and accelerated atherosclerosis. There is an increased risk, as well, for other autoimmune endocrine diseases. However, even elderly individuals can develop IDDM. On the other hand, IDDM patients (especially the older ones) do not always have acute ketoacidosis when they first come to medical attention. They may go through a period of weeks or months when their insulin need is not absolute, but eventual absolute insulin deficiency supervenes.

NIDDM generally presents in an overweight person, later in life, and much less dramatically than IDDM. More often than not, NIDDM is noted in connection with one of its vascular complications, or asymptomatically, on a routine glucose tolerance test. However, NIDDM can occur at any age and can present with ketoacidosis related to the stress of a superimposed infection, or with classic polyuria, polydipsia, and weight loss. The complications are similar to those of IDDM, but they occur somewhat less frequently. No increased risk for other autoimmune disorders has been noted. There is a continuum between impaired glucose tolerance and NIDDM. The current diagnostic criteria for NIDDM are an unequivocal and gross elevation of plasma glucose, a fasting venous plasma glucose of at least 140 mg/dl, or an elevated glucose level at two time points after a 75 g glucose load. Since early diagnosis of the full-blown disorder can promote early treatment, which may prevent complications, periodic monitoring of individuals known to be at risk may be beneficial.

MODY must be distinguished from IDDM, because it is frequently diagnosed in adolescence or young adulthood, and from NIDDM, since it presents similarly. On the whole, MODY is a milder disorder than either of the others and, in some families at least, is relatively complication free.

MATERNAL DIABETES IN PREGNANCY: CONGENITAL MALFORMATIONS

Aside from other potential complications, the risk for a major congenital malformation in the infant of a diabetic mother is estimated to be two to three times as great as in the general population. In the wide range of reported teratogenic effects, the strongest association is with caudal regression, an otherwise rare malformation, combining sacral agenesis and hypoplastic femors. Other aberrations found with increased frequency include renal and cardiovascular abnormalities; for instance, transposition of the great vessels; and CNS anomalies, such as neural-tube defects.

Women with gestational diabetes do not appear to be at risk, but women using oral hypoglycemic agents may be. Insulin dependent women are clearly at highest risk. The overall incidence of birth defects in offspring of diabetic mothers has been estimated at 6% or higher. Many studies report that earlier onset, longer duration, and the presence of complications are associated with higher malformation rates.

The cause of the teratogenic effect is not clear, but preliminary evidence, such as the aforementioned risk figures, suggests that poor diabetic control increases the risk of congenital malformations. Also, high first-trimester hemoglobin A_1c values, indicating high glucose levels over the preceding weeks to months, have been associated with high malformation rates. Major malformation rates of more than 20% have been reported in offspring of women with evidence of poor control in the first trimester. Whether hyperglycemia itself, or some related factor such as acidosis or ketosis is responsible, is not known.

The incidence of other complications of diabetic pregnancy—such as neonatal hypoglycemia, macrosomia, and respiratory distress—has been reduced through improved perinatal management. However, attempts to prevent the malformations are more complicated. They require identification of patients well before pregnancy, because organogenesis occurs early in gestation, and it takes time to achieve good diabetic control. The difficulty of managing insulin-dependent diabetics is compounded by pregnancy and normalization of blood sugar may entail a risk for frequent hypoglycemia, which may also have adverse effects on the fetus. Nevertheless, it is hoped that good metabolic control, both before and during pregnancy, can reduce the teratogenic risk of maternal diabetes.

PROCEDURES FOR DIAGNOSTIC CONFIRMATION

- Blood-glucose determination
- Urinalysis
- Glucose tolerance test
- Insulin assays

CONSIDERATIONS IN MANAGEMENT

- Genetic
  - Pedigree analysis for evidence of a specific inheritance pattern in the family
  - Determination of diabetic subtype and exclusion of potential underlying condition with predisposition for diabetes
  - Risk counseling
  - Evaluation of asymptomatic at-risk family members
  - Discussion of teratogenic potential of diabetes with insulin dependent women of reproductive age, including postulated benefits of achieving good control prior to conception
  - Discussion of risks and consideration of evaluating fetus by available prenatal diagnostic methods, especially when a woman presents after the seventh week of pregnancy with a history of poor diabetic control (e.g., AFP analysis to rule out neural-tube defects, ultrasound to rule out cardiac or limb defects)
  - Discussion of sharing information with other family members of NIDDM patients to encourage counseling and screening for individuals at risk
- Psychosocial, educational, familial
  - Family counseling
  - Education/guidance for nutritional management, use of insulin, appropriate exercise, planning pregnancies, etc.
  - Assessment of community resources and voluntary support groups, as indicated
- Medical
  - Treatment of diabetes and complications, as indicated
  - Screening of first-degree relatives of NIDDM patients by periodic glucose tolerance testing
  - Periodic screening of IDDM patients:
    - For other endocrine autoimmune disorders for which easy noninvasive tests are available (e.g., T_4 and TSH levels for thyroid dysfunction; future use of new pepsinogen I/pepsinogen II ratio for atrophic gastritis)
    - For complications (e.g., yearly retinal examination and renal-function studies)
  - In diabetic pregnancy:
    - Achievement of good diabetic control, preferably prior to pregnancy
    - Hgb A_1c levels for a measure of glucose levels in prior weeks
    - Consideration of prenatal diagnosis, as available, for defects of increased frequency
    - Consideration of referral to specialist for close surveillance

SUGGESTED READING

Cahill GF, Jr, McDevitt HO: Insulin dependent diabetes mellitus: The initial lesion. *N Engl J Med* 1981;304:1454.

Chung CS, Myrianthopoulos NC: Factors affecting risks of congenital malformations: II. Effects of maternal diabetes. (Birth Defects Original Article Series), vol 11, no. 10. White Plains, NY, March-of-Dimes Birth Defects Foundation, 1975.

Ellenberg M, Rifkin H: *Diabetes Mellitus, Theory and Practice,* 3rd ed. New York, Medical Examination Publishing, 1982.

Fajans SS, Cloutier MC, Crowther RL: Clinical and etiologic heterogeneity of idiopathic diabetes mellitus. *Diabetes* 1978;27:1112.

Horii K, Watanabe G, Ingalls TH: Experimental diabetes in pregnant mice: Prevention of congenital malformations in offspring by insulin. *Diabetes* 1966;15:194.

Kucera J: Rate and type of congenital anomalies among offspring of diabetic women. *J Reprod Med* 1971;7:61.

Miller E, Hare JW, Cloherty JP, et al: Elevated maternal hemoglobin A in early pregnancy and major congenital anomalies in infants of diabetic mothers. *N Engl J Med* 1981;304:1331.

Mills JL, Baker L, Goldman AS: Malformations in infants of diabetic mothers occur before the seventh gestational week, *Diabetes* 1979;28:292.

Mills J: Malformations in infants of diabetic mothers. *Teratology* 1982;25:384.

National Diabetic Data Group International Workgroup. Classification of diabetes mellitus and other categories of glucose intolerance. *Diabetes* 1979;28:1039.

Pedersen LM, Tygstrup I, Pederson J: Congenital malformations in newborn infants of diabetic women. *The Lancet* 1964;1:1124.

Riley WJ, Maclaren NK, Lezotte DC, et al: Thyroid autoimmunity in insulin dependent diabetes mellitus: The case for routine screening. *J Pediatr* 1981;98:350.

Rotter JI: The modes of inheritance of insulin dependent Diabetes. *Am J Hum Genet* 1981;34:835.

Rotter JI, Rimoin DL: Etiology—Genetics, in Brownlee M, (ed): *Handbook of Diabetes Mellitus*, vol 1. New York, Garland Publishing, 1981, pp 3-93.

Rotwein PS, Chirgwin J, Province M, et al: Polymorphism in the 5′ flanking region of the human insulin gene: A genetic marker for non-insulin-dependent diabetes. *N Engl J Med* 1983;308:65, 1983.

Samloff IM, Varis K, Ihamaki T, et al: Relationships among serum pepsinogen I, serum pepsinogen II, and gastric mucosal histology: A study in relatives of patients with pernicious anemia. *Gastroenterology* 1982;83:204.

Chapter 37

Phenylketonuria (PKU)

Until it was discovered that strict adherence to a low phenylalanine diet prevents the development of significant symptoms, PKU was a major cause of severe mental deficit, affecting an estimated 0.64% of individuals institutionalized for mental retardation. One of the more common inborn errors of metabolism, classic PKU is estimated to occur in approximately 1/11,000 whites or Orientals in the United States; incidence among blacks is lower. Most affected individuals are now identified through neonatal screening and can be expected to develop normally. However, a secondary problem is the high risk for mental retardation in the offspring of women treated in childhood for PKU.

ETIOLOGY AND PATHOGENESIS

Classic PKU is the consequence of a defect in the hepatic enzyme, phenylalanine hydroxylase, which catalyzes the conversion of phenylalanine to tyrosine. As a result, phenylalanine accumulates in plasma, abnormal metabolites are excreted in the urine, and insufficient tyrosine is produced. The exact pathogenesis is not clear, but failure of myelination and of brain development is thought to underlie the mental retardation; impaired melanin synthesis is believed to be responsible for the lighter-than-expected pigmentation of PKU patients.

Other variants of hyperphenylalaninemia detected in the neonatal screen include transient and milder conditions. Recent recombinant DNA studies suggest that multiple distinct mutations in the phenylalanine hydroxylase gene confer different levels of clinical severity, and in various combinations result in different phenotypes. A small percentage of the variants (1% to 3%) leading to much more malignant disorders, are thought to result from defects in the coenzyme tetrahydrobiopterin.

GENETIC CHARACTERISTICS AND MODE OF INHERITANCE

Classic PKU is an autosomal recessive disorder, as are other variants of hyperphenylalaninemia (with the possible exception of one of the very rare malignant forms, which may be X-linked recessive).

- The carrier rate among U.S. whites and Orientals is about 1/50.
- When both parents are carriers, PKU has a 25% occurrence risk.
- Unaffected siblings of individuals with PKU have a 67% risk of being carriers.

- For treated PKU patients, reproduction has become a realistic goal. The theoretical risk for an affected person to have a child with PKU is about 1%. If the spouse is a carrier, the risk for an affected child is 50% for each pregnancy.
- In addition, for any female PKU patient not on a diet during pregnancy, the risk is high for mental retardation in children who do not have PKU (see Clinical notes).

Carrier testing is available and becoming increasingly reliable. Prenatal diagnosis by recombinant DNA techniques is expected to be available for informative families in selected centers in the very near future.

CLINICAL NOTES

Untreated classic PKU is rarely evident in the newborn, whose prenatal environment has served as a protection from the effects of the abnormal fetal metabolism. Clinical features appear gradually after protein intake has begun. They include a ''musty'' odor, eczema, reduced pigmentation, neurologic deterioration, seizures, and severe mental retardation (96% to 98% ultimately have an IQ below 50, although an occasional patient functions in the normal range). Without treatment, 50 IQ points may be lost by the end of the first year of life. Treated patients can develop normally.

Clinical expression of the other variants of hyperphenylalaninemia ranges from normal, to mild, to lethal neurologic abnormalities.

Identification of new cases of PKU is accomplished mainly through state-mandated *neonatal screening*, mostly by the Guthrie bacterial inhibition test, after the infant has had sufficient protein intake for a significant accumulation of phenylalanine in the blood (some controversy still exists about the timing of the test to avoid false negatives). Positive results are confirmed by a second assay. Since a large proportion of false-positive results reflect the transient, benign, or other forms of hyperphenylalaninemia, further evaluation may be necessary for accurate diagnosis and treatment.

With prompt institution of the *low phenylalanine diet*, treatment of PKU has become a model for dietary restriction of a precursor in management of a metabolic error. Careful monitoring of the diet is crucial, since a fine balance is necessary to assure sufficient phenylalanine for normal growth, and not enough for toxic accumulation. Periodic diagnostic reevaluation is indicated, because the diet, best instituted quickly for most infants with hyperphenylalaninemia, may, in the long run, be inappropriate or ineffective for cases that turn out not to be classic PKU. Because the risk for further brain damage decreases as the brain matures, the rather unpalatable diet has in the past been discontinued approximately when the child reached school age. The current inclination, however, is to begin relaxing it only at ages 8 to 10 years', or even later.

Maternal PKU. In adulthood the restricted diet may not be necessary for normal function of an affected individual. However, as more such children grow up to lead normal lives, it becomes evident that the women among them are at increased risk for spontaneous abortion and at a high risk for retarded offspring because of the maternal PKU. Such children often have microcephaly, congenital heart defects, and low birth weight, even though they themselves do not have PKU. The degree of risk for this pregnancy outcome depends on the maternal blood levels of phenylalanine, having been estimated to be as high as 95%, when the blood concentration is 20 mg/dL or higher. Reinstitution of the phenylalanine- restricted diet before conception and throughout the pregnancy is recommended to reduce the risk. Whether it can be entirely eliminated is not yet clear.

Conversely, the birth of one or more retarded children with microcephaly may raise suspicion that the mother may be one of the rare individuals with unexpressed PKU.

PROCEDURES FOR DIAGNOSTIC CONFIRMATION

Guthrie bacterial inhibition test
Fluorometric determination of phenylalanine
Urinary chromatography
Phenylalanine loading test (also used for carrier testing)

CONSIDERATIONS IN MANAGEMENT

Genetic
- Risk counseling
- Consideration of carrier testing for close relatives and their mates to evaluate potential increased risk
- Consideration of prenatal diagnosis for high-risk pregnancies when test becomes available at selected centers (available now for dihydropteridine reductase deficiency)
- Discussion with successfully treated female PKU patients, regarding the importance of reinstating the low phenylalanine diet prior to pregnancy

Psychosocial, educational, familial
- Family counseling
- Education and support for strict adherence to the unnatural diet required to prevent mental retardation
- Consideration of referral for psychologic help in dealing with problems of the diagnosis and restricted life-style, e.g., depression, denial, guilt, anxiety
- Assessment of community/family resources, such as support groups, financial assistance for dietary supplements

Medical
- Phenylalanine restricted diet, as soon as possible after birth, through childhood, and during pregnancy, beginning before conception
- Periodic re-evaluation of diet by monitoring phenylalanine levels (elimination of diet in cases of transient or mild hyperphenylalaninemia; consideration of other therapy for malignant forms)

SUGGESTED READING

Committee on Genetics. More on newborn screening for phenylketonuria: Recommendations of the Committee on Genetics. *Pediatrics* 1983;71:139.

Ledley FD, Levy HL, Woo SLC: Molecular analysis of the inheritance of phenylketonuria and mild hyperphenylalaninemia in familes with both disorders. *N Engl J Med* 1986;314:1276.

Schneider AJ: Newborn phenylalanine/tyrosine metabolism: Implications for screening for phenylketonuria. *Am J Dis Child* 1983;137:427.

Scriver CR, Clow CL: Phenylketonuria: Epitome of human biochemical genetics. *N Engl J Med* 1980;303:1336, 1394.

Tourian A, Sidbury JB: Phenylketonuria and hyper- phenylalaninemia, in: Stanbury JB, Wyngaarden JB, Fredrickson DS, et al, (eds): *The Metabolic Basis of Inherited Disease*, 5th ed. New York, McGraw-Hill, 1983.

Woo SLC, Lidsky AS, Gûttler F, et al: Prenatal diagnosis of classical phenulketonuria by gene mapping. *JAMA* 1984;251:1998.

Chapter 38

Familial Forms of Breast Cancer

Breast cancer has the highest incidence of any cancer in women. The cumulative lifetime risk that a woman will develop breast cancer has been estimated to be 7%. About 100,000 new cases, nearly all in women, are diagnosed each year in the United States and more than 30,000 deaths result. Studies of clinical and family history have identified populations at relatively high risk, for whom increased surveillance is prudent and justified.

ETIOLOGY AND PATHOGENESIS

The etiology of breast cancer is not known. An association with hormones is widely recognized, particularly in the proportion of primary breast cancers that require estrogenic activity for tumor growth. Reproductive risk factors suggest that hormones may be important in the susceptibility to malignant change. However, no clear relationship has been established between estrogen levels and breast cancer. Studies on excretion rates of the different estrogen fractions may provide information on this point. Other research strategies are being applied to locate a potential susceptibility gene (or genes) and to develop methods for identifying high-risk individuals in family studies. Among them are linkage studies, using DNA probes and enzyme markers; high resolution chromosome analysis; and investigations of the role of growth factors in malignant transformation.

GENETIC CHARACTERISTICS AND MODE OF INHERITANCE

Familial aggregation in breast cancer has been noted for many years.

- Early studies showed that when one woman in the family had breast cancer, the risk for the same cancer in a female first-degree relative was increased two to three times, compared with a woman lacking such a family history.
- The risk estimate can be refined considerably on the basis of disease characteristics and family history (Table 38-1).

Positive family history, bilaterality, and age at diagnosis are interrelated. Patients with a positive family history are more likely to have bilateral breast cancer and to have breast cancer at an earlier age than others. The pedigree may resemble an autosomal dominant pattern of inheritance in some families, but it is not clear whether a single gene is often responsible.

Breast Cancer in Males. Although breast cancer in men is a rarity, reported cases appear to have a strong genetic component. Families have been noted in

TABLE 38-1
Cancer in First-Degree Relative vs. Age-Matched Control

FAMILY HISTORY	RISK
Mother or sister had breast cancer	2 to 3-fold higher
Mother and sister had breast cancer	6-fold higher
Mother had breast cancer; sister had premenopausal bilateral breast cancer	40-fold higher

which father and son, or father and daughter both had breast cancer. Also, in breast cancer families, predisposition apparently can be passed to the next generation through a clinically unaffected male.

Klinefelter's syndrome, a chromosomal anomaly in which males have an additional X chromosome (an XXY complement), carries a risk of breast cancer which approaches that of normal women. Klinefelter's syndrome is not inherited, and the possible role played by the additional X chromosome in the development of breast cancer is not known.

Other Cancers in Breast Cancer Families. There are breast cancer families in which a seemingly disproportionate incidence of other cancers is also observed. The descriptive terms "cancer families" and "familial adenocarcinomatosis" have been applied. The cancers most frequently associated with breast cancer are as follows:

Colon cancer (less often, other GI tract cancers)
Endometrial cancer
Ovarian cancer
Soft-tissue sarcomas, acute leukemias

It has been proposed that in these families the same genotype which predisposes to breast cancer predisposes other tissues to malignant change. Characteristically, these specific cancers tend to be observed at an earlier age than their sporadic counterparts, and multiple primary malignant lesions occur in the same individual. Transmission of the predisposing trait appears to be vertical—in successive generations.

- Risk for first-degree relatives of an affected individual may be as high as 50% in some of these families.

Breast cancer is also associated with Cowden's syndrome, a disease characterized by multiple hamartomatous lesions of skin and mucous membranes. Fibroadenomatous enlargement of the breasts is observed, apparently predisposing to malignant change in some cases.

- Cowden's syndrome is inherited as an autosomal dominant disease with a recurrence risk of 50% for the offspring of an affected individual.

- About 1/3 women with Cowden's syndrome develops breast cancer.

CLINICAL NOTES

Age at menarche, at menopause, and at first full term of pregnancy all have been related to risk for breast cancer. The highest risk conferred by these factors would be for a nulliparous woman who had early menarche and late menopause. Full-term delivery of a child before the age of 26 has a protective effect, apparently unrelated to the number of subsequent pregnancies. Other empiric data modulate the risk as indicated in Table 38-2.

The incidence of breast cancer with regard to age has a bimodal distribution. There is a steep increase with increasing age until menopause and a second, though shallower, rise following menopause.

Breast cancers generally grow slowly. The likelihood of intralobular or intraductal carcinoma in situ progressing to invasive carcinoma has been difficult to predict.

Benign cystic mastopathy is a condition that may, but does not always, predispose to malignant change. The critical factor appears to be the severity of the disease, especially when there is a family history of breast cancer.

Paget's disease of the nipple is an inflammatory, eczemalike carcinoma that generally has multiple points of origin within the duct system. It is a disease of middle age, with no apparent genetic basis.

PROCEDURES FOR DIAGNOSTIC CONFIRMATION

Mammography
Xeroradiography
Thermography
Biopsy

CONSIDERATIONS IN MANAGEMENT

Genetic
- Extended family history, including other cancers
- Risk counseling: determination of estimated risk on the basis of empiric data, family history, age at onset, laterality, and reproductive history (Table 38-1)
- Discussion of sharing information with other family members to encourage evaluation and risk counseling

Psychosocial, educational, familial
- Education of at-risk patient in breast self-examination
- Education about need for periodic (medical) breast examinations or other indicated diagnostic procedures
- Education about use of occult-blood smear in rare families also at risk for GI cancer

Medical
- Surveillance program for at-risk family members
- Evaluation of at-risk family members for associated tumors in specific families
- Consideration of prophylactic mastectomy in specific high-risk cases (has been used in special situations to prevent breast cancer)

TABLE 38-2
Risk Factors for Breast Cancer in Females

FACTOR	HIGH RISK	LOW RISK	RISK DIFFERENTIAL*
Age	Old age	Young age	+ + +
At menarche	Early	Late	+
At first birth	Older than 30	Younger than 20	+ +
At menopause	Late	Early	+
Country of residence	North America, northern Europe	Asia, Africa	+ + +
Socioeconomic class	Upper	Lower	+ +
Marital status	Never married	Ever married	+
Place of residence	Urban Northern United States	Rural Southern United States	+ +
Race	White	Black	+
Oophorectomy	No	Yes	+ +
Body build	Obese	Thin	+ +
Family history of pre-menopausal bilateral breast cancer	Yes	No	+ + +
History of cancer in one breast	Yes	No	+ + +
History of fibrocystic disease	Yes	No	+ +
Any first-degree relative with breast cancer	Yes	No	+ +
History of primary cancer in ovary or endometrium	Yes	No	+ +
Radiation to chest	Large doses	Minimal exposure	+ +

*+ + + = relative risk of greater than 4.0;
+ + = relative risk of 2.0-4.0;
+ = relative risk of 1.1-1.9

Sources: Kelsey JL: A review of the epidemiology of human breast cancer. *Epidemiol Rev 1979; 1:98.*

SUGGESTED READING

American Cancer Society, 1981 Cancer Facts and Figures.

Anderson DE: A genetic study of human breast cancer. *Journal of the National Cancer Institute* 1972;48:1029.

Anderson DE: Genetic study of breast cancer: Identification of a high risk group. *Cancer* 1974;34:1090.

Anderson DE: Genetic predisposition to breast cancer. *Recent Results in Cancer Research* 1974;57:10.

Anderson DE: Breast cancer in families. *Cancer* 1977;40:1855.

Everson RB, Li FP, Fraumeni JF, JR, et al: Familial male breast cancer. *The Lancet* 1976;i:9.

Fishman J, Fukushima DK, O'Connor J, et al: Low urinary estrogen glucuronides in women at risk for familial breast cancer. *Science* 1979;204:1089.

Lemon HM: Genetic predisposition to carcinoma of the breast: multiple human genotypes for estrogen 16 alpha hydroxylase activity in Caucasians, *J Surg Onocol* 1972;4:255.

Li F, Fraumeni JF, Jr: Familial breast cancer soft tissue sarcomas, and other neoplasms, *Ann Intern Med* 1975;83:833.

Lippman ME, Allegra JC: Receptors in breast cancer. *N Engl J Med* 1978;299:930.

Lynch HT, (ed.): *Genetics and Breast Cancer*, New York, Van Nostrand-Reinhold, 1981.

Lynch HT, Albano WA, Danes, BS, et al: Genetic predisposition to breast cancer, *Cancer* 1984;53:612.

Miller AB, Bulbrook RD: The epidemiology and etiology of breast Cancer. *N Engl J Med* 1980;303:1246.

Mulvihill J, Miller RW, Fraumeni JF, Jr, (eds): *Genetics of Human Cancer*, vol. 3. (Progress in Cancer Research and Therapy). New York, Raven Press, 1977.

Mulvihill J: Cancer control through genetics, in Arrighi, FE, Stubblefield E, Rao, N, (eds): *Genes, Chromosomes and Neoplasia*. New York, Raven Press, 1981.

Chapter 39

Genetic Forms of Colon Cancer

Colorectal cancers are among the most common in the United States and it is widely recognized that a genetic predisposition plays a significant role in their development.

ETIOLOGY AND PATHOGENESIS

The etiology of most colorectal cancers is still poorly understood. The familial polyposis syndromes have been the most thoroughly studied as genetic disorders that predispose cells to malignant transformation. In addition, in certain families colon cancer can also confer a risk of cancer in multiple primary sites in the index case as well as in first-degree relatives.

Approximately 95% of colorectal cancers are adenocarcinomas, but only a small percentage of adenomas become cancers. However, with or without a family history, individuals with adenomas of the colon are at an increased risk for colorectal cancer. Hyperplastic and inflammatory polyps do not seem to confer this risk.

A good deal of current research is devoted to the search for markers—molecular, biochemical, chromosomal—to clarify the etiology of colon cancer and provide methods for premorbid identification of high-risk individuals.

GENETIC CHARACTERISTICS AND MODE OF INHERITANCE

A number of distinct genetic syndromes are known to predispose to colon cancer. They may or may not be characterized by pre-existing polyposis. Colon cancer can exist as the only clinical entity, or can be associated with other tumors.

- Several syndromes are inherited in a straightforward mendelian fashion with risks as high as 50% for first-degree relatives. Others have an apparent familial aggregation which is less clear-cut.
- The occurrence of an isolated case of colon cancer in a family does not preclude genetic etiology.

Table 39-1 summarizes clinical characteristics and mode of inheritance for each of the known familial colon cancer syndromes discussed below.

CLINICAL NOTES

Polyposis Syndromes Predisposing to Colon Cancer comprise approximately 5% of all colon cancers. Although malignant potential varies from one syndrome to another, they are all prudently regarded as premalignant conditions. Although generally gradual, malignant change may be extremely variable. Approximately 75% of the carcinomas included in hereditary polyposis syndromes are in the rectosigmoid area.

Familial polyposis of the colon (FPC, multiple adenomatosis of the colon and rectum, autosomal dominant). Multiple polyps invariably undergo malignant transformation, usually by the fourth decade. The colon epithelium is typically covered by hundreds to thousands of polyps, which can form in childhood and are generally limited to the colon. Reported incidence is estimated at approximately 1/8,000 births.

TABLE 39-1
Summary: Genetic Forms of Colon Cancer

	INHERITANCE PATTERN	ASSOCIATED MALIGNANCIES AND OTHER LESIONS
Syndromes With Preexisting Polyposis		
Familial polyposis of the colon (FPC)	Autosomal dominant	Limited to colon
Gardner's syndrome	Autosomal dominant	GI malignancies, cutaneous fibromas, osteomas, epidermoid cysts, dental anomalies
Peutz-Jeghers syndrome	Autosomal dominant	Jejunal polyps; pigmented lips, buccal mucosa; ovarian tumors
Scattered, discrete polyps	Familial aggregation	Limited to colon
Juvenile polyposis	Autosomal dominant	Limited to colon
Cowden's syndrome	Autosomal dominant	Breast/thyroid cancer mucocutaneous/facial/ gingival hamartomas
Familial gastrointestinal adenomatosis	Familial aggregation	Polyps throughout GI tract
Turcot syndrome	(?) Autosomal recessive	CNS malignancies
Syndromes Without Preexisting Polyposis		
Familial colonic adenocarcinomatosis	Familial aggregation	Limited to colon
Cancer family syndrome	Familial aggregation	Endometrial/ovarian/ lung/breast/gastric cancers, etc., in various combinations (see Chapter 40)
Ulcerative colitis	Multifactorial;	Limited to colon

Gardner's syndrome (autosomal dominant). Adenomatous colon polyps and colon carcinomas are consistent in the disease. Other clinical manifestations vary considerably and can include epidermoid cysts, cutaneous fibromas, osteomas of the mandible and other bones, and dental anomalies. Desmoid tumors and retroperitoneal fibrosis are common sequelae of surgery, some developing into fibrosarcomas. Colorectal malignancies develop somewhat later than those of FPC, and predisposing polyps are not limited to the colon. The incidence of Gardner's syndrome is approximately half that of familial polyposis (1/16,000).

Peutz-Jeghers syndrome (autosomal dominant). Diagnostic signs for the heterogeneous Peutz-Jeghers syndrome include abnormal patterns of pigmentation, particularly on the lips and buccal mucosa. Hamartomatous polyps can occur throughout the GI tract, but are most probable in the jejunum. They have a considerably lower malignant potential than adenomatous polyps. An estimated 2% to 3% of patients develop malignant lesions of the colon, stomach, and duodenum. An estimated 5% of patients develop benign granulosa cell ovarian tumors. Their risk of cervical cancer may be increased.

Scattered, discrete polyps (familial aggregation). Solitary polyps are a common finding in adults, but a familial occurrence has been associated with adenocarcinoma of the colon. The polyps in this syndrome are colonic, and their status as benign, precancerous, or malignant is best determined by histologic studies.

Juvenile polyposis (autosomal dominant). Cystic hamartomatous polyps typically appear in childhood and can be sparse or numerous. Although these polyps may not predispose to cancer, adenomatous polyps which have a significantly higher malignant potential are thought to be a secondary complication of the syndrome. Reports of relatives of index patients with multiple polyposis and/or colonic carcinoma have been noted in the literature.

Cowden's syndrome (multiple hamartoma syndrome, autosomal dominant). Cowden's syndrome consists of multiple hamartomatous lesions, primarily in mucocutaneous tissue of the face and gingivae, and fibroadenomatous breast enlargement. A variety of colon polyps can be found, including adenomatous polyps, which undergo malignant transformation. The syndrome also predisposes to thyroid and breast cancer.

Familial gastrointestinal adenomatosis (familial aggregation). This condition differs from familial polyposis of the colon in that the distribution of polyps in familial gastrointestinal adenomatosis is throughout the entire GI tract. Rectal carcinoma has been documented.

Turcot syndrome (?autosomal recessive). Adenomatous polyps of the colon in association with gliomas characterize this rare syndrome.

Nonpolyposis Syndromes Predisposing to Colon Cancer. Although colon cancer is common and more than one family member can be affected by chance alone, certain families do exist who are predisposed to bowel cancer (and malignancy in general). Factors that raise suspicion of this familial predisposition include:

- Multiple family members with colon cancers

- Multiple sites of primary malignancies in any one family member
- Malignancies in different body sites in first-degree relatives
- Generally earlier onset of symptoms than the population average

The following types of familial aggregations have been reported:

Familial colonic adenocarcinoma. Malignant lesions are more randomly distributed throughout the colon than are those in patients with sporadic colon cancer. Carcinomas are limited to the colon, and adenomas and carcinomas are present concurrently. The likelihood of developing a secondary primary adenocarcinoma in remaining bowel tissue is high.

Ulcerative colitis. This inflammatory bowel disorder has a well-established risk of colon cancer, with a strong correlation between the risk and the extent and duration of the bowel disorder.

Cancer Families. Colon cancer is also found in families that seem to have a predisposition to malignant disease in general and in others where it may occur in association with specific adenocarcinomas, e.g., gastric or endometrial (see Chapter 40).

PROCEDURES FOR DIAGNOSTIC CONFIRMATION

Fecal occult blood
Sigmoidoscopy
Colonoscopy
Air-contrast barium enema
Polypectomy with histologic assessment of polyps
Carcinoembryonic antigen (CEA)
Upper GI series

CONSIDERATION IN MANAGEMENT

Genetic
- Determination of mode of transmission or probable risk
- Evaluation of at-risk family members for possible associated findings (e.g., epidermoid cysts in Gardner's syndrome, mucosal pigment in Peutz-Jeghers syndrome)
- Risk counseling
- Consideration of reproductive options, such as, donor insemination, adoption
- Discussion of sharing information with other family members to encourage evaluation and risk counseling

Psychosocial, educational, familial
- Education for compliance with patient-conducted surveillance procedures, for example, occult-blood smears, breast self-examinations
- Education about need for regular physical evaluations
- Psychologic support if colostomy is indicated or performed
- Consideration of psychotherapy for physical disfigurement (e.g., Cowden's syndrome, Gardner's syndrome)
- Guidance in informing minors of family medical background in view of necessity for surveillance

Medical
- Monitoring for colon cancer in at-risk family members, and for additional primary tumors in index case

Screening for associated neoplasia in at-risk family members (e.g., breast examination in families with breast/colon cancer association)

Prophylactic colectomy in selected high-risk cases (e.g., familial polyposis of the colon)

SUGGESTED READING

Fraumeni JF Jr, (ed): *Persons at High Risk of Cancer,* New York, Academic Press, 1975.

Lynch HT, (ed): *Cancer Genetics*, Springfield, Ill., Charles C. Thomas, 1976.

Lynch HT, Krush AJ, Guirgis J: Genetic factors in families with combined gastrointestinal and breast cancer. *Am J Gastroenterol* 1973;59:31.

McAllister AJ, Kent FR: Peutz-Jeghers Syndrome—experience with twenty patients in five generations, *Am J Surgery* 1977;134:717.

Mulvihill JJ, Miller RW, Fraumeni JF, Jr, (eds): *Genetics of Human Cancer*, vol 3. (Progress in Cancer Research and Therapy) New York, Raven Press, 1977.

Naylor EW, Gardner EJ: Penetrance and expressivity of the gene responsible for the Gardner syndrome. *Clin Genet* 1977;11:381.

Pavlides GP: Hereditary polyposis Coli I. The diagnositc value of colonoscopy, barium enema, and fecal occult blood. *Cancer* 1977;40:2632.

Schimke NR: *Genetics and Cancer in Man*, Edinburgh, Churchill Livingstone, 1978.

Sherlock P, Winawer SJ: Are there markers for the risk of colorectal cancer? (editorial) *N Engl J Med* 1984;311:118.

Weary PE, Gorlin, RJ, Gentry WC, Jr, et al: Multiple hamartoma syndrome (Cowden's disease). *Arch Dermatol* 1972;106:682.

Williams W: Management of malignancy in "cancer families." *The Lancet* 1978;I:198.

Winawer SJ, Schuttenfeld D, Sherlock P, (eds): *Colorectal Cancer: Prevention, Epidemiology and Screening*, New York, Raven Press, 1980.

Chapter 40

Other Heritable Cancers

Genetic predisposition to breast cancer and to cancer of the colon has been discussed in other chapters, as has neurofibromatosis, one of many single-gene disorders which may have malignancies or benign tumors as manifestations. This section catalogues a number of mendelian disorders with a propensity for tumors and emphasizes the counseling for retinoblastoma and Wilms' tumor. Other genetic disorders with neoplastic features will be presented primarily in tabular form, (Table 40-3).

While cancer at any age may have a genetic component, cancer in children, although much less frequent, is more likely to have a genetic etiology. This is well recognized for such childhood cancers as Wilms' tumor and retinoblastoma, but is also thought to apply to other malignancies. Clearly genetic, of course, are those that occur in association with a predisposing genetic disorder, such as, Bloom's syndrome and xeroderma pigmentosum.

ETIOLOGY AND PATHOGENESIS

Despite a wealth of research, relatively little is known about the etiology of cancer in humans. From analogies in the mouse, we can surmise that somatic mutations in the affected organ and latent viral infections in susceptible individuals may be important in man. The H2 locus in the mouse is especially important in influencing susceptibility to tumor viruses; a counterpart in humans is still uncertain, since no strong association between the human leukocyte antigen (HLA) system and a particular tumor has been discovered. The best evidence for a viral origin of human tumors is the Epstein-Barr viral lymphoma or Burkitt's lymphoma; HLA association is not involved. Epstein-Barr virus is also associated with cancer of the nasopharynx; hepatitus B virus infections that lead to a persistent carrier state are highly associated with primary cancer of the liver; and human papilloma viruses of certain subtypes are associated with cancer of the uterine cervix.

Cells with chromosomal imbalance may be predisposed to viral or chemical transformation. Basic research has indicated that transformability, whether by viruses or chemicals, seems to follow a "common pathway" involving predisposing "oncogenes". Patients with chromosome instability syndromes (ataxia-telangiectasia, Bloom's syndrome, Fanconi's anemia) or trisomy 21 have a higher incidence of leukemia and other tumors, as well as a predisposition to viral transformation of cells. For the rare recessive disorders in this list, heterozygous carriers will be fairly common in the general population and *may* be represented disproportionately among those with leukemias or other malignancies; however this hypothesis is still speculative. Tumors are also more common in patients with inborn or acquired immunodeficiency, possibly reflecting defects in immune surveillance mechanisms.

Knudson has proposed a two-hit model to explain occurrences of hereditary and nonhereditary tumors of identical form, including retinoblastoma, Wilms' tumor, neuroblastoma, and pheochromocytoma. In this model, it is hypothesized that all tumors arise from at least two mutations, the second always being a random hit of a somatic cell. When the first mutation occurs in a germ cell, it will be present in all cells of the individual and will be hereditary (transmitted to 50% of the offspring as an autosomal dominant trait). The gene carrier may develop none, one, or multiple tumors, and the number will follow the so-called Poisson distribution. If the first mutation, on the other hand, is somatic, it will be confined to a single cell and the cancer will be nonhereditary in that family.

Thus, all tumors require two mutational events. Only a single second event is necessary for tumor development in hereditary cases; hereditary tumors will occur at an earlier average age than nonhereditary ones and will be more likely to be multiple. The nonhereditary type will occur later and singly, since tumor development in a single cell requires two infrequent mutational events.

Research on genetic cancers suggests that the mutational events are not necessarily point mutations, but may take the form of chromosome defects such as breakage with resultant deletion or rearrangement, or somatic loss of an entire chromosome. The hereditary predisposition is characterized by heterozygosity in all cells (i.e., only one of two homologous chromosomes has undergone such a change). In the cancer cell, the normal chromosome has also become abnormal at the relevant locus, making the cell homozygous, or was lost, producing a hemizygous cell that also lacks the homologous normal gene.

GENETIC CHARACTERISTICS AND MODE OF INHERITANCE

The disorders listed in the tables are autosomal dominant, autosomal recessive, or X-linked mendelian traits. Except in those disorders in which the tumors themselves are the basis for the diagnosis, only a minority of gene carriers develop tumors, the proportions depending on the disorder. The risk for tumors is thus lower than the risk for inheriting the gene, even though the mode of inheritance for the predisposing condition may be straightforward.

In general, information about specific conditions can be found in standard textbooks or the references to this section of the handbook. Special emphasis will be given to a few conditions, however.

Wilms' Tumor, a nephroblastoma, occurs in 1/25,000 to 1/10,000 livebirths throughout the world, without any racial or ethnic predilection. It accounts for nearly 15% of all tumors diagnosed before age 15 years; about 50% occur before age 3 and 90% before age 10. The tumor is probably derived from the metanephric blastema. Familial Wilms' tumor was seldom noted before effective therapy became common. Now descriptions of siblings and parent-child affected family members are appearing. As in other heritable tumors of paired organs, familial cases are more often bilateral and occur at an earlier age. One third of all cases are thought to be familial.

When hereditary, the tumor is transmitted as an autosomal dominant trait with about 63% penetrance, meaning that 37% of the gene carriers manifest no overt malignancy, while 15% have bilateral and 40% to 50% have unilateral tumors. Roughly 30% of all unilateral sporadic cases actually carry the genetic trait. Risks for future children are given in Table 40-1.

TABLE 40-1
Risk of Subsequent Child Developing Embryonal Neoplasm

FAMILY HISTORY	WILMS' TUMOR (%)*	RETINOBLASTOMA (%)†
Parent with bilateral tumors; parent with unilateral and positive family history; unaffected parent with either two children affected, or child and near collateral relative affected	22	48
Parent with unilateral tumor and negative family history	9.5	7.5
Single child with bilateral tumors	10	1.2
Single child with unilateral tumor	<5	1.2

*Assumes 63% penetrance
†Assumes 95% for penetrance
Source: Jackson LG, Anderson DE, Schimke RN: Genetics and cancer, in Jackson LG, Schimke RN, (eds): *Clinical Genetics, A Source Book for Physicians.* New York, John Wiley & Sons, 1979

Other congenital anomalies can be associated with Wilms' tumor. One third of patients with nonfamilial congenital aniridia (absence of the iris) have Wilms' tumor, though only 2% of the tumor patients have the eye lesion. The dominant gene form of aniridia is not associated with Wilms' tumor. There is also a syndrome of Wilms' tumor, aniridia, pseudohermaphroditism (male), and nephrotic syndrome. Others may have Wilms'tumor with hemihypertrophy, sometimes also with hepatoblastoma or adrenocortical carcinoma, or the Beckwith-Wiedemann syndrome. After a five- to 15-year latent period, leukemia may occur in survivors of Wilms' tumor.

The finding, in some cases, of an interstitial chromosome deletion of the short arm of chromosome 11 (involving band 11p13), first suggested that the locus for Wilms' tumor may be in this area. Further research supporting this hypothesis includes the identification of a linkage between Wilms' tumor and the catalase gene, which is mapped on chromosome 11p.

Retinoblastoma develops in about 1/18,000 liveborn infants. About 30% are hereditary and transmitted as an autosomal dominant trait with penetrance in excess of 90%, including all bilateral cases and about 15% of unilateral cases. A parent with a history of bilateral or multifocal unilateral retinoblastoma has a 50% risk of passing the gene on to each offspring. Since penetrance is incomplete, up to 10% of gene carriers will not develop the tumor, but may pass on the gene. By the same token, a seemingly sporadic case may actually be inherited from a non-expressing parent. Tumors may have occurred and spontaneously regressed in such a parent, without coming to medical attention. A careful ophthalmologic evaluation may reveal retinal scars that document a familial nature. Recurrence risks for various family-history patterns are given in Table 40-1.

Some children with sporadic retinoblastoma and associated anomalies, which may include mental retardation, have an abnormal chromosome karyotype, with a deletion of part of the long arm of chromosome 13 (involving band 13q14). While only about 5% of cases have been shown to have the chromosome deletion, it has been suggested that a submicroscopic deletion or a mutation at the same chromosome location may be necessary for the development of retinoblastoma. This hypothesis, supported by other findings, has led to the identification of markers for the susceptibility to retinoblastoma. A recent report describes the use of recombinant DNA probes and assays for the closely linked enzyme, esterase D, for prenatal and postnatal tests in informative families with the hereditary form of the disorder.

Follow-up of children with retinoblastoma reveals that some develop a secondary malignancy, particularly in the head and neck regions, thought to be a direct complication of radiotherapy (or intracarotid chemotherapy). However, an excess of tumors has been noted outside the cranial area in patients treated only surgically, particularly in those with bilateral, familial retinoblastoma. These second malignancies are usually long-bone sarcomas and may appear many years later.

Neuroblastoma produces an entirely different clinical picture. Very few familial cases have been reported; associated congenital anomalies are not frequent; and second malignancies have not been observed. However, chromosome abnormalities have also been noted in tumor cells. About 70% of those studied revealed a deletion of the terminal portion of chromosome 1. Survival is poor (less than 25% survived three years in one large series of 500 patients). Neuroblastomas are derived from neural crest tissue; they occur most often in the paravertebral area in conjunction with autonomic ganglia and present with abdominal mass. Many unusual presentations have been noted, including Horner's syndrome, polymyoclonia, and hypoglycemia. If the patient survives, these tumors have an unusal propensity to undergo spontaneous regression or else maturation to ganglioneuromas or ganglioneurofibromas, especially in women.

Multiple Endocrine Gland Neoplasms are recognized clinically because of hormonal effects: Sensitive hormone assays now permit early diagnosis, which can be life-saving. There are several types.

Multiple endocrine neoplasia, type I (MEN I). MEN I includes tumors of the parathyroids, pancreas, pituitary, adrenal cortex and thyroid (in that order, as shown in Table 40-2). Parathyroid involvement is usually symptomless but readily detected from serum calcium and phosphorus levels. Pancreatic involvement

TABLE 40-2
Screening Patients at Risk for MEN 1

ENDOCRINE GLAND	FREQUENCY OF INVOLVEMENT (%)	SCREENING STUDIES*
Parathyroid	90	Serum calcium and phosphorus
Pancreas	80	Glucose tolerance test with plasma insulin levels if possible; fasting serum gastrin
Pituitary	65	Plasma growth hormone response to insulin-induced hypoglycemia; metapyrone for ACTH, plasma (or urinary) gonadotropin levels
Adrenal cortex	35	Plasma cortisol (a.m. and p.m.); urinary steroids, less reliable
Thyroid	20	True thyroxine; I^{131}-uptake and scan if gland palpably abnormal

*Other screening studies include: urinary 5-HIAA, gastric analysis with measurement of basal and maximal acid content; upper-GI series, biopsy if ulcer symptoms are present; X-ray films of sella turcica; visual-field examination; and selective stimulatory and suppression tests (dexamethasone and calcium infusion).

Source: Jackson LG, Anderson DE, Schimke RN: Genetics and cancer, in Jackson LG, Schimke RN, (eds): *Clinical Genetics, A Source Book for Physicians.* New York, John Wiley & Sons, 1979

can produce the Zollinger-Ellison syndrome due to excessive gastrin secretion and hypoglycemic episodes due to insulinomas. A combination of necrotizing skin rash, stomatitis, and diabetes mellitus is a clue to glucagonoma. Pancreatic delta cell tumor causes the pancreatic cholera syndrome of watery diarrhea, due to vasoactive intestinal peptide. Thyroid lesions are usually simple adenomas, never medullary carcinomas. The genetic predisposition is straightforward autosomal dominant with high penetrance. First-degree relatives should be tested every year.

Multiple endocrine neoplasia, type II (MEN II). MEN II, also autosomal dominant, comprises medullary carcinoma of the thyroid, pheochromocytomas, and parathyroid hyperplasia (probably secondary to secretion of calcitonin by the medullary carcinoma of the thyroid). Gliomas, glioblastomas, and meningiomas may occur at increased frequency in this syndrome. Plasma calcitonin assay is an important diagnostic and screening tool.

Multiple endocrine neoplasia, type III (MEN III). MEN III is characterized by pheochromocytoma and medullary carcinoma of the thyroid, plus an array of other findings, including some striking physical features: neuromas of the conjunctival, labial, and buccal mucosa (hence, the synonym of multiple mucosal neuroma syndrome), the tongue, the larynx, and the GI tract; enlarged corneal nerves; soft, ''blubbery'' lips; pseudoprognathism due to soft-tissue hypertrophy of the chin; marfanoid habitus; café-au-lait spots; and megacolon. The life-threatening feature is the thyroid malignancy; once the phenotype is recognized, early thyroidectomy is indicated. Like MEN II, this disorder appears to be autosomal dominant with high penetrance.

Skin Cancers. Cutaneous tumors are particularly informative from a genetic point of view because they are readily detected, diagnosed, treated, and traced through families. Hereditary skin neoplasms are included in Table 40-3. In other disorders, including MEN III and ataxia-telangiectasia, skin lesions may be a clue to internal malignancies.

Xeroderma pigmentosum, a rare skin disease, is of great importance because of its mechanism. The patient's skin is extremely sensitive to ultraviolet light. After early erythema and freckling, basal cell carcinomas, squamous cell carcinomas, melanomas, angiosarcomas, fibrosarcomas, and keratocanthomas occur. The defect is transmitted as an autosomal recessive trait. It is known to be due to one of a family of mutations that interfere with repair of ultraviolet-damaged DNA, showing that normal people, as well as heterozygotes, have considerable protective repair capacity.

Cancer Family Syndrome. Other tumors at multiple sites cluster in families, suggesting a genetic etiologic factor. The term cancer family syndrome has been used to reflect the indisputable occurrence of families with an inordinately high incidence of multiple primary tumors in multiple organ sites. Prominent in the group are families with familial adenocarcinomatosis, arising from consistent sites. The tumors occur at an earlier age than observed in sporadic cases at the same sites. Within families, association of specific tumors appears, e.g., gastric and colon adenocarcinomas or adenocarcinoma of the endometrium and colon. Familial association of other types of neoplasms have been observed, as well, e.g., ovarian, lung, breast, renal, pancreatic, brain tumors, melanomas, and acute leukemia.

PROCEDURE FOR DIAGNOSTIC CONFIRMATION

Procedures depend upon detection and pathologic examination of tumors, plus clinical and laboratory recognition of associated features in hereditary syndromes. When the tumors listed in Table 40-3 are encountered, it is wise to consider these often rare genetic disorders in the differential diagnosis, to seek medical histories on family members, and to examine those members available.

Laboratory tests are available for certain specific conditions, such as the endocrine neoplasia syndromes. More tests are expected to become available through research for markers, to be used for diagnostic and predictive analysis. See Table 40-2 for screening studies on individals at risk for MEN I.

TABLE 40-3
Selected Hereditary Neoplasia and Genetic Conditions That Feature Neoplasia as a Complication

DISORDER	NEOPLASTIC FEATURES	OTHER CHARACTERISTICS	INHERITANCE
Neurofibromatosis	Neurofibroma, multiple iris nevi, fibrosarcoma,	Multiple café-au-lait spots, axillary freckling; scoliosis	AD

(Continued)

DISORDER	NEOPLASTIC FEATURES	OTHER CHARACTERISTICS	INHERITANCE
	neuroma, schwannoma, meningioma, polyps, optic glioma, pheochromocytoma		
Tuberous sclerosis	Angiofibroma, periungual fibroma, glial tumors, renal tumors, lung cysts; rhabdomyoma of heart	Seizures, mental retardation; depigmented skin patches, shagreen patches	AD
von Hippel-Lindau syndrome	Retinal angioma, cerebellar and other hemangioblastoma, pheochromocytoma, renal carcinoma	Onset most common in 4th decade; retinal detachment, blindness	AD
Sturge-Weber syndrome	Angioma in a variety of organs	Intracranial anomaly behind facial angioma; seizures (90%) mental retardation (30%); ocular defects	Sporadic
Nevoid basal cell carcinoma syndrome	Basal cell carcinomas—face and upper trunk; medulloblastoma, ovarian fibroma, and carcinoma	Jaw cysts, prognathism, congenital ocular and skeletal anomalies, frontal/parietal bossing	AD
Multiple hamartoma (Cowden's) syndrome	Mucosal and cutaneous papillomatosis and fibrosis, thyroid adenoma and carcinoma, lipoma, meningioma	Fibrocystic breast disease, bone and liver cysts, polyps; hypertrichosis; mild mental retardation (50%)	AD
Multiple lentigines (LEOPARD) syndrome	Multiple lentigines	Congenital heart disease; abnormal genitalia; growth retardation; deafness (15%); triangular face, parietal bossing, epicanthic folds, ptosis, skeletal anomalies	AD
Multiple leiomyomata	Angiomyomata of skin, uterus, and/or esophagus		AD
Multiple lipomatosis	Skin cancer		AD

TABLE 40-3 *Continued*

DISORDER	NEOPLASTIC FEATURES	OTHER CHARACTERISTICS	INHERITANCE
Beckwith-Wiedemann syndrome	5% risk of Wilms' tumor, hepatoma or adrenal cortical neoplasia	Macroglossia, neonatal hypoglycemia (30-50%), omphalocele, visceromegaly, earlobe grooves and pits, microcephaly (50%), facial birthmarks (90%)	Most sporadic, 5% recurrence risk for sibs
Werner's syndrome	Sarcoma, hepatoma, breast carcinoma	Premature aging—onset age 15-30; growth arrested at puberty; beaked nose, soft-tissue calcifications	AR
Rothmund-Thomson syndrome	Squamous cell carcinoma	Progressive hyperpigmented skin lesions (beginning as erythema of cheeks); juvenile cataracts; sparse eyebrows, eyelashes, hair; short stature; photosensitivity	AR
Focal dermal hypoplasia (Goltz's syndrome)	Mucocutaneous papillomas	Atrophy/linear hyperpigmentation of skin, eye anomalies, digital syndactyly, herniation of subcutaneous fat, usually lethal in males	?XD
Fanconi's anemia	Acute myelogenous leukemia, squamous cell carcinoma of mucocutaneous junctions, adenoma, hepatic carcinoma	Chromosomal breaks; skeletal anomalies, skin hyperpigmentation, café-au-lait spots, pancytopenia, urinary tract anomalies	AR
Bloom's syndrome	Leukemia, intestinal cancer, squamous cell carcinoma of the tongue, reticulum cell sarcoma	Chromosomal breaks; photosensitivity; erythematous, telangiectasic facial lesions, proportionate short stature, dolichocephaly, narrow face, café-au-lait spots	AR

(Continued)

DISORDER	NEOPLASTIC FEATURES	OTHER CHARACTERISTICS	INHERITANCE
Multiple endocrine neoplasias (MEN)			
Type I	Endocrine adenomas (parathyroid 80%, pituitary, thyroid, pancreas, adrenal cortex), carcinomas	Hormonal hypersecretion, diagnosis not usually made before 2nd decade	AD
Type II	Thyroid medullary carcinoma, parathyroid adenoma, pheochromocytoma	Hormonal hypersecretion	AD
Type III	Pheochromocytoma, thyroid medullary carcinoma in over 85%, neurofibroma; submucosal neuromas of tongue, lips, eyelids, and GI tract	Asthenic habitus; muscle wasting, lumbar lordosis, café-au-lait spots or diffuse freckling; hormonal hypersecretion; usually diagnosed before 4th decade	AD
Agamma-globulinemia (Bruton's)	Leukemia, lymphoreticular malignancies	Recurrent infections; malabsorption, rheumatoid-like arthritis, dermatomyositis, hypoplastic lymph nodes; lethal in infancy if untreated	XR
Wiskott-Aldrich syndrome	Lymphoreticular malignancies	Thrombocytopenia; inability to form antibodies to polysaccharide antigens	XR
Ataxia-telangiectasia	Lymphoreticular malignancies, leukemia, stomach carcinoma, brain tumors	Chromosomal breaks; facial cutaneous telangiectasia, thin face, dull expression, cerebellar ataxia, premature greying of scalp hair	AR
Multiple polyposis of the colon	Colon carcinoma	Multiple intestinal polyps; diagnosis usually in early adulthood	AD
Gardner's syndrome	Multiple osteomas of facial bones, epidermoid or	Intestinal polyps, rectal bleeding, diagnosis usual in	AD

TABLE 40-3 *Continued*

DISORDER	NEOPLASTIC FEATURES	OTHER CHARACTERISTICS	INHERITANCE
	sebaceous cysts, skin fibromas or desmoids, colon adenomas	childhood or early adulthood	
Peutz-Jeghers syndrome	Ovarian tumors (granulosa cell), colon carcinoma	Colon polyps, lip pigmentations, rectal bleeding	AD
Hemochromatosis	Hepatoma	Iron overload, diabetes mellitus, weakness	AR
Gonadal dysgenesis XX type	Gonadoblastoma, dysgerminoma	Dysgenic internal genitalia, bilateral streak ovaries, lack of secondary sexual development, infertility	AR
Gonadal dysgenesis XY type	Gonadoblastoma, dysgerminoma	Clinical female with male chromosomes, streak ovaries, no secondary sexual development, infertility	XR
Testicular feminization syndrome	Gonadoblastoma, dysgerminoma	Female external genitalia, male chromosomes, androgen insensitivity, presence of testicles, inguinal hernia (in 50%), primary amenorrhea, infertility	XR
Wilms' tumor	Wilms' tumor	Thought to result from an abnormality at or near band p13 of chromosome 11	AD—all bilateral cases, 30% of unilateral cases Sporadic—70% of unilateral cases
Aniridia, Wilms' tumor syndrome	Wilms' tumor	11p- chromosome deletion, mental retardation, long face, high nasal root, eyelid ptosis, lowset poorly lobulated ears	Sporadic
Retinoblastoma	Retinoblastoma, sarcoma	Other malformations and/or mental retardation may be present; associated	AD—all bilateral cases, 15% of unilateral cases

(Continued)

DISORDER	NEOPLASTIC FEATURES	OTHER CHARACTERISTICS	INHERITANCE
		with 13q14 chromosome defect (e.g., deletion, rearrangement, point mutation)	Sporadic—85% of unilateral cases
Acoustic neuroma	Meningiomas, schwannomas	Deafness in early adulthood	AD
Multiple exostoses	Osteosarcoma, chondrosarcoma	Multiple cartilaginous exostoses	AD
Familial malignant melanoma	Multiple primary malignant melanomas	Early age of onset	?AD
Cutaneous albinisms	Skin cancers	Decreased/absent cutaneous pigmentation, photosensitivity; visual deficits (some forms)	Varied, most AR
Xeroderma pigmentosum	Skin cancers	Photosensitivity; skin lesions (keratoses), corneal scarring, hyperpigmentation/ hypopigmentation	AR

Key: AD = autosomal dominant (50% risk to each offspring of an affected individual); AR = autosomal recessive (25% risk to each offspring of unaffected carrier parents); XR = sex-linked recessive (50% risk to each male offspring of a carrier female, 50% risk for each daughter of a carrier female to also be a carrier, all daughters of affected males are carriers, no male-to-male transmission); XD = sex-linked dominant (50% risk to each offspring of an affected female, 100% risk to daughters of an affected male, no risk to sons); Sporadic = a single occurrence in a family is usual

CONSIDERATIONS IN MANAGEMENT

Genetic
- Recognition of the hereditary types of tumors
- Risk counseling
- Consideration of reproductive options, if indicated by condition
- Investigation of current availability of predictive tests

Psychosocial, educational, familial
- Early and periodic examination of relatives
- Explanation, supportive care, and counseling

Medical
- Appropriate diagnosis
- Appropriate surgery, radiation, or chemotherapy, including prophylatic surgery in special settings
- Follow-up, especially for second malignancies in paired organs or in other sites
- In xeroderma pigmentosum: rigorous protection from sunlight from infancy on
- For children at risk for retinoblastoma: regular ophthalmologic examinations beginning shortly after birth

SUGGESTED READING

Abramson DH: Retinoblastoma: Diagnosis and management. *CA* 1982;32:130.

Cavenee WK, Murphree AL, Shull MM, et al: Prediction of familial predisposition to retinoblastoma. *N Engl J Med* 1986;314:1201.

Dryja TP, Cavenee W, White R, et al: Homozygosity of chromosome 13 in retinoblastoma. *N Engl J Med* 1984;310:550.

Goustin AS, Leof EB, Shipley GD, et al: Growth factors and cancer. *Cancer Res* 1985;46:1015.

Knudson AG, Jr, Strong LC: Mutation and cancer: Neuroblastoma and pheochromocytoma. *Am J Hum Genet* 1972;24:514.

Knudson AG: Hereditary cancer, oncogenes, and antioncogenes. *Cancer Res* 1985;45:1437.

Koufos A, Hansen MF, Copeland NG, et al: Loss of heterozygosity in three embryonal tumours suggests a common pathogenetic mechanism. *Nature* 1985;316:330.

Kraemer KH: Cancer-prone genodermatoses and DNA repair. *Prog Dermatol* 1981;15(2):1.

Mulvihill JJ: Cancer control through genetics, In Arrighi FE, Stubblefield E, Rao PN, (eds): *Genes, Chromosomes and Neoplasia*. New York, Raven Press, 1981.

Riccardi VM, Sujanski E, Smith AC, et al: Chromsomal imbalance in the aniridia-Wilms tumor association: 11p interstitial deletion. *Pediatrics* 1978;61:416.

Seeger RC, Brodeur GM, Sather H, et al: Association of multiple copies of the N-*myc* oncogene with rapid progression of neuroblastomas. *N Engl J Med* 1985;313:1111.

Strong LC, Herson J, Haas C, et al: Cancer mortality in relatives of retinoblastoma patients. *JNCI* 1984;73:303.

Yunis J: Chromosomal rearrangement, genes, and fragile sites in cancer: Clinical and biological implications, In De Vita VT (ed.): *Important Advances in Oncology*. Philadelphia, JB Lippincott Co., 1986.

Chapter 41

Affective Disorders

Manic-depressive illness, or bipolar disorder, is a distinctive type of affective disorder. This psychosis tends to be episodic, with a propensity for depression, but also exhibits one or more striking periods of enhanced alertness, great energy, flights of fancy, and other typical symptoms of the manic phase. Major depression, or unipolar disorder, is marked by recurrent depressive episodes without symptoms of mania.

ETIOLOGY AND PATHOGENESIS

Pharmacologic treatments of affective disorders have opened many fruitful avenues for investigation of pathogenetic mechanisms. Effective antidepressants include tricyclics, which inhibit presynaptic re-uptake of serotonin, depamine, and norepinephrine, thereby increasing their concentrations in the synapses and on postsynaptic receptors. Another class of antidepressants consists of the monoamine oxidase (MAO) inhibitors, which block mitochondrial metabolism of the same biogenic amines, with similar effect on their levels. These findings have given rise to the biogenic amine hypothesis of affective disorders.

For manic-depressive illness, this hypothesis predicts that brain levels of the amines will be low during depressive periods and high during manic periods. Many biochemical and pharmacologic observations are consistent with this interpretation. In addition, reserpine—a drug used to treat high blood pressure—can precipitate depression in predisposed persons, presumably through its known action of depleting the biogenic amines from their nerve ending vesicles. L-DOPA, in turn, can trigger hypomania, presumably by supplying more dopamine and possibly norepinephrine to the brain. Various pharmacologic manipulations have led to the view that dopamine and norepinephrine concentrations may mediate the locomotor and energy-level signs of affective disorders, while serotonin levels may be reflected primarily in mood changes.

Other researchers have placed emphasis on changes in levels of adrenal glucocorticoids in depression and manic-depressive illness.

Finally, in a search for the mechanism of the effectiveness of lithium treatment for manic-depressive illness, abormalities in the transport of sodium (Na) and lithium (Li) across RBC membranes have received much attention.

There is also some biochemical evidence for heterogeneity among unipolar patients. The metabolite MHPG (3-methoxy-4-hydroxyphenylglycol) can cross the blood-brain barrier and be measured in urine as a marker for norepinephrine turnover in the brain. Unipolar depressed patients with low urinary MHPG tend to respond to the tricyclics, imipramine and desipramine, as well as to maprotiline and dextroamphetamine; patients with high urinary MHPG tend to respond better to amitriptyline. It is possible that such research will lead to better ways of choosing therapy for individual patients, replacing the current practice of waiting for weeks to determine the clinical response to the drug.

GENETIC CHARACTERISTICS AND MODE OF INHERITANCE

Except, perhaps, in cases where other organic disorders underlie the depression (see Clinical notes), the mode for inheritance is unknown. However, the evidence for strong genetic predisposition is overwhelming. Family studies have differentiated the bipolar and unipolar types:

- Relatives of bipolar probands have a higher risk of affective disorders and some of them will be bipolar, though most seem to be unipolar; relatives of unipolar probands are almost always unipolar
- Relatives of early-onset probands are at greater risk than relatives of late-onset probands
- Female first-degree relatives of bipolar probands are one and a half to two times more frequently affected than male first-degree relatives of the same probands; there is no such sex difference among relatives of unipolar probands
- 30% of bipolar patients have a defect in transport of Na and Li across RBC membranes (high ratio of Li in RBC to Li in plasma); their relatives with affective illness also have higher Li ratios, while unaffected relatives have normal ones

Attempts have been made to demonstrate X-linked dominant inheritance in certain families by showing linkage of manic-depressive illness to genetic markers known to be coded for by X-chromosome genes, namely color blindness and glucose-6-phosphate dehydrogenase (G6PD). These claims have been disputed by others who have failed to demonstrate such linkage.

- Analysis of all the data suggest heterogeneity in patterns of inheritance
- Generally, empiric risk figures are used to estimate recurrence risks (see Table 41-1)

A recent report has suggested a possible marker for affective disorders in studies looking at muscarinic cholinergic receptors on fibroblasts. It is suggested that it may help clarify inheritance patterns, at least in some families.

CLINICAL NOTES

Descriptions of the clinical characteristics and natural history of affective illness are in standard textbooks. However, it is important to note here that at least 10% to 20% of patients commit suicide.

Also, it is essential to consider organic causes of bipolar or unipolar affective disorders, especially those that require specific therapies. Among potential underlying genetic diagnoses are hyperparathyroidism with hypercalcemia, Wilson's disease, porphyria, and Huntington's disease.

TABLE 41-1
Empiric Risks for Affective Disorders

	Risk for general population (%)		
	Males	1-2	
	Females	2-3	
	Risks for first-degree relatives*		
	RELATIVES (%)		
PATIENTS	BIPOLAR	UNIPOLAR	TOTAL
Bipolar	4-18	6-28	11-42
Unipolar	0.3-2	7-19	8-23
	Risks for twins		
		RISK(%)	RANGE/ SEVERAL STUDIES
Identical		80	50-92
Fraternal		20	16-38
	Influence of age of onset in bipolar and unipolar probands†		
AGE	RISK IN BIPOLAR AND UNIPOLAR RELATIVES (%)		
Early onset (<40)		32	
Late onset (>40)		23	

*Parent, sibling, child

†Source: Baron M, Mendlewicz J, Klotz, J: Age of onset and genetic transmission in affective disorders. *Acta Psychiatr Scand* 1981;64:373.

PROCEDURES FOR DIAGNOSTIC CONFIRMATION

There are no current specific laboratory tests in routine use. Assays for MHPG in urine and for lithium concentration in RBCs compared with plasma after incubation in vitro are being developed and evaluated by research groups.

Serum calcium and electrolytes, plasma ceruloplasmin, and urinary porphyrins may be indicated to rule out organic causes mentioned above.

CONSIDERATIONS IN MANAGEMENT

- Genetic
  - Exclusion of other causes
  - Risk counseling for offspring, siblings
- Psychosocial, educational, familial
  - Evaluation of family for related disorders
  - Supportive counseling for relatives to cope with patient
  - Assessment of community mental health resources
- Medical
  - Psychiatric treatment as indicated, including consideration of carefully selected drugs
  - Precautions against suicidal tendencies

SUGGESTED READING

Baron M, Mendlewicz J, Klotz J: Age of onset and genetic transmission in affective disorders. *Acta Psychiatr Scand* 1981;64:373.

Gershon ES, Targum SD, Kessler LR, et al: Genetic studies and biologic strategies in the affective disorders. *Prog Med Genet* 1978;2:101.

Mendlewicz J, Fleiss JL: Linkage studies with X-chromosome markers in bipolar (manic-depressive) and unipolar (depressive) illness. *Biol Psychiatry* 1974;9:261.

Mass JW: Clinical and biochemical heterogeneity of depressive disorders. *Ann Intern Med* 1978;88:556.

Nadi NS, Nurnberge JI, Gershon ES: Muscarinic cholinergic receptors on skin fibroblasts in familial affective disorder. *N Engl J Med* 1984;311:225.

Tosteson DC: Lithium and mania. *Sci Am* 1981;244:164.

Chapter 42

Schizophrenia

Schizophrenia is a category of major psychoses characterized by a collection of disorders affecting thought processes and mood. A number of clinical "types" have been described in standard texts, but these descriptions are not always consistent over the lifetime of the patient or, often, among the affected members of the same family.

ETIOLOGY AND PATHOGENESIS

Essentially nothing is clearly understood about the causes or mechanisms of the schizophrenias. Certain psychoanalytic assumptions of disturbed early childhood development are untestable. Family studies of adopted children have given more emphasis to biologic factors, but the absolute risks are low, suggesting that nongenetic factors must be important in triggering or preventing expression of the disorder. Since the drugs which are effective antipsychotic agents in schizophrenic patients have a primary site of action in competing for dopamine receptors, an abnormality in dopamine metabolism or dopamine receptors sensitivity has been postulated.

Many studies of platelet monoamine oxidase (MAO) activity show lower mean activity in schizophrenics, compared with controls. The significance of this finding is uncertain and of little value in diagnosing individual patients. It has also been claimed that MAO is low in individuals with paranoia and auditory hallucinations, normal in those without auditory hallucinations, and high in some depressed patients with chronic, asocial, eccentric, or bizarre behavior.

GENETIC CHARACTERISTICS AND MODE OF INHERITANCE

The mode of inheritance is unknown, except in some instances where another disorder may be responsible for schizophrenic behavior (see Clinical notes). The category of schizophrenia is almost surely heterogeneous, although a model for autosomal dominant genetic predisposition has been proposed, in which some of those with the gene would be frankly schizophrenic, while others were classified as schizoid.

Family and twin studies point strongly to a genetic predisposition. However, intrafamilial social and environmental factors could account for such observations. This nature/nurture tangle was separated through adoption studies. The first study examined 47 adults who had been put up for adoption early in life by mothers who were hospitalized with schizophrenia. Five of the offspring were diagnosed as schizophrenic, an incidence similar to that expected for offspring who lived with their schizophrenic parent (see Table). None of 50 individuals in a matched control group of adopted persons with nonschizophrenic parents became schizophrenic. Many of the other children of the schizophrenic women also had psychosocial problems.

Studies conducted in Denmark began with individuals who were diagnosed as schizophrenic and had been adopted early in life. Then information about the biologic and adoptive relatives was obtained, to test whether the expected increased incidence in relatives would be found in the families who reared the schizophrenic probands, or in the families who passed on the genes. The results were unequivocal that the increase in risk for relatives was among the biologic family members, not the adoptive ones. Some of these affected relatives were related through the fathers, thus excluding the possibility of effects from the uterine environment.

In the absence of a clear mode of inheritance and useful diagnostic tests, counseling must be based upon empiric-risk estimates derived from numerous studies with varying results. Approximate risk figures are shown in Table 42-1. Evaluation of intervening relatives prior to counseling may improve precision (for example, when the risks are estimated for offspring of an individual with an affected brother, the prospective parent should be evaluated to rule out any evidence of psychopathology in him or her before risks are given based only on an affected uncle).

TABLE 42-1
Empiric Risks for Schizophrenia

Risk for general population	About 1%
RELATIONSHIP TO AFFECTED INDIVIDUAL	**RISK (%)**
Sibling	
Neither parent affected	7-8
One parent affected	14
Twin	
Identical	40-60
Fraternal	6-12
Child	
One parent affected	10-15
Both parents affected	40
Parent	4
Uncle, aunt, nephew, niece, grandchild, first cousin	2-3

CLINICAL NOTES

Descriptions of clinical characteristics and the generally chronic course of schizophrenic illness are in standard textbooks. Some support exists for clinical subtypes from analysis of clinical manifestations within families or twin pairs, but there is also much overlap.

It is essential to consider and to rule out other specific causes of "schizophrenia," especially those which are treatable or preventable, such as Wilson's disease or porphyria. Also, Huntington's disease is commonly misdiagnosed as schizophrenia, if neurologic symptoms are not yet apparent. Temporal lobe seizure disorders are another possibility. Rare inborn errors of metabolism may be diagnosed initially as schizophrenia (e.g. metachromatic leukodystrophy, before the accumulation of sulfatides produces neurologic impairment).

PROCEDURES FOR DIAGNOSTIC CONFIRMATION

No specific laboratory tests are available. The platelet MAO assay is of no value for individual diagnosis, which rests upon clinical criteria and personal and family histories.

Diagnostic tests, such as plasma ceruloplasmin determination, urinary porphyrin assays, and EEG should be considered to rule out treatable and preventable conditions.

CONSIDERATIONS IN MANAGEMENT

- Genetic
  - Pedigree analysis
  - Exclusion of other causes
  - Evaluation of intervening relatives
  - Risk counseling
- Psychosocial, educational, familial
  - Evaluation of family for related disorders
  - Supportive counseling for relatives to cope with patient and appreciate likely clinical course
  - Assessment of community mental health resources
- Medical
  - Treatment, as indicated, including consideration of antipsychotic medications

SUGGESTED READING

Gottesman II, Shields J: *Schizophrenia and Genetics*. New York, Academic Press, 1972.

Heston LL: The genetics of schizophrenia and schizoid disease. *Science* 1970;167:249.

Heston LL: Psychiatric disorders in foster home reared children of schizophrenic mothers. *Br J Psychiatry* 1966;112:819.

Tsuang M: *Genes and the Mind; Inheritance of Mental Illness*. New York, Oxford University Press, 1980.

Wender PH, Rosenthal D, Kety SS, et al: Crosstesting: A research strategy for clarifying the role of genetic and environmental factors in the etiology of schizophrenia. *Arch Gen Psychiatry* 1976;30:121.

Wyatt RJ, Potkin SG, Murphy DL: Platelet monoamine oxidase activity in schizophrenia: A review of the data. *Am J Psychiatry* 1979;136(4A):377.

Chapter 43

Alcoholism

Roughly 90 million people in the United States are estimated to use or abuse alcohol. About 10 million are problem drinkers with difficulties in personal and/or social adjustment, and 6 million of these are thought to have severe alcohol dependence with associated medical, psychologic, and social impairment. How many are considered true alcoholics, rather than heavy drinkers, depends on the variable criteria used to define the term. Population surveys have indicated conservatively that among adults 3% to 5% of men and 0.1% to 1% of women are alcoholics.

Alcoholism has long been recognized to run in families, but the reasons for this are not fully understood. Obviously, environmental factors, such as the availability of alcohol and the acceptability and encouragement of its use foster the development of alcoholism. However, a growing body of evidence indicates that, at least in some cases, a genetic susceptibility may underlie the addiction. Other individuals may become alcoholic through environmental stimuli.

A known teratogen, alcohol consumed during pregnancy has been implicated in the etiology of birth defects and mental retardation (see Clinical notes—Fetal alcohol syndrome).

ETIOLOGY AND PATHOGENESIS

The biochemical reasons for alcohol dependence are not known. None of the evidence accumulated to date has provided a satisfactory explanation for genetic mechanisms in alcohol addiction.

GENETIC CHARACTERISTICS AND MODE OF INHERITANCE

Most family studies show a risk of 25%, or more, for first-degree male relatives of alcoholics and 5% to 10% for first-degree female relatives to become alcoholic, too.Identical twins are more likely to be concordant for alcoholism than fraternal twins, especially when onset is early in life. Some studies have suggested that a genetic vulnerability for alcoholism may be associated with a family history of depressive illness, other psychoses, or hyperactivity. However, others have noted a risk for alcoholism independent of other psychopathology.

Using stringent diagnostic criteria, studies of adopted persons showed that biologic sons of hospitalized alcoholics were four times as likely to become alcoholic as biologic sons of nonalcoholics, when both groups were raised by nonalcoholic foster parents. The increased risk did not differ from that for sons raised by their biologic alcoholic parent. The risk for alcoholism was not increased when it was the adoptive parent that was considered alcoholic. Risks to daughters of alcoholics were found to be lower than those for sons, but the findings were not as conclusive. A threefold increased risk was reported for biologic daughters of alcoholic mothers and only a trend, but not a significant increase, for daughters of alcoholic fathers. Daughters of alcoholics were also shown to have an increased risk for depression, but only when raised by the alcoholic parent.

CLINICAL NOTES

The clinical definition of alcoholism varies from study to study. Criteria that have been used to distinguish alcoholism from heavy drinking include hospitalization; marital, job, and police-related problems; morning drinking; repeated blackouts; tremors; withdrawal symptoms; and delirium tremens.

Since alcoholics typically deny abuse of alcohol, diagnosis in primary practice often hinges on a high suspicion, which may be aroused by a finding of alcohol-related illness, reports of personal or job-related problems, or concerns expressed by a relative. A careful history may then confirm the presence of problem drinking, perhaps severe enough to constitute alcoholism.

Fetal Alcohol Syndrome is a specific constellation of anomalies related to fetal exposure to alcohol. About one third of children born to severely chronic alcoholic mothers who drank heavily during pregnancy, have been reported to have some of the alcohol-related congenital anomalies and mental retardation. The syndrome includes the following features to varying degrees: prenatal-onset growth deficiency; typical facies (short palpebral fissures, epicanthic folds, upturned nose, hypoplastic philtrum, flat midface, thin upper lip); microcephaly; mild to moderate mental retardation; joint anomalies; and CNS, cardiac, and ocular defects.

It is unclear whether, or to what extent, lower alcohol consumption or binge drinking can produce abnormalities, particularly developmental ones. Thus, no safe level of alcohol consumption in pregnancy has been established.

PROCEDURES FOR DIAGNOSTIC CONFIRMATION

Alcoholism is generally diagnosed on the basis of a history of personal and social problems and medical complications resulting from chronic alcohol abuse.

CONSIDERATIONS IN MANAGEMENT

- Genetic
  - Counseling about risks associated with alcohol consumption in families of alcoholics
  - Discussion with women of reproductive age about prevention of fetal alcohol syndrome by avoiding alcohol consumption in pregnancy
- Psychosocial, educational, familial
  - Family counseling; support for patient and other family members
  - Education about physical, psychologic, and social rehabilitation
  - Education about coping mechanisms for family members
  - Consideration of referral for long-term psychotherapy

Assessment of community resources/support, such as Alcoholics Anonymous, Al-Anon

Medical

Pharmacologic, psychologic, and supportive measures to achieve and maintain abstinence

Treatment as indicated, for medical complications

SUGGESTED READING

Bohman M, Sigvardsson S, Cloninger CR: Maternal inheritance of alcohol abuse: Cross-fostering analysis of adopted women. *Arch Gen Psychiatry* 1981;38:965.

Cloninger CR, Bohman M, Sigvardsson S: Inheritance of alcohol abuse: Cross-fostering analysis of adopted men. *Arch Gen Psychiatry* 1981;38:861.

Goodwin DW: Alcoholism and heredity. *Arch Gen Psychiatry* 1979;36:57.

Grouse LD: Recognition of fetal alcohol syndrome. *JAMA* 1981;245:2436.

Winokur G, Reich T, Rimmer J, *et al:* Alcoholism. III. Diagnosis and familial psychiatric illness in 259 alcoholic probands. *Arch Gen Psychiatry* 1970;23:104.

Chapter 44

Syndromes/Symptom Complexes

About 2% to 3% of infants are born with major malformations, among them are many with multiple malformation syndromes. Some of these are well characterized genetic and nongenetic disorders; others constitute more complex diagnostic and etiologic challenges and may not yield to precise diagnosis, pending further progress in syndrome identification.

This chapter will tabulate known genetic implications of selected common syndromes (Table 44-1) and provide some clinical guidelines for evaluation of the dysmorphic infant or child. Some common syndromes are described in more detail in Chapters 4, 5, 6, 21, and 23. A general classification is given herein.

ETIOLOGY AND PATHOGENESIS

Birth defects can be classified into categories. *Malformations* refer to birth defects that are due to abnormal or incomplete organ development, e.g., cleft lip. *Disruptions* are caused by an extrinsic interference with an otherwise normal development, e.g., teratogen exposure or amniotic banding. *Deformations,* e.g., clubfoot, are caused by mechanical forces such as intrauterine constraint. *Dysplasias* are disorders with abnormal organization of cells into tissues, e.g., connective-tissue disorders or ectodermal dysplasias. Several etiologic categories have been defined for multiple congenital anomalies. *Multiple malformation syndromes* are patterns of symptoms believed to be related pathogenetically, e.g., Down's syndrome or de Lange's syndrome. Multiple anomalies, which occur together more often than would be expected by chance alone but which have not been shown to be related pathogenetically, are described as *associations,* e.g., VATER association. A cause and effect relationship is not implied in an association. A *sequence* describes a pattern of anomalies secondary to a primary malformation, e.g., Pierre Robin sequence.

GENETIC CHARACTERISTICS AND MODE OF INHERITANCE

The distinction between sporadic, fetal teratogenic, and genetic (single gene, multifactorial, or chromosomal) syndromes is vital to recurrence risk assessment. It is also important to distinguish between normal population variation and a hereditary syndrome with multiple mildly affected family members, because there may be a risk for more severely affected offspring. See Table 44-1 for modes of inheritance of selected syndromes.

TABLE 44-1
Selected Disorders With Multiple Congenital Anomalies

DISORDER	TYPICAL CHARACTERISTICS	DIFFERENTIAL DIAGNOSTIC CONSIDERATIONS	MODE OF INHERITANCE
Potter's syndrome (oligohydramnios sequence)	Bilateral renal agenesis, oligohydramnios, fetal compression, characteristic facies (wrinkled skin, hypertelorism, skin fold under eye, flat nose, dysplastic ears), limb positioning defect; consider ultrasonography of first-degree relatives for possible renal/genito-urinary anomalies, fatal	Renal adysplasia with kidney remnants, branchiootorenal syndrome, chromosomal syndromes	Usually sporadic, recurrence risk 1% to 3%
Meckel's syndrome	Microcephaly, encephalocele, cleft palate, polycystic kidneys, polydactyly; fatal	Trisomy 13, holoprosencephaly, Smith-Lemli-Opitz syndrome	Heterogeneous, AR
Branchiootorenal syndrome	Small, thick ears, preauricular pits, hearing loss, cervical fistulae, renal anomalies	Cup-shaped ears without hearing loss	AD, reduced penetrance
Smith-Lemli-Opitz syndrome	Microcephaly, ptosis, micrognathia, hypospadias, cryptorchidism, occasional stippled epiphyses	Trisomy 13, Meckel's syndrome, Zellweger's syndrome	AR
Holoprosen-cephaly	Facial involvement predicts brain involvement, cyclopia most severe form; hypotelorism, nasal and eye anomalies; mental retardation; less severe forms can include midfacial hypoplasia and single central incisor	Chromosomal syndromes or Meckel's syndrome	Heterogeneous, can be associated with chromosomal syndromes
Prune-belly syndrome	Protuberant abdomen due to hypoplastic or absent abdominal muscles, urinary tract anomalies, bilateral cryptorchidism, hip dysplasia, clubfeet, cardiac defects	Congenital ascites, congenital urethral obstruction	Unknown

(Continued)

DISORDER	TYPICAL CHARACTERISTICS	DIFFERENTIAL DIAGNOSTIC CONSIDERATIONS	MODE OF INHERITANCE
Crouzon's syndrome	Craniosynostosis, midface hypoplasia, ocular proptosis, conductive hearing loss	Simple craniosynostosis, Apert's syndrome, Saethre-Chotzen syndrome	AD, 50% new mutations
Hemihypertrophy	Can be isolated or associated with other syndromes; usually unilateral enlargement of digits, limbs, or entire body; variable mental retardation (in 15%); associated with neoplasms, renal dysplasia	Neurofibromatosis, Russell-Silver syndrome, Beckwith-Wiedemann syndrome, Klippel-Trenaunay-Weber syndrome	Usually sporadic
Klippel-Trenaunay-Weber syndrome	Unilateral hypertrophy; vascular anomalies including nevus flammeus, cutaneous and subcutaneous hemangiomas	Neurofibromatosis, hemihypertrophy, Beckwith-Wiedemann syndrome	Unknown, sporadic
Russell-Silver syndrome	Short stature, variable body/limb/facial asymmetry, triangular face with long eyelashes, frontal bossing, micrognathia, café-au-lait spots	Neurofibromatosis, hemihypertrophy, Klippel-Trenaunay-Weber syndrome	Usually sporadic
Beckwith-Wiedemann syndrome	Omphalocele, macroglossia, increased birth weight and size, visceromegaly, neonatal hypoglycemia, mild mental retardation, earlobe grooves, risk for renal tumor	Hypothyroidism, mucopolysaccharidoses, mucolipidoses, maternal diabetes, hemihypertrophy	Sporadic, familial in 5%
Sturge-Weber anomaly	Unilateral angiomatosis on face, chest, neck, back, upper extremities, seizures, intracranial calcification, mental retardation, glaucoma	Neurofibromatosis, Klippel-Trenaunay-Weber syndrome, Beckwith-Wiedemann syndrome	Sporadic
Klippel-Feil anomaly	Fusion of cervical vertebrae, short neck, low posterior hairline, scoliosis	Turner's syndrome, Noonan's syndrome, multiple pterygium syndrome	Most sporadic, more common in females

TABLE 44-1 *Continued*

DISORDER	TYPICAL CHARACTERISTICS	DIFFERENTIAL DIAGNOSTIC CONSIDERATIONS	MODE OF INHERITANCE
Noonan's syndrome	Short stature, broad forehead, webbed neck, pectus excavatum, cryptorchidism, pulmonic stenosis, mental retardation	Turner's leopard, multiple pterygium, or Aarskog's syndromes	Sporadic, AD in some families
Multiple pterygium syndrome	Popliteal, neck, axillary, antecubital, digital, and intercrural webbing; cleft palate, flexion of fingers, short stature	Popliteal pterygium syndrome, Turner's or Noonan's syndromes	AR, AD
G syndrome	Asymmetric skull, hypertelorism, esophageal abnormality, dysphagia, stridorous cry, hypospadias	Aarskog's syndrome	AD with male sex limitation or ? XR
Aarskog's syndrome	Short stature, rounded facies, hypertelorism, hypodontia, shawl scrotum, inguinal hernias, brachyclinodactyly, syndactyly	Noonan's or Robinow's syndromes	XR
Robinow's syndrome (fetal-face syndrome)	Large cranium, bulging forehead, hypertelorism, flat face, short forearms, dental anomalies, hypoplastic genitalia	Aarskog's syndrome	AD, AR
Asymmetric crying facies	Unilateral weakness or agenesis of lower lip muscle, cardiac anomalies, minor skeletal or genitourinary anomalies in some	Goldenhar's syndrome, facial palsy from birth trauma	AD, reduced penetrance
Moebius syndrome	VI-VII cranial-nerve palsy, masklike facies, eye anomalies, limb reduction defect, Poland's anomaly; mild mental retardation (10%), normal life span	Hypoglossia-hypodactylia spectrum, amniotic banding	Most sporadic

(Continued)

DISORDER	TYPICAL CHARACTERISTICS	DIFFERENTIAL DIAGNOSTIC CONSIDERATIONS	MODE OF INHERITANCE
Amniotic band complex (ADAM complex)	Tissue bands, unilateral amputation and constriction limb anomalies, distal syndactyly, facial clefting, encephaloceles	Genetic limb-reduction defects—usually bilaterally symmetric with rudimentary digits	Sporadic
Goldenhar's syndrome (hemifacial microsomia)	1st and 2nd branchial-arch anomalies, facial asymmetry, dysplastic ears, ear tags, epibulbar dermoid, eye anomalies	Pierre Robin anomaly, Moebius syndrome, Treacher Collins syndrome	Unknown, 1% to 2% recurrence risk
Pierre Robin sequence or anomaly	Micrognathia, U-shaped cleft palate, glossoptosis, heart murmurs/defects, congenital glaucoma, respiratory difficulties, choking	T-E fistula, Stickler's syndrome, fetal trimethadione, alcohol and hydantoin syndromes	Unknown
Treacher Collins syndrome (mandibular dysostosis)	Downslanting palpebral fissures, colobomas of lower eyelid, micrognathia, ear anomalies, ear tags, conductive hearing loss in some, facial asymmetry	Nager's syndrome, Goldenhar's syndrome	AD, high penetrance, variable expression
VATER association	Vertebral or vascular anomalies, anal malformations, tracheoesophageal fistula, esophageal atresia, renal or radial defects	Chromosomal syndromes or thrombocytopenia-radial aplasia syndrome	Sporadic
CHARGE association	Colobomas-ocular, heart, atresia-choanal, mental and somatic retardation, genital, ear anomalies/deafness	DiGeorge's syndrome	Sporadic
Cockayne's syndrome	Loss of subcutaneous fat, senile appearance, microcephaly, thin face, beaked nose, short stature, cataracts, nystagmus, photosensitive skin, death in early adulthood	Seckel's syndrome, Bloom's syndrome, progeria, erythropoietic porphyria	AR

TABLE 44-1 *Continued*

DISORDER	TYPICAL CHARACTERISTICS	DIFFERENTIAL DIAGNOSTIC CONSIDERATIONS	MODE OF INHERITANCE
Dubowitz's syndrome	Short stature, microcephaly, facial asymmetry, sparse hair and eyebrows, prominent low-set ears	Bloom's syndrome, fetal alcohol syndrome	AR
de Lange's syndrome	Microcephaly, short stature with digit anomalies, mental retardation, facial hirsutism, with characteristic synophrys, "carpmouth"	3q+ syndrome	Usually sporadic, 2%-recurrence risk for sibs
Rubinstein-Taybi syndrome	Microcephaly, nevus flammeus, broad thumbs and toes, heavy eyebrows and long eyelashes, beaked nose, grimacing smile	Trisomy 13, de Lange's syndrome, Apert's syndrome	Unknown
Cerebral gigantism (Sotos' syndrome)	Increased birth weight, large cranium, dolichocephaly, enlarged cerebral ventricles, frontal bossing, receding hairline, anteverted nares, high arched palate	Hypophyseal tumor, XYY syndrome, neurofibromatosis, aqueductal stenosis, Arnold-Chiari malformation, Dandy-Walker syndrome	Most sporadic
Williams' syndrome (Elfin facies)	Full cheeks, open mouth with thick lips, long philtrum, heart murmur, hoarse voice, stellate iris pattern	Supra valvular aortic stenosis, idiopathic hypercalcemia	Sporadic
Leprechaunism	Grotesque elfin facies, thick lips, hirsutism, emaciated appearance, abnormal endocrine functions, enlarged breasts, and external genitalia	Marasmus, congenital lipodystrophy	AR
Bardet-Biedl syndrome	Mental retardation, pigmentary retinopathy, obesity, polydactyly, hypogenitalism, renal problems	Prader-Willi syndrome, Laurence-Moon syndrome (no polydactyly or obesity)	AR

(Continued)

DISORDER	TYPICAL CHARACTERISTICS	DIFFERENTIAL DIAGNOSTIC CONSIDERATIONS	MODE OF INHERITANCE
Prader-Willi syndrome	Hypotonia, obesity, small hands and feet, hypogonadism, voracious appetite, association with interstitial deletion of chromosome 15 in many cases	Congenital muscular dystrophy, Bardet-Biedl, Laurence-Moon, and Alstrom's syndromes	Sporadic

Key: AD = autosomal dominant; AR = autosomal recessive; XR = X-linked recessive; Sporadic = single occurrence in a family; Heterogeneous = several different modes of inheritance have been described

CLINICAL NOTES

Valuable diagnostic information can be gathered from careful prenatal and family histories; physical examination; and metabolic, chromosomal, and other laboratory screening. Comparison of the patient's physical measurements to standard charts can alert the clinician to deviations from normal development and help differentiate similar syndromes (most entries in the reading list have standard charts useful for syndrome identification). The table lists other conditions with similar manifestations for consideration in the differential diagnosis of a syndrome.

Examination of relatives or family photographs may be useful to detect subtle manifestations of a syndrome in other family members. Often a diagnosis remains elusive, and a substantial proportion of cases remain undiagnosed even after referral to a geneticist, dysmorphologist, or other specialist. In these cases, comparison of clinical findings with case reports in the literature of unidentified syndromes may prove useful, and increasingly is resulting in the delineation of new syndromes.

PROCEDURES FOR DIAGNOSTIC CONFIRMATION

Antibody screens (bacterial, viral, TORCH)
Metabolic screening
Chromosome analysis
Thyroid-function studies
Radiograph
Ultrasonography

CONSIDERATIONS IN MANAGEMENT

Genetic
- Establish diagnosis
- Pedigree analysis, recurrence risk assessment

- Risk counseling
- Consideration of prenatal diagnosis, when available and indicated

Psychosocial, educational, familial

- Family counseling and support for acceptance of child with serious malformations and retardation
- Assessment of psychologic impact on family, e.g., parental guilt, shame, grief
- Cosmetic considerations
- Consideration of referral for psychiatric therapy and/or support
- Assessment of community resources and support groups

Medical

- As indicated by diagnosis and severity, and may include:
  - Supportive therapy
  - Corrective or cosmetic therapy
  - Orthodontic and/or speech therapy
  - Periodic surveillance for early detection of associated problems

SUGGESTED READING

Bergsma D, (ed): *Birth Defects Atlas and Compendium,* 2nd ed. New York, Alan R. Liss, 1979.

Feingold M, Bossert WH: Normal values for selected physical parameters: An aid to syndrome delineation, vol 10, no 13. (Birth Defects Original Article Series). White Plains, N.Y., March-of-Dimes Birth Defects Foundation, 1974.

Goodman RM, Gorlin RJ: *The Malformed Infant and Child: An Illustrated Guide,* New York, Oxford University Press, 1983.

Kalter H, Warkany J: Congenital malformations—etiologic factors and their role in prevention. *N Engl J Medicine* 1983;308(8,9):424.

Smith DW: *Recognizable Patterns of Human Deformation,* Philadelphia, WB Saunders, 1981.

Smith DW: *Recognizable Patterns of Human Malformation*, 3rd ed, Philadelphia, WB Saunders, 1982.

Chapter 45

Teratogenesis

A teratogen is an environmental agent capable of causing abnormal prenatal variation in form or function. Ever since the recognition of the teratogenic potential of thalidomide, many substances as well as other environmental factors have become suspect.

Given the general population risk of 2% to 5% for any pregnancy to result in a child with a birth defect, the hazard imposed by a specific teratogen must be viewed as an increase over this baseline, even though such agents may be responsible for some of the malformations of unknown origin. While the contribution of teratogens to the total number of birth defects remains ill-defined and much is yet to be learned about their mechanisms of action, this summary of current knowledge can provide an overview of recognized and alleged teratogenic agents, as well as a framework for the identification of birth defects caused by them.

In appropriate cases this chapter may serve to rule out suspected genetic etiology and its associated high recurrence risks. In addition, it categorizes the substances most prudently avoided during pregnancy and describes risk estimates when exposure has already occurred or cannot be avoided.

ETIOLOGY AND PATHOGENESIS

Teratogens interfere with basic functions of developing cells and organisms in several ways, such as:

Alteration of tissue growth resulting in hyperplasia, hypoplasia, asynchronous growth
Interference with cellular differentiation of fetal morphogenesis
Direct causation of cell death

These few pathogenic pathways can be induced by a wide range of environmental agents, and, as a corollary, agents which have a known or suspected role in such disruptive processes should be viewed as potentially teratogenic. Known carcinogens or mutagens should be suspect, as well, as likely teratogens, since the potential for the three types of insult may be found in the same agents.

Teratogens are rarely restrictive in their disruptive action, as they are likely to affect more than one tissue and to produce more than one aberrant process. The resultant damage may be expressed in one or more of the following categories:

Infertility or fetal wastage
Prenatal-onset growth deficiency

Birth defects
Alterations in nervous system function

For given agents, characteristic patterns of clinical consequences have been described. However, the effects of a particular teratogen can vary considerably, with the recognized syndromes more generally observed at the severe end of the spectrum. The nature of the more frequent milder manifestations may be shared by various sources of disruption and therefore may be less specific to any given teratogen.

The sources of variability are not always clearly evident in individual cases, but are known to include a number of critical factors. The time of prenatal exposure to a teratogen is a crucial variable in assessing risk and outcome. In general, organs are most vulnerable to adverse influences during their formation and period of rapid differentiation. Malformations are most likely to be produced in the first trimester. Noxious agents may be lethal in the earliest days of pregnancy, or have no observable effect at all. They have comparatively mild effects late in gestation. Results of later exposures are more likely to approximate postnatal damage.

Another important variable is the degree of exposure generated by the dose of the noxious agent, as there is evidence for increased risk with increased dosage. However, safe levels of exposure have, in most cases, not been established for recognized teratogens.

TERATOGENS

Five general groups of hazardous agents will be considered:

1. Infectious agents
2. Physical agents
3. Drugs and chemical agents
4. Maternal metabolic and genetic factors
5. Paternal factors

Infectious Agents. Organisms that cause only a glancing infection in a normal adult can be catastrophic in a fetus. Although the placenta frequently is a barrier to infectious organisms, most fetal infections, nonetheless, result from an admixture of fetal and maternal blood across the placenta. Also, the placenta itself can become a focus of infection which may subsequently spread into the fetal circulation. Infection can enter the amniotic space through the vagina and the cervix, as well, suggesting the possibility of direct transmission through sexual intercourse.

A number of viruses are known teratogens, but the spirochete which causes syphilis (*Treponema pallidum*) and mycoplasmas are the only bacterial agents believed to be of teratogenic significance. Among parasitic organisms, *Toxoplasma,* often transmitted by cats, is recognized as a major infectious teratogen. It is one of the best documented of such agents, along with rubella virus and cytomegalovirus.

Infectious agents generally induce teratogenic effects through direct invasion and inflammation of fetal tissues. Another potential hazard, possibly precipitating abortion or premature birth, is disruption of placental function as a consequence of maternal infection. Inflammation or thrombosis of fetal blood vessels has been proposed to account for the consistency in manifestations attributable to several organisms. In addition to interference with normal cellular functions, some pathogens, e.g., lytic viruses, directly induce cell death, and thereby inflict severe intrauterine damage.

The fetal CNS is particularly vulnerable to pathogenic effects of infectious organisms: disordered movement and muscle tone, seizures, mental retardation, and CNS-mediated auditory and visual defects are typical sequelae. Prematurity, low birth weight, failure to thrive, and evidence of general sepsis (e.g., hepatitis, pneumonitis) also suggest prenatal infection. Several organisms, in addition, produce characteristic signs, e.g., pathognomonic skin lesions. Tables 45-1 and 45-2 list the most common teratogenic infectious agents with their characteristic adverse effects.

Physical Agents. A diverse number of agents, grouped for convenience as physical factors, are capable of causing teratogenic effects. These include radiation, heat, and abnormalities of the intrauterine environment, such as compression. Of these, radiation has occasioned the most concern.

Radiation. Among the most frequently expressed concerns about potential hazards in pregnancy, are those relating to the effects of exposure to radiation, either before the pregnancy was recognized, or from unavoidable sources later in gestation (e.g., diagnostic, therapeutic, occupational).

Very high radiation has, indeed, been implicated, not only as a teratogen, but as a mutagen and carcinogen as well. Teratogenic effects have been inferred primarily from animal studies and post-World War II studies of survivors of nuclear explosions. These studies indicate that the lower the dose rate and the longer the period over which the dose is absorbed, the less severe the effect.

In assessing the risk, precise calculation of dosage absorbed by the fetus, and time of gestation are critical factors.

- High doses of irradiation (well above 50 rads to the fetus) are associated with lethal effects very early in pregnancy (if any effects are observed at all), and with growth retardation somewhat later. Malformations of multiple systems (particularly microcephaly and ocular defects) are likeliest between two and four weeks of gestation. The CNS remains sensitive to radiation damage throughout gestation, however, and indeed into postnatal life.
- If exposure is limited to 2 rads or less, anytime during the pregnancy, the increased risk is considered extremely low or negligible.
- For doses of 2-10 rads no clear association with birth defects has been shown, but some concern has been expressed, if exposure was between two and four weeks' gestation.

- In general, increased rates of congenital malformations or growth retardation have not been observed following common diagnostic procedures of 5 rads or less.
- In any case, even an extremely low risk is best avoided, unless a significant benefit is expected from the procedure.
- In spite of widespread anxiety, present evidence does not support any causal relationship between birth defects and such low energy radiation sources as ultrasonography or video display terminals.

TABLE 45-1
Bacterial and Parasitic Teratogens

INFECTIOUS AGENT	ADVERSE EFFECTS	COMMENTS
Treponema pallidum (syphilis)	Early congenital syphilis (presents in first 2 years): evidence of widespread infection: hepatosplenomegaly, jaundice, inflammation of lymph nodes; cutaneous and mucosal lesions; nephrosis; CNS manifestations; ocular abnormalities Later manifestations (after age 2): Hutchinson's teeth; interstitial keratitis; sensorineural deafness; Clutton's joints; characteristic craniofacial anomalies; mental retardation	Clinical signs related to time of infection; infections late in pregnancy may not manifest; half of infants infected early in pregnancy may be premature, stillborn, or die soon after birth. Serologic test available
Mycoplasma	Maternal infection has been related to spontaneous abortion	Possible association with anencephaly, based on anecdotal evidence; *Mycoplasma* common in female genital tract
Listeria monocytogenes	Strong association with fetal wastage, prematurity, neonatal death	Data on congenital malformations unavailable
Toxoplasma gondii	CNS abnormalities predominate: microcephaly, intracranial calcification; seizures; mental retardation; ocular manifestations; generalized sepsis	Effects observed in cases of acute maternal infection and are related to stage of gestation; early pregnancy carries lower risk of infection spreading to the fetus, but higher risk of severe anomalies; drug therapy for maternal toxoplasmosis may be teratogenic. Exposure can be through raw meat, eggs, and cat feces

TABLE 45-2 Continued
Viral Teratogens

INFECTIOUS AGENT	ADVERSE EFFECTS	COMMENTS
Rubella	Intrauterine growth retardation; failure to thrive; ocular defects; cardiovascular defects; CNS abnormalities, e.g., microcephaly, mental retardation; sensorineural deafness; abnormal ossification of long bones; endocrine disturbances	Abnormalities vary with month of gestational infection. Neurologic signs, endocrine malfunctions, and deafness may not be apparent in neonatal period. Vaccine available
Cytomegalovirus (CMV)	Increased neonatal mortality; CNS abnormalities, (microcephaly: periventricular calcification); ocular/auricular defects; GI malformations; epilepsy; retardation	Abnormalities vary with month of gestational infection. Most infections in newborn infants are clinically silent. Vaccine currently unavailable
Herpes simplex (Types I and II)	Primary infection before 20 weeks can lead to increased fetal mortality, congenital malformations (hydranencephalus, chorioretinitis)	Most fetal infections acquired at or close to birth; manifestations variable in severity. Vaccine currently unavailable
Varicella zoster	Growth retardation; microcephaly with cortical atrophy; muscle weakness/palsy/paralysis; limb anomalies; positional limb deformities; ocular defects; cutaneous lesions	Risk to fetus estimated as less than 10% for severe teratogenic effects. Critical period is three to four months of pregnancy
Venezuelan equine encephalitis virus	CNS anomalies; hydrocephalus; cataracts	Findings based on limited clinical reports and animal studies
Coxsackieviruses	Myocarditis characteristic; disseminated infection; urogenital and GI anomalies	Defects tend to be associated with specific viruses
Influenza, measles, mumps	Increased risk of fetal wastage. Neural-tube defects and hydrocephaly have been proposed for influenza; endocardial fibroelastosis and diabetes mellitus for mumps	Reported associations with congenital anomalies need confirmation for evaluation of increased risk, if any

Heat. A teratogenic potential of fetal exposure to excessive heat, due to maternal febrile illness or deliberate inducement (e.g., saunas), has been postulated, but findings are inconclusive. A Canadian retrospective study reported a significant association between maternal febrile episodes and microphthalmia, with some cases accompanied by microcephaly and mental retardation. Animal studies have suggested a cause/effect relationship between excessive heat and fetal malformation; and in humans, neural-tube defects and neuromigrational errors have been reported anecdotally. However, prospective analyses did not document such associations in humans.

However inferential the evidence, medical management of febrile illnesses, particularly in the first trimester of pregnancy, might prudently take the possibility of teratogenicity into account. In addition, avoidance of deliberate exposure to excess heat would seem a routine caveat for pregnant women.

Mechanical (intrauterine) factors. The configuration of the intrauterine environment can itself adversely affect fetal morphogenesis. Fibrous bands, for example, can be the cause of human limb abnormalities, and intrauterine head constraint is probably the most common cause of cranial synostosis. A number of intrauterine mechanical influences of diverse or unknown cause can result in fetal deformations. Uterine constraint, in general, is more severe in the breech position, and a high percentage of related deformations are attributable to this position. Additional factors predisposing to fetal malformations are oligohydramnios, uterine malformations, large myomas, early amnion rupture, placental defects, and multiple gestation.

It is important to distinguish such deformations, which by definition are caused by mechanical factors, from malformations, which are due to an error of morphogenesis, because the prognosis is much better for the former.

Drugs and Chemical Agents. The role of drugs and chemical agents as potential teratogens has been the subject of much public concern. Unfortunately, conclusive evidence is often not available to confirm or rule out postulated risks, which may have been based on subjective reports. Clear association of ill effects with prenatal exposure has been shown for only a few of the accused substances.

Attempts to delineate the teratogenic risk of one or another chemical agent are confounded by a number of factors. First, when birth defects are anecdotally associated with a reportedly hazardous substance, the fact that the general population risk may account for them is not always considered. Even in controlled, usually retrospective studies, the number may be small and findings contradictory, so that conclusive evidence cannot be gained. Although animal studies may be informative in many instances, direct extrapolation to effects on humans is not always warranted, especially if the doses used are not comparable to likely human exposure. In any case, a human embryo may have a very different sensitivity from a test animal. In humans, it is difficult to study one substance alone, since exposure is rarely limited to just one (e.g., the usual combination of alcohol, caffeine, and smoking). Finally, genetic factors have an undoubted but unknown influence on susceptibility to teratogenic action and its expression.

Suspicion of teratogenic potential has been cast on a great many substances in the environment and on drugs and medications, both prescribed and self-selected. Of these substances, some have been established as proven teratogens (e.g., mercury, alcohol, anticonvulsants, anticoagulants) or have been shown to have a clear association with increased fetal wastage or reduced birth weight (e.g., smoking, alcohol, addictive drugs). For most of the suspected agents no consistent association with teratogenic effects has been documented, and some, for which hazards had been proposed and generally accepted, are no longer considered teratogenic in customary use. Table 45-3 provides an overview of current information about the most common recognized chemical teratogens and about other suspect substances that have raised public and medical concern.

For many of the common pharmaceutical products the available evidence (or lack thereof) can provide reassurance when exposure has already occurred. However, except in cases of medical necessity, suspect drugs, as well as alcohol and smoking, are clearly best avoided during pregnancy.

TABLE 45-3
Teratogenic Drugs/Chemicals

TERATOGEN	ADVERSE EFFECTS	COMMENTS
Environmental Chemicals		
Organic mercury compounds	CNS damage; microcephaly; dental anomalies; mental retardation	Exposure generally through food chain or industrial contamination
Polychlorinated biphenyls	Intrauterine growth retardation; brown-stained skin (cola-colored-baby syndrome); exophthalmos; natal teeth	Evidence anecdotal
Industrial fat solvents* (e.g., xylene)	Caudal regression malformations	Evidence based on animal data and scattered reports
Agent orange and dioxin*	Unspecified increased anomalies	Evidence generally unreliable
Pesticides*	Little information available	
Heavy metals*	Little information available	
Occupational exposure to anesthetic gases	?Increased spontaneous abortion	Data inadequate to assess risk
Drugs—Nonprescription		
Thalidomide	Limb-reduction malformations; cardiac, renal, GI, ocular, and aural anomalies	Documentation of direct relationship to birth defects spurred investigation of other substances
Ethyl alcohol	Fetal alcohol syndrome at severe end of range of anomalies: characteristic facial appearance; intrauterine growth retardation; developmental and mental retardation; cleft palate; CNS, cardiac, and ocular defects; fetal wastage	Spectrum of abnormalities and fetal wastage. Offspring of chronic alcoholic mothers at high risk for the syndrome; lower consumption or binge drinking may produce recognizable abnormalities; no threshold has been established for a safe level of consumption

TABLE 45-3 *Continued*

TERATOGEN	ADVERSE EFFECTS	COMMENTS
Tobacco	Reduced birth weight; fetal wastage; no clear association with human malformations	Dose-related
Caffeine*	Reduced birthweight; no clear association with human malformations	Dose-related
Salicylates	Fetal hemorrhage; intrauterine growth retardation; patent ductus arteriosus when use is in late pregnancy	Effects have been related to *excessive* use, low-dose effects not clearly identified
Addictive drugs (hallucinogens, opium, heroin, methadone, amphetamines, etc.)*	Reduced birth weight, neonatal withdrawal symptoms, perinatal mortality; no clear link to human malformations	Little objective evidence available
Drugs—Prescription		
Anticancer agents Folic acid antagonists (aminopterin, methotrexate)	Intrauterine growth retardation; early abortion and stillbirth; neonatal death; craniofacial anomalies; neural-tube defects; mental retardation; skeletal anomalies	Methotrexate also prescribed for psoriasis; safety of topical administration not demonstrated
Alkylating agents (busulfan, chlorambucil, nitrogen mustard	Intrauterine growth retardation; fetal death; microphthalmia; cleft palate; genitourinary anomalies; limb reduction malformations	Risk for severe defects estimated at 10% to 35%, particularly if exposure occurs early in pregnancy
Other cytotoxic agents	Insufficient information, probably hazardous	
Anticoagulants Coumarin derivatives (Warfarin)	Intrauterine growth retardation; characteristic syndrome includes neurologic damage, mental retardation, microcephaly, optic atrophy, skeletal anomalies	Exposure, at six to nine weeks postconception carries the greatest risk of severe anomalies
Heparin derivatives	Stillbirth; prematurity; no known defects or malformation syndrome	Risk of stillbirth estimated at 10% to 15%, risk of prematurity at 20%
Antibiotics Tetracyclines	Dental abnormalities; enamel hypoplasia	Second and third trimester are vulnerable periods
Streptomycin (and related compounds)	Eighth-cranial-nerve damage: sensorineural hearing loss	Risk has been reported as high as 10% to 15%, but evidence is not yet conclusive

(Continued)

TERATOGEN	ADVERSE EFFECTS	COMMENTS
Chloroquine (antimalarial)	Optic/auditory abnormalities: chorioretinitis, optic nerve hypoplasia/atrophy, deafness	Overall risk for severe abnormalities relatively low with standard doses
Pyrimethamine, penicillins, sulfas*	No consistent association with human malformations	Single report of a malformed infant subsequent to use of pyrimethamine as an antitoxoplasmosis agent
Anticonvulsants (see Chapter 9) Oxazolidine derivatives (trimethadione, paramethadione)	Fetal wastage; intrauterine growth retardation; major abnormalities: characteristic facial appearance; CNS damage resulting in delayed speech and mental deficiency; cardiovascular/urogenital abnormalities; limb defects; cleft palate, T-E fistula	Risk is high; most exposures are medically unnecessary or inappropriate
Hydantoin derivatives	Fetal hydantoin syndrome: intrauterine growth retardation; characteristic facial appearance; CNS damage resulting in mental deficiency; cleft lip/palate; limb-reduction malformations; cardiovascular/skeletal abnormalities	Risk of recognizable effects is estimated at 7%; risk of subtler effects may be higher; possible association with neuroblastoma
Barbiturates and Primidone*	?Possible syndrome of unusual facial appearance and mental deficiency	Few cases are reported and evidence is inconclusive
Diazepam*	?Possible increased risk for cleft lip/palate	Absolute risk relatively low; one of most frequently prescribed drugs (Valium)
Valproic acid	?Increased risk for neural-tube defects	A risk of about 1% has been suggested by some studies, but others have not been able to confirm it
Hormonal Agents/Antagonists Female sex hormones: Diethylstlibestrol	Malformation of vaginal epithelium, cervix, and uterus; testicular abnormality	Increased risk of vaginal adenocarcinoma and testicular malignancy
Other estrogens	Possible "feminization" of male fetuses	
Progestins	Masculinization of external genitalia in female fetuses	Risk estimated at 1% to 2% before 12th week of gestation, uncertain after 12th week

TABLE 45-3 *Continued*

TERATOGEN	ADVERSE EFFECTS	COMMENTS
Estrogen/progestin combinations (birth control pills)	Possible congenital cyanotic heart disease; caudal defects; VATER association	Risk debatable, but low if it exists
Male sex hormones: Androgens	Masculinization of female fetuses	
Clomiphene*	Multiple gestation may predispose to positional limb deformations	About 6% of treated pregnancies result in multiple births; relationship to other defects not clearly identified
Adrenal corticosteroids*	No consistent association with human malformations	
Antithyroid agents (thiouracil, propylthiouracil, radioactive iodine, methimazole)	Possible hypothyroidism, mental retardation; limb/scalp defects	Physical defects are associated with methimazole, and risk is probably low
Iodides	Hypothyroidism; neonatal goiter	High-risk period after first trimester
Psychotropic drugs		
Lithium	Lithium toxicity; increased risk of congenital heart disease	Data inadequate to confirm risk
Amphetamines*	(See addictive drugs)	
Benzodiazepine*	Possible increased risk for cleft lip/palate	Absolute risk relatively low
Meprobamate*	Variety of anomalies reported, particularly cardiovascular	Epidemiologic studies have not born out reports
Tricyclic antidepressants*	Upper-limb deformities	Anecdotal evidence
Phenothiazines*	Sporadic reports of birth defects	Chlorpromazine has been associated with congenital retinopathy
Butyrophenones* (haloperidol)	?Limb-reduction defects	Based on two reports
Barbiturates	?Cardiac and brain defects	Evidence is inconclusive
Diazepam*	?Possible increased risk for cleft lip/palate	Absolute risk relatively low; one of most frequently prescribed drugs (Valium)
Other drugs		
Bendectin and antihistamines*	No consistent association with human malformations	Reports of diverse anomalies aroused public concern about Bendectin, probably the most commonly used drug in early pregnancy,

(Continued)

TERATOGEN	ADVERSE EFFECTS	COMMENTS
		but no evidence has been obtained to confirm them
Thiazide diuretics*	?Hyponatremia; fetal hematopoietic system, e.g., thrombocytopenia	No consistent reports
Disulfiram (Antabuse)*	?Limb-reduction defects	Based on four cases; drug is antialcoholic
Vitamin A or retinoids (Accutane, 13-cisretinoic acid) in therapeutic doses	Spontaneous abortion, absence or hypoplasia of external ears, CNS anomalies, cortical blindness, congenital heart disease	Relative risk of 9.2 for any major malformation; 25.6 for these particular malformations

*Association with congenital defects has been proposed but available information does not provide conclusive evidence; increased risk, if any, is estimated to be very low.

Maternal Metabolic, Genetic, and Related Conditions. The maternal biochemical environment, when disordered, can have serious consequences for the fetus, as listed in Table 45-4 and discussed in relevant chapters. Two diseases, in particular, merit detailed mention in this context.

Diabetes. This condition has significant teratogenic potential in terms of both severity and prevalence. However, the increased risk for maldevelopment of the conceptus does not appear to be associated with prediabetes or gestational diabetes. It is insulin dependent diabetes that carries a potential for birth defects. Overall risk estimates for offspring of diabetic mothers have been as high as 9%, or three times the population risk. The rate of malformation and mortality appears to be associated with the degree of severity of the disease in the mother. The suggestion that good metabolic control of the maternal diabetes, both before and during the pregnancy, may reduce the risk, argues for vigilant monitoring and surveillance of the diabetic mother.(See Chapter 36.)

Maternal phenylketonuria (PKU). The teratogenic effects of maternal PKU are unfortunate consequences of dietary control of the disorder among affected children. Phenylketonuric females are rarely still on a low phenylalanine diet when they reach childbearing age. If they become pregnant without returning to the diet, they subject their unborn children to elevated maternal levels of phenylalanine and phenylpyruvic acid. Such a prenatal environment has been shown to produce mentally retarded, though biochemically unaffected, offspring with intrauterine growth retardation, microcephaly, and other congenital malformations. The high rate of microcephaly in these children suggests consideration of unreported or unrecognized maternal PKU as a cause in cases of otherwise unexplained microcephalus. A strict dietary regimen is recommended for women with PKU who are planning a family. It should begin before conception and be maintained throughout the duration of the pregnancy, to minimize the risk of damage

TABLE 45-4
Teratogenic Maternal Metabolic/Genetic Conditions*

MATERNAL DISORDER/CONDITION	ADVERSE EFFECTS	COMMENTS
Diabetes	Caudal regression malformation complex; transposition of great vessels; renal anomalies; neural-tube defects; cleft lip/palate; perinatal mortality	Severe risks relate to diabetes and have been reported at 7% to 13%. Good metabolic control of maternal diabetes is reported to decrease the risk
Phenylketonuria (PKU)	Mental retardation; intrauterine growth retardation; microcephaly; congenital heart defects; vertebral anomalies; strabismus; increased fetal wastage	Risk of mental retardation has been estimated as high as 95% with maternal blood phenylalanine concentration of 20 mg/dL or higher
Myotonic dystrophy	Polyhydramnios; decreased fetal movement; hypotonia; joint deformities, e.g., arthrogryposis; inguinal hernia; cryptorchidism; mental retardation	Anomalies are congenital when mother *and* child carry the gene for myotonic dystrophy; mental retardation occurs in approximately 85% of surviving children
Malnutrition	Intrauterine growth retardation	Growth retardation may be irreversible if malnutrition is severe enough
Myasthenia gravis	Transient myasthenia gravis; hypotonia, ptosis; sucking, swallowing, breathing difficulties	The disorder responds to treatment and disappears within a few weeks. It is reported in about 15% of babies born to myasthenic mothers
Systemic lupus erythematosus	Increased risk of first trimester miscarriage; congenital heart block	Magnitude of risk for heart block is not known

*Also see chapters on diabetes, PKU, and genetic neuromuscular disorders (36,37,15).

to the fetus. Even so, complete avoidance of the risk cannot be assured. (Also see Chapter 37).

Paternal Factors. The question of teratogenic capability of preconception paternal exposure to hazardous agents has caused considerable anxiety and public discussion. Available evidence indicates that such effects are unlikely, although a decrease in fertility may result under some conditions (e.g., heat). With exposure to noxious agents, one might expect an increase in genetic mutation leading to dominant or X-linked disorders, but no consistent pattern of this type of increased risk has been observed. It is believed, however, that lifelong exposure to various environmental agents may contribute to the documented paternal age-related increase in new mutations.

CLINICAL NOTES

The common clinical consequences of recognized teratogens are discussed above and listed in the accompanying tables. These can be kept in mind for diagnosis of unexplained birth defects or functional abnormalities. Conversely, the presence of one or more of certain general signs and symptoms can raise a high index of suspicion, suggesting careful examination of the maternal prenatal history for exposure to teratogenic agents and/or evidence of maternal disease. Consideration of teratogenic influence may thus help to exclude genetic etiologies and their potentially high recurrence risks. Table 45-5 lists the most common of these indications.

This discussion and the accompanying tables are intended to serve as an orientation to the area of teratology, a field in which continuing investigation is steadily providing new information. The physician is directed to the bibliography for more detailed information on a possible or known teratogen, and to the journal *Teratology*, for current reports in the literature.

PROCEDURES FOR DIAGNOSTIC CONFIRMATION

Pregnancy history
Examination of child for subtle abnormalities associated with teratogenic exposures
Consultation with a medical microbiologist
Virologic isolation from urine/body tissues
Serologic studies
Hemagglutination studies
Antibody inhibition
Radioimmunoassays
Virus culture of vesicular lesions (Varicella zoster)

TABLE 45-5
Possible Indications of Teratogenesis

Fetal death
Prematurity
Intrauterine growth retardation
Microcephaly
Cleft palate
Mental retardation
Seizures
Craniofacial anomalies
Central auditory and visual defects (e.g., chorioretinitis, cataracts, microphthalmia)
Contractures and/or limb deformations (generally not malformations like polydactyly and syndactyly)
Cardiac anomalies
Widespread sepsis
Chronic skin rashes
Failure to thrive

Dark-field microscopic examination (syphilis)

Sabin-Feldman dye test (toxoplasmosis)

Examination of mother for maternal factors

Long-term follow-up for further signs and symptoms or exclusion of other etiologies

CONSIDERATIONS IN MANAGEMENT

Genetic
- Explanation of nongenetic nature of teratogen
- Risk counseling to clarify reduced recurrence risks, if teratogen can be avoided

Psychosocial, educational, familial
- Family counseling; support for acceptance of birth of an affected child
- Education regarding use of potentially hazardous agents during future pregnancies, especially use of alcohol and tobacco
- Consideration of referral for psychosocial therapy and/or support for families with a problem of teratogenic origin
- Assessment of family/community resources relevant to specific abnormalities
- Discussion of risks and benefits in cases involving a clear need for maternal therapy with potentially teratogenic agent

Medical
- Dependent on anomaly

At the time of this writing, telephone *Teratogen Information Services* are being set up in various parts of the country. Most serve a proscribed geographic area. For information on a Teratogen Line in your area call your nearest genetics clinic or write to: Science Information Division, The March of Dimes—National Foundation, 1275 Mamaroneck Avenue, White Plains, New York, 10605.

SUGGESTED READING

Brent RL: Radiation teratogenesis. *Teratology* 1980;21:281.

Brent RL, Harris MI, (eds): *Prevention of Embryonic, Fetal, and Perinatal Disease,* John E. Fogarty International Center for Advanced Study in the Health Sciences, DHEW Publication No. (NIH) 76-853, 1976.

Briggs GG, Bodendorfer TW, Freeman RK, et al: *Drugs in Pregnancy and Lactation,* Baltimore, Waverly Press, 1983.

Clarren SK, Smith DW: The fetal alcohol syndrome. *N Engl J Med,* 1978;298:1063.

Hanson JW: Teratogenic agents, in Rudolph AM, (ed): *Pediatrics,* 17th ed. New York, Appleton-Century-Croft, 1983.

Kalter H, Warkany J: Congenital malformations, etiologic factors and their role in prevention. *N Engl J Med* 1983;308:425.

Lammer EJ, Chen DT, Hoar RM, et al: Retinoic acid embryopathy, *N Engl J Med* 1985;313:837.

Layde PM, Edmonds LD, Erickson J: Maternal fever and neural- tube defects. *Teratology* 1980;21:105.

Physicians' Desk Reference®, 40th ed. New Jersey, Medical Economics Co, Inc., 1986.

Physicians' Desk Reference for Nonprescription Drugs®, 7th ed. New Jersey, Medical Economics Co., Inc., 1986.

Schardein JL: *Drugs as Teratogens*. Cleveland, CRC Press, 1976.

Shepard TH: *Catalog of Teratogenic Agents,* 5th ed. Baltimore, The Johns Hopkins University Press, 1986.

Stagnos S, Whitley RJ: Herpesvirus infections in pregnancy, Part I: Cytomegalovirus and Epstein-Barr virus infections; Part II: Herpes simplex virus and varicella zoster virus infections, *N Engl J Med* 1985;313:1270, 1327.

Chapter 46

Cleft Lip and/or Cleft Palate

Cleft lip with or without cleft palate (CL/P) and isolated cleft palate (CP) should be considered as separate genetic entities. They differ etiologically, occur at different times in embryonic development, and carry different genetic risks.

- CL/P is estimated to occur in 0.1% of births
- CP is estimated to occur in 0.04% of births

Both conditions are believed to be under-reported because of possible failure in reporting them in stillbirths, in mild defects, and when associated with other conditions or syndromes.

ETIOLOGY AND PATHOGENESIS

CL/P and CP. Etiology of both CL/P and CP is uncertain and diverse. Some cases are caused by single-gene disorders, chromosomal abnormalities, or environmental agents. The majority are transmitted as a multifactorial condition.

CL/P. In the developing embryo, closure of the lip is dependent on the merging in the midline of three mesodermal processes (lateral nasal, median nasal, and maxillary) and is generally completed by 45 days. Failure of fusion of these processes results in lip clefting, which can be complete or partial, bilateral or unilateral. Extension of the cleft into the palate may occur as a secondary phenomenon which is different from the mechanism causing CP.

CP. The secondary palate is formed by the union in the midline, of paired palatal shelves (derived from the maxillary processes) and occurs by nine weeks' gestation. Failure of fusion of these shelves results in cleft palate, which can be complete or partial (e.g., submucous cleft or just a high arched palate), with variation in the size and location of the cleft.

GENETIC CHARACTERISTICS AND MODE OF TRANSMISSION

Precise diagnosis and classification are necessary because CL/P and CP each can carry different recurrence risks:

- They can be a part of other syndromes, with risks based on the syndrome. An estimated 184 of these have been identified. Most are rare. However, it is important to consider the possibility of syndromes, since recurrence risk can be as high as 25% or 50%. Tables 46-1 and 46-2 list some syndromes that may include CL/P or CP.

- Environmental insults can cause clefting, in which case risk of recurrence is very low, unless the insult is still operative.
- When syndromes and environmental insults are ruled out, CL/P and CP are considered to be inherited as multifactorial conditions.
- For the isolated clefts, the estimated general population risk and recurrence risks based on empiric data are summarized in Table 46-1.

TABLE 46-1
Estimated Risks for Isolated Clefts

	CL WITH OR WITHOUT CP (%)	ISOLATED CP (%)
General population	0.1	0.04
One affected sibling	4-7	2-5
One affected parent	2-4	7
One affected parent and one affected sibling	11-14	14-17
Two affected siblings	9-10	10

A careful check of other family members may reveal or rule out subtle signs, such as a submucous cleft or a high arched palate, allowing more precise risk estimation.

Key: CL = cleft lip; CP = cleft palate

TABLE 46-2
Selected Genetic Disorders With Cleft Lip and/or Cleft Palate

SYNDROME	OTHER CHARACTERISTICS	MODE OF INHERITANCE
EEC (ectrodactyly-ectodermal dysplasia-clefting)	Ectrodactyly, absent lacrimal puncta, sparse hair	AD
Robert's	Phocomelia, facial hemangioma	AR
Cleft lip and lip-pit	Pits of lower lip in patient and/or relatives; syndrome is 80% penetrant with highly variable expression	AD
Orofaciodigital, I and II	Multiple hyperplastic frenula, alopecia, cleft tongue, digital abnormalities; mental retardation	Varied
Popliteal pterygium	Popliteal pterygium lip pits, genital anomalies, syndactyly	AD
Various chromosomal anomalies, particularly trisomy 13 and 4 p- syndrome	Developmental retardation, multiple anomalies	CH

Key: AD = autosomal dominant; AR = autosomal recessive; CH = chromosomal

TABLE 46-3
Selected Genetic Disorders With Cleft Palate

DISORDER	OTHER CHARACTERISTICS	MODE OF INHERITANCE
Stickler's syndrome	High myopia, epiphyseal changes, Pierre Robin anomaly	AD
Cerebrocostomandibular syndrome	Microcephaly, micrognathia, posterior rib gap defects	AR
Pierre Robin anomaly	Micrognathia, glossoptosis; occurs alone or with a variety of syndromes; mandibular growth usually catches up with maxillary growth	Varied depending on associated defects
Treacher Collins syndrome	Downward slanting palpebral fissures, displastic ears micrognathia, lower eyelid coloboma, conductive hearing loss	AD
Meckel's syndrome	Microcephaly, neural-tube defect, polydactyly, polycystic kidneys, genital ambiguity in males	AR
Orofaciodigital syndrome	Multiple hyperplastic frenula, alopecia, cleft tongue, digital abnormalities; lethal in males	XD
Otopalatodigital syndrome I	Growth retardation, pugilistic facies, bone dysplasia, conductive deafness	XR
Various chromosomal syndromes	Multiple congenital anomalies	CH

Key: AD = autosomal dominant; AR = autosomal recessive; XR = X-linked recessive; XD = X-linked dominant; CH = chromosomal

CLINICAL NOTES

- Cleft lip alone occurs in 25% of cases, cleft lip with cleft palate in 50%, cleft palate alone in 25%.
- Cleft lip is unilateral in 80% of cases, bilateral in 20%.
- Clefting extends to the palate in 70% of unilateral cases, in 80% of bilateral cases.
- Possible complications include: tooth defects in the area of cleft, speech problems, increased susceptibility to ear and sinus infections, and hearing loss.
- About 7% to 13% of patients with cleft lip alone are born with associated defects, as are 11% to 14% of patients with cleft lip and cleft palate. About 35% to 50% of patients with cleft palate are born with other defects.
- The prevalence of patients with associated defects drops after the neonatal period because of the lethal nature of some of the defects. Other associated defects, however, may be so inconspicuous as to be missed. Perhaps the easiest to detect and easiest to neglect is the association with lip pits, an autosomal dominant disorder.

PROCEDURES FOR DIAGNOSTIC CONFIRMATION

Condition visible on examination

Additional procedures, including chromosome analysis, may be indicated to rule out associated defects.

CONSIDERATIONS IN MANAGEMENT

Genetic
- Risk counseling—dependent on classification and etiology of cleft
- Consideration of prenatal diagnosis in pregnancies at risk:
  - For associated syndrome, if available and indicated
  - For cleft palate by ultrasound (now available in some centers)

Psychosocial, educational, familial
- Education for handling feeding difficulties associated with cleft
- Psychologic preparation of child for surgical procedures and anesthesia
- Preparation of child for dental work
- Consideration of psychotherapy when facial deformity is present

Medical
- Surgical repair
- Speech therapy
- Orthodontic and dental prosthetic procedures
- Control of ear and sinus infections
- Audiologic monitoring
- Management of associated defects

SUGGESTED READING

Cohen MM: Syndromes with cleft lip and cleft palate. *Cleft Palate J* 1978;15:306.

Fraser FC: The genetics of cleft lip and cleft palate. *Am J Hum Genet* 1970;22:336.

Gorlin RJ, Pindborg JJ, Cohen MM: *Syndromes of the Head and Neck*, New York, McGraw-Hill, 1976.

Shields ED, Bixler D, Fogh-Andersen P: Cleft palate: A genetic and epidemiological investigation. *Clin Genet* 1981;20:13.

Chapter 47

Other Isolated Birth Defects

An estimated 2% to 3% of infants are born with multiple malformations or a major isolated birth defect that has medical, surgical, or cosmetic significance. Such congenital malformations account for an estimated 10% of neonatal deaths and are the most common cause of pediatric hospitalizations in North America. Minor malformations, of little importance other than cosmetic appearance, occur in approximately 4% of the population, and commonly involve the face, ears, hands, and feet.

While some birth defects are identified as nongenetic, or have an unknown etiology, it is important to identify those which do have a genetic component in light of the possibly significant risk implications to the family.

Selected, localized birth defects, e.g., cleft lip and/or palate, heart defects, neural-tube defects, hydrocephalus, and limb defects are described in separate chapters. This chapter will briefly review the genetic implications of other common, isolated birth defects. Table 47-1 tabulates these birth defects and provides available recurrence risks, as well as other pertinent information.

ETIOLOGY AND PATHOGENESIS

Four major etiologic categories delineated for birth defects may all be mechanisms for isolated anomalies. A *malformation* refers to a birth defect which results from abnormal or incomplete organ development, e.g., cleft lip and/or palate. Malformations are often familial. A *disruption,* e.g., amniotic band syndrome, is caused by external forces (trauma, teratogen, infection) interfering with development which was originally normal. Although these are usually not genetic, it has been suggested that a fetal genetic predisposition may influence the disruptive effect. A *deformation* is an abnormality caused by mechanical prenatal forces and is usually not considered genetic, e.g., clubfoot. However, genetic factors may control the maternal intrauterine environment, or the deformation may be secondary to an underlying genetic disorder, e.g. Potter's syndrome and infantile polycystic kidney disease. *Dysplasias* are characterized by an abnormal organization of cells into tissues and are usually not confined to single organs, e.g., connective-tissue disorders.

Twinning results in an increased risk for structural defects, particularly in monozygotic twins. These include malformations, vascular disruptions, and deformations due to intrauterine crowding.

TABLE 47-1 Continued
Isolated Birth Defects

BIRTH DEFECTS	INCIDENCE	COMMENTS	RECURRENCE RISK: ONE CASE, NO OTHER FAMILY HISTORY		MODE OF INHERITANCE
			SIBLING (%)	OFFSPRING (%)	
Craniofacial					
Cleft lip +/− palate (see Chapter 46)	1/1,000	Can be associated with chromosomal or other syndromes	4	2-4	Most MF
Cleft palate (see Chapter 46)	1/2,500	Can be associated with chromosomal or other syndromes	2	7	Most MF
Craniosynostosis	0.4-1/1,000 live births	Most cases sporadic; can be associated with Crouzon's, Apert's, and other syndromes	Depends on etiology		Sporadic, MF, AR, AD
Microcephaly	1/10,000 for all types	Can be associated with teratogen exposure, maternal PKU, de Lange's, Rubenstein-Taybi, and chromosomal syndromes	Depends on etiology		Heterogeneous
Neural-tube defects (see Chapter 7)	1-2/1,000 births	Can be associated with genetic syndromes	3-5	3-5	MF
Hydrocephalus (see Chapter 8)	2/1,000 births	Can be associated with genetic syndromes	Depends on etiology		Heterogeneous
Cardio-Respiratory					
Congenital heart defects (see Chapter 21)	8/1,000	Can be associated with genetic or teratogenic syndromes	1-4	1-4	Most MF

TABLE 47-1 *Continued*

BIRTH DEFECTS	INCIDENCE	COMMENTS	RECURRENCE RISK: ONE CASE, NO OTHER FAMILY HISTORY		MODE OF INHERITANCE
			SIBLING (%)	OFFSPRING (%)	
Limb/Skeletal					
Simian crease	Unilateral—4% of normal infants; bilateral—1% of normal infants	Twice as common in males; can be found in Down's, fetal alcohol, hydantoin, trimethadione, Aarskog's, de Lange's, and other syndromes			
Clubfoot	1-4/1,000	Can occur with some genetic syndromes, e.g., neural-tube defects, arthrogryposis	3	3	Most MF
Congenital hip dislocation	1-5/1,000	Occurs more commonly in females; can occur in bone and connective-tissue disorders; see text	4-6	12	MF
Scoliosis (idiopathic)	4-10% of adolescent population	More common in females; can be associated with genetic syndromes	5-7, up to 30 in first-degree relatives, depending on criteria for diagnosis		MF
Limb defects (see Chapter 23)	Most prevalent localized malformations		Depends on etiology		Heterogeneous, many due to single gene
Gastrointestinal (GI)					
Pyloric stenosis	1/330	More common in male Caucasians, rare in Chinese	(See Table 47-2)		MF
Hirschsprung's disease	1/8,000 live births	Higher recurrence risk in more severely affected index case (those with longer segment of bowel affected); common in Caucasians	6	6	MF

Omphalocele (exomphalocele, includes umbilical hernia)	1/6,000 live births	Most cases sporadic; can be associated with Beckwith-Wiedemann, Hunter's, Hurler's, trisomy 13 and 18 syndromes; may cause increased AFP levels, often detected incidentally at prenatal diagnosis	Depends on etiology	Heterogeneous
Gastroschisis	1/20,000-1/30,000 live births	Umbilicus intact, may cause increased AFP levels, often detected incidentally at prenatal diagnosis		Sporadic
Tracheoesophageal fistula	1/100,000 live births	One of the components of VATER association; also can be associated with many other anomalies	Not thought to be increased over general population risk, unless part of a syndrome	Sporadic
Duodenal atresia	1/10,000 live births	Many patients have associated malformations (Down's syndrome, colon malrotation, heart defects, T-E fistula, renal defects); has been reported in rare families as AR	Not thought to be increased over general population risk, unless part of a syndrome	Sporadic
Esophageal atresia	1/3,000 live births	Associated anomalies in approximately half the cases, including heart, GI, genitourinary, skeletal, limb, facial clefts, CNS lesions, Down's syndrome	Not thought to be increased over general population risk, unless part of a syndrome	Sporadic
Inguinal hernia	1/100 children under age 12	Much more common in males; chromosome analysis desirable in females to rule out testicular feminization	Depends on etiology	Varied
Esophageal atresia with tracheoesophageal fistula	1/3,000 live births		Not thought to be increased over general population risk	Sporadic

TABLE 47-1 *Continued*

BIRTH DEFECTS	INCIDENCE	COMMENTS	RECURRENCE RISK: ONE CASE, NO OTHER FAMILY HISTORY SIBLING (%)	OFFSPRING (%)	MODE OF INHERITANCE
Imperforate anus	1/5,000 live births	48% of patients have associated anomalies (skeletal, renal, esophageal, cardiac, other)	Depends on etiology		Sporadic, AR in some families
Diaphragmatic hernia	1/2,200 births	May have associated defects, commonly nervous system; also seen in congenital myotonic dystrophy	<1 if increased; when familial, appears to be MF	Probably not increased	Sporadic
Genitourinary					
Hypospadias	1/186 live births	May be associated with other anomalies or can occur in genetic syndromes (Smith-Lemli-Opitz, Russell-Silver, pterygium, Klinefelter's, Down's, and 4p) or after fetal exposure to progestins	7-12		Heterogeneous
Cryptorchidism	3% of newborns (higher in premature infants)	May be associated with genetic syndromes (cryptophthalmos, Aarskog's, Noonan's . . .)			

Upper urinary tract malformations (hydroureter, nephrosis, absent or hypoplastic kidney, duplication of renal pelvis or ureter, horseshoe kidney)	Individually rare, collectively very common	Can be associated with other anomalies, deafness, CV, skeletal, or genetic syndromes	Depends on etiology		Most sporadic; some are hereditary
Renal agenesis and/or dysgenesis (when bilateral, can lead to Potter's syndrome)	1/2,000 live births	Absence of left kidney more common; can be associated with extrarenal anomalies (heart, CNS, skeletal, GI, genital, ear); ultrasonography of first-degree relatives may reveal less severe renal malformations with medical and genetic implications	Depends on etiology		
Vaginal atresia	Rare	Normal external genitalia and fertility, cytogenetic studies normal; can be associated with genetic syndromes (Laurence-Moon-Biedl, and a rare AR syndrome of renal, genital, and middle-ear anomalies)	2-5	2-5	MF
Bicornuate uterus/double vagina	?	Can be associated with renal agenesis, Beckwith-Wiedemann syndrome, trisomy 13	2-5	2-5	?MF
Absent uterus (müllerian aplasia)		Normal external genitalia; can be associated with renal and vertebral anomalies; cytogenetic studies normal	5 for female sibs	Patients are infertile	MF

TABLE 47-1 *Continued*

BIRTH DEFECTS	INCIDENCE	COMMENTS	RECURRENCE RISK: ONE CASE, NO OTHER FAMILY HISTORY SIBLING (%)	OFFSPRING (%)	MODE OF INHERITANCE
Ear (see Chapter 30)					
Anotia/microtia/atretic ear canal	1/20,)00-1/30,000 births	More common in right ear; bilateral in 17% of patients; conductive hearing loss usually present	Depends on etiology		AR, AD, CH, teratogens
Cup-shaped ears		Usually bilateral, with normal hearing; can be associated with Pierre Robin and conductive hearing loss	Low-50	50	AD
Eye (also see Chapter 29)					
Aniridia (includes simple iris colobomas-AD)	1/100,000-1/200,000	Usually bilateral; sporadic cases can be associated with chromosome 11q deletion syndrome and Wilms' tumor	Low-50	50	AD, 85% penetrance
Anophthalmia	Rare	Can be associated with chromosome abnormalities, fetal teratogens, or other syndromes	25	Negligible	AR
Microphthalmia	Rare	Vision good if no other ocular defects; can be associated with teratogen exposure or chromosome abnormalities	Depends on etiology		AR, AD

Key: , AR = autosomal recessive, AD = autosomal dominant, MF = multifactorial, Sporadic = single occurrence within a family, Heterogeneous = several different modes of inheritance likely, CH = chromosomal

GENETIC CHARACTERISTICS AND MODE OF INHERITANCE

The majority of single, localized defects (e.g., pyloric stenosis, duodenal atresias, hip dislocations) can be attributed to the interaction between genetic and environmental factors, i.e., a multifactorial inheritance. Only a small proportion of them, including some forms of microcephaly, craniosynostosis, and many limb defects are due to a single mutant gene.

Some common observations on multifactorial inheritance include:

- Recurrence risks decrease rapidly as the degree of relationship goes beyond first-degree relatives
- Recurrence risks increase after the birth of a second affected child
- Incidence may vary with sex, ethnicity, or other demographic features
- Recurrence risks are higher for relatives of an affected individual who is of the more rarely affected sex (these affected individuals are thought to carry more of the predisposing genes and therefore have a greater chance of passing them on)

Genetic characteristics of some common birth defects are summarized below.

Pyloric Stenosis is five times more common in males than in females. The frequency in the general population is 1/330 (both sexes combined). The frequency in males is 5/1,000, and in females 1/1,000.

- If a male is affected, the risk for an affected son is 5%, and for an affected daughter, 3%.
- However, if a female is affected, the risk to her son is 20%, and to her daughter, 8%.

Table 47-2 summarizes empiric recurrence risks for various relatives.

Omphalocele (exomphalocele) is more common in males than in females, with a fequency of 1/600 live births. Controversy surrounds whether or not a ruptured omphalocele and gastroschisis are in fact one entity. An omphalocele is the protrusion of the intestine through an abdominal-wall defect at the umbilicus. *Gastroschisis* is a fissure in the abdominal wall not involving the site of insertion of the umbilical cord. Gastroschisis is lateral and more often right-sided. Most cases of omphalocele or gastroschisis are sporadic. Some cases can

TABLE-47-2
Pyloric Stenosis: Risks to Relatives

RELATIVE AT RISK	MALE AFFECTED	FEMALE AFFECTED
Brother	4%	10%
Sister	3%	4%
Son	5%	20%
Daughter	3%	8%
Nephew	2%	5%
Niece	0.4%	—

Source: Adapted from Harper PS: *Practical Genetic Counseling,* Baltimore, University Park Press, 1982.

be detected in utero by elevated alpha-fetoprotein levels, although this occurs most often as an incidental finding when testing is performed for another reason. Similarly, ultrasonography may, on occasion, reveal the defect. Omphalocele can be associated with macroglossia, macrosomia, and hypoglycemia in the Beckwith-Wiedemann syndrome, which is usually not considered familial.

Atresias. Most bowel atresias are sporadic. Duodenal atresia is found in 30% of infants with Down's syndrome, and, rarely, as an isolated autosomal recessive trait. Esophageal atresia is associated with other anomalies 50% of the time. Malrotation of the colon is also usually sporadic, although it can occur in the "apple peel syndrome" with jejunal atresia, inherited as an autosomal recessive trait. Isolated anal atresia (imperforate anus) tends to be sporadic, but autosomal recessive inheritance has been reported in some families. It is associated with other anomalies in 50% of cases.

Hypospadias is a very common malformation, occurring in 1/186 live births, and is marked by genetic heterogeneity. Multifactorial inheritance probably accounts for many cases. It is important to identify those with intersexuality, either pseudohermaphrodites, or true hermaphrodites (chromosome analysis is desirable to establish chromosomal sex and to identify possible sex chromosome abnormalities). Hypospadias is a diagnostic feature in the autosomal recessive hypertelorism-hypospadias syndrome, which may also include cranial asymmetry, mental retardation, and cleft lip and palate. It has also been reported in association with fetal exposure to progestins.

Cryptorchidism occurs with a 3% (higher in premature infants) frequency in newborns, and is usually a benign condition, which resolves spontaneously. It can be associated with a variety of genetic syndromes—including Aarskog's, Noonan's ("male-Turner"), Prader-Willi, and Reifenstein's. The risk for development of testicular tumor is increased when the testes do not descend. Generally, when females have an inguinal hernia, chromosome analysis is desirable to rule out testicular feminization, in which unsuspected undescended testes may be subject to such malignant transformation.

Macroorchidism can occur in normal newborns, although it is generally noted after puberty. It is clinically significant for suspicion of fragile X syndrome in a mentally retarded male.

Vaginal Atresia affects the lower third of the vagina. Multifactorial inheritance is assumed, with 2% to 5% recurrence risks for female first-degree relatives. This condition is often not detected until puberty when hydrometrocolpos occurs. Surgical correction allows for normal fertility.

Clubfoot is twice as common in males as in females. Usually it is caused by uterine constraint, often due to oligohydramnios.

Overall recurrence risk to siblings is about 3%

2% for siblings of an affected male

5% for siblings of an affected female

Recurrence risks may approach 10% to 15% when an affected parent already has an affected child

Clubfoot can be a part of other syndromes, and is often a finding in individuals with neural-tube defects or other neurologic anomalies.

Congenital Hip Dislocation is approximately three to four times more common in females than in males, and is often noted after breech delivery. Multifactorial inheritance is postulated, with genetic factors thought to affect shape and laxity of the hip joint. Empiric recurrence risks for the isolated condition are given in Table 47-3. Dislocations can occur with some syndromes, such as neural-tube defects or Larsen's syndrome (cleft palate, unusual facies, multiple dis locations, with both autosomal recessive and autosomal dominant inheritance reported).

TABLE 47-3
Hip Dislocation: Risks to Relatives

AFFECTED RELATIVE	MALE	FEMALE	OVERALL
Sibling	1%	11%	6%
Parent	6%	17%	12%
Parent and one child	—	—	36%
Second degree relative	—	—	1%

Source: Adapted from Harper PS: *Practical Genetic Counseling,* Baltimore, University Park Press, 1982.

Idiopathic Scoliosis is more common in females than in males. Onset may be infantile, juvenile, or adolescent. Polygenic multifactorial inheritance is thought to account for the estimated 5% to 7% recurrence risk for first-degree relatives. Estimates as high as 30% have been given, but these are believed to reflect lax diagnostic criteria. Scoliosis may also result from vertebral anomalies or neurologic problems. It is a common finding in some genetic syndromes, including various forms of dwarfism and Marfan's syndrome.

Isolated Craniosynostosis is generally considered to be sporadic or multifactorial, although both autosomal recessive and autosomal dominant inheritance have been reported. An estimated 8% of coronal and 2% of sagittal synostoses are thought to be familial. Mental retardation and other associated anomalies have been reported more frequently in bilateral coronal synostosis than in the sagittal form. Many genetic syndromes include craniosynostosis, for example, Crouzon's and Apert's syndromes.

Microcephaly. True isolated hereditary microcephaly is reported to be autosomal recessive. Microcephaly is also a component of various genetic syndromes, or can be the result of prenatal teratogen exposure or maternal PKU. Mental retardation is common in individuals with microcephaly.

CLINICAL NOTES

A thorough diagnostic evaluation for a birth defect may include family and pregnancy histories, physical examination, chromosome analysis, laboratory studies, and radiographic studies, as well as consultation with references and/or specialists when a syndrome is suspected. Various physical measurements can alert the clinician to early deviations from normal development. In addition to the commonly measured head circumference, height, and weight, the following can be compared with standard charts (see Suggested reading list for references with appropriate charts): interpupillary distance; inner and outer canthal distance; palpebral length and width; ear measurements; nasolabial distance (for philtrum length); hand measurements; dermatoglyphics; internipple distance and chest circumference; erruption time of primary teeth; stretched penile length; and testicular volume.

A negative family history does not preclude a genetic etiology, nor does it necessarily implicate an environmental factor. By the same token, a positive family history does not preclude a teratogen as the cause if the teratogenic agent has operated in multiple pregnancies (e.g., maternal medication for seizures, maternal alcohol abuse).

PROCEDURES FOR DIAGNOSTIC CONFIRMATION

- Antibody screens (bacterial, viral)
- Radiography
- Ultrasonography
- Chromosome analysis
- Other procedures indicated by specific birth defect or suspected syndrome

CONSIDERATIONS IN MANAGEMENT

- Genetic
  - Examination of relatives, family photographs, for subtle findings
  - Establish diagnosis
  - Pedigree analysis, recurrence risk assessment
  - Risk counseling
  - Consideration of prenatal diagnosis, if available and indicated
- Psychosocial, educational, familial
  - Family counseling and support for acceptance of child with serious or minor abnormalities, and/or with problem requiring surgery
  - Cosmetic considerations
  - Occupational rehabilitation, as indicated
  - Consideration of referral for psychotherapy for families reacting strongly
- Medical
  - As indicated by diagnosis and severity, may include:
    - Supportive therapy
    - Corrective and/or cosmetic surgery
    - Orthodontic and/or speech therapy

SUGGESTED READING

Bergsma D, (ed): *Birth Defects Atlas and Compendium,* 2nd ed, New York, Alan R. Liss, 1979.

Feingold M, Bossert WH: Normal values for selected physical parameters: An aid to syndrome delineation, vol 10, no 13. (Birth Defects Original Article Series). White Plains, N.Y., March-of-Dimes Birth Defects Foundation, 1974.

Goodman RM, Gorlin RJ: *The Malformed Infant and Child—An Illustrated Guide*. New York, Oxford University Press, 1983.

Kalter H, Warkany J: Congenital malformations: Etiologic factors and their role in prevention. *N Engl J Med* 1983;308(8,9):424, 491.

Smith DW: *Recognizable Patterns of Human Deformation*. Philadelphia, WB Saunders, 1981.

Smith DW: *Recognizable Patterns in Human Malformation*, 3rd ed. Philadelphia, WB Saunders, 1982.

Chapter 48

Malignant Hyperthermia

Malignant hyperthermia is an often lethal hereditary disorder. It is characterized by attacks of rapidly rising temperature, associated with muscular rigidity, and induced by succinylcholine or by anesthetics (notably halothane), as well as by diversified stress situations. The incidence has been reported as approximately 1/20,000 cases of general anesthesia, with about 67% of those affected dying as a consequence. The disorder occurs as an isolated entity and is also reported in association with several well-defined neuromuscular diseases, such as central core disease, myotonia congenita, and myotonic dystrophy.

ETIOLOGY AND PATHOGENESIS

Varying degrees of hyperthermia and rigidity, as well as diverse expression of other symptoms in different patients, suggest a heterogeneity of malignant hyperthermia, which may involve more than one kind of mutant gene. There is variation in the reaction to different exposures by individuals at risk. Why this is so is not clear, as the pathophysiology of the disorder is not fully understood. Various theories have been proposed; a lesion of the excitation-contraction coupling mechanism in the affected muscle is regarded as a plausible explanation. Studies of muscle taken during an attack or post mortem show severe pathologic changes of rhabdomyolysis. Between attacks, when the patient is asymptomatic, microscopic and electronmicroscopic examinations of muscle may show a nonspecific myopathy with such changes as increased centralization of nuclei, variation of fiber size, fiber splitting, subsarcolemmal mitochondrial aggregates (ragged red fibers), streaming of Z lines, and expansions of the neuromuscular junction beyond the synaptic trough.

GENETIC CHARACTERISTICS AND MODE OF INHERITANCE

Although genetic heterogeneity has been proposed, malignant hyperthermia as an isolated disorder is, as a rule, inherited as an autosomal dominant trait.

- Children of an affected parent have a 50% risk of inheriting the mutant gene.
- The chance of the gene being present in parents or siblings of an affected individual is also 50%.
- A special variant, associated with dysmorphic features and mental retardation is thought to be inherited in an autosomal recessive pattern, with 25% recurrence risk for siblings.

When malignant hyperthermia is found in association with a clearly recognized neuromuscular disease, it is, of course, inherited in the same mode as the underlying disorder.

CLINICAL NOTES

The acute attack of malignant hyperthermia appears either during surgery or within hours of its termination. The rapidly developing hyperthermia may be accompanied by varying degrees of muscular rigidity, tachycardia, tachypnea, cyanosis, metabolic and respiratory acidosis, rhabdomyolysis, highly elevated serum creatine kinase (CK), myoglobinuria, and renal shutdown. The mortality is high, especially when the temperature rises to 41°C (106°F) or more.

Between attacks, the usual patient does not show clinical signs of myopathy. An otherwise, unexplained elevation of serum CK is found in 50% of the cases. In vitro contraction of muscle fibers after addition of caffeine and halothane identifies 80% of the gene carriers.

Since no test is 100% reliable for gene carrier detection, and exposure to an anesthetic may carry catastrophic consequences for a carrier, all offspring and other first-degree relatives of an affected individual should be considered potential candidates for malignant hyperthermia; appropriate precautions should be taken when surgery is required. Similarly, suspicion of potential hyperthermia in individuals with a family history of death from anesthesia can lead to testing and precautions. The fact that an individual at risk has had prior surgery without an attack of malignant hyperthermia does not preclude the presence of the gene, since it has been found that the syndrome does not necessarily develop every time a gene carrier is anesthetized (50% of patients have had previous anesthesia without a recognized attack).

PROCEDURES FOR DIAGNOSTIC CONFIRMATION

- Serum CK determination
- Inorganic pyrophosphate levels
- Muscle biopsy for contracture studies

CONSIDERATIONS IN MANAGEMENT

- Genetic
  - Risk counseling
  - CK and other studies of relatives at-risk
  - Discussion of sharing information with other family members to encourage evaluation for identification of gene carriers
- Psychosocial, educational, familial
  - Education of affected or at-risk individuals and their families as to the nature of the disorder and the importance of informing health professionals about the risk
  - Consideration of using an identifying device (bracelet, necklace, etc.) for individuals at risk
- Medical
  - Preoperative serum CK studies
  - Use of local or spinal anesthesia, when possible

Agents other than halothane or succinylcholine, if general anesthesia cannot be avoided
Constant monitoring of temperature during surgery
Use of dantrolene for relaxation of muscles, cooling, as indicated

SUGGESTED READING

Aldrete JA, Britt BA: *Malignant Hyperthermia; Second International Symposium*. New York, Grune & Stratton, 1978.

Nelson TE, Flewellen EH: The malignant hyperthermia syndrome. *N Engl J Med* 1983;309:416.

Chapter 49

Pseudocholinesterase Deficiency

The plasma enzyme pseudocholinesterase is responsible for the rapid hydrolysis of succinylcholine (suxamethonium), a muscle relaxant widely used in surgery. The drug is a potent agent, causing immediate paralysis of respiratory and other muscles. If the enzyme fails to terminate the succinylcholine action, severe consequences may ensue from lack of resumption of respiration.

ETIOLOGY AND PATHOGENESIS

Suxamethonium sensitivity is due to a mutation affecting the gene for plasma pseudocholinesterase. The result is an abnormal enzyme which is incapable of rapid hydrolysis of the drug. Several mutations affect the enzyme structure and function. The most common mutant is the ''atypical'' allele; its enzyme product is not really deficient when tested with the usual laboratory substrate, benzoylcholine. However, against a succinylcholine substrate, it fails to work. The use of dibucaine, an inhibitor of pseudocholinesterase, is an inexpensive and efficient means for distinguishing patients with homozygous atypical, heterozygous, and homozygous normal enzymes. Atypical homozygotes are abnormally resistant to dibucaine. Another allele, termed ''silent'', is common among Alaskan Eskimos, who have no enzyme activity at all in the homozygous state.

About 3% to 4% of Caucasians are heterozygous for the atypical allele and about 1/3,000 to 1/2,000 is homozygous and sensitive to suxamethonium.

GENETIC CHARACTERISTICS AND MODE OF INHERITANCE

Succinylcholine sensitivity is inherited as a classic autosomal recessive trait.

- Risk to siblings is 25%.
- The sensitivity will not be passed on to children unless the unaffected parent is a heterozygote, in which case the risk is 50%.

Genetic heterogeneity is readily demonstrated, with several variants of the enzyme differentiated by their electrophoretic properties, their residual enzyme activity against certain substrates, and their susceptibilty to dibucaine or fluoride ion as inhibitors.

CLINICAL NOTES

It is essential that all physicians, especially anesthesiologists and surgeons, be alert to this genetic predisposition. The patient is entirely normal until treated with succinylcholine. The condition may be suspected if the patient has a family history of unexplained "surgical" or "anesthetic" death. If the problem is recognized, the patient can be supported with artificial respiration until the effects of the drug wear off (sometimes hours) or can be given plasma or enzyme concentrate to terminate the action of the succinylcholine.

This condition is unrelated to malignant hyperthermia, another hazard of anesthesia (see Chapter 48).

PROCEDURES FOR DIAGNOSTIC CONFIRMATION

Assay of plasma pseudocholinesterase, with and without dibucaine, is usually sufficient to make the diagnosis. A bedside test is available.

Confirmatory tests include electrophoresis and staining for enzyme activity, and assay with acetylcholine or succinylcholine.

CONSIDERATIONS IN MANAGEMENT

- Genetic
  - Risk counseling
  - Test siblings and then other first degree relatives
- Psychosocial, educational, familial
  - Educate family in regard to deficiency and need to avoid succinylcholine or take precautionary measures
- Medical
  - Avoid succinylcholine, or be prepared with plasma or enzyme preparation for immediate treatment.

SUGGESTED READING

Goedde HW, Agarwal DP: Pseudocholinesterase variation. *Hum Genet Suppl* 1978;1:45.

Morrow AC, Motulsky AG: Rapid screening method for the common atypical pseudocholinesterase variant. *J Lab Clin Med* 1968;71:350.

Chapter 50

G6PD Deficiency
Drug-Induced Hemolytic Anemia

Glucose-6-phosphate dehydrogenase (G6PD) deficiency is the most common inherited enzyme deficiency. An X-linked recessive trait, it is determined by a variety of different mutations at a single gene locus and affects males almost exclusively. Depending on the severity of the deficiency, affected persons may have a chronic mild-to-moderate hemolytic nonspherocytic anemia, but they are more likely to be entirely normal in the absence of particular stresses. The stresses which trigger hemolytic crises include certain oxidizing drugs, infections, diabetic acidosis, ingestion of fava beans, and occupational exposures. Over the years many agents have been reported to precipitate the adverse effects of G6PD deficiency. Like some of the hemoglobinopathies, G6PD deficiency provides protection against malaria and, thus, has become common in populations from malaria-ridden parts of the world, such as the Mediterranean countries and parts of Africa.

ETIOLOGY AND PATHOGENESIS

G6PD is the first step in the pentose-phosphate shunt pathway, which generates NADPH, which in turn keeps glutathione in the -SH form. This sequence is essential to maintaining the integrity of the RBC membrane, especially in the presence of oxidizing agents.

Multiple mutations produce structural and functional changes in the enzyme G6PD; 140 such G6PD variants have been distinguished and categorized. Different G6PD variants are common in different populations and cause different degrees of abnormality. Table 50-1 lists some of the common G6PD types and their distribution patterns and enzyme activities.

Certain common G6PD deficient variants, such as G6PD A-, are relatively unstable, so that G6PD activity is normal or near-normal in young red cells and falls with aging of the cells. If hemolysis occurs, the older cells will be destroyed, leaving behind the younger cells with normal G6PD activity. An assay for G6PD at that time will fail to discover the G6PD deficiency, of course.

TABLE 50-1
Common G6PD Variants

TYPE	POPULATION DISTRIBUTION	ENZYME ACTIVITY (%)
B	Normal variant in American Caucasian and black populations	100
A+	Nearly normal variant in about 20% of American black males	90
A−	Deficient variant in about 12% of American black males	10-20
B−	Deficient variant in 2% to 5% of American males of Mediterranean ancestry	0-5
G6PD-Chinese; G6PD-Canton	Common variants in Chinese populations	4-25

GENETIC CHARACTERISTICS AND MODE OF INHERITANCE

G6PD deficiency is inherited as a classic X-linked recessive trait.

- Only males are expected to be G6PD deficient.
- All the daughters of an affected male are obligate carriers.
- The sons of an affected male cannot inherit the trait, and cannot transmit it to their children.
- Women with one X chromosome coding for a G6PD variant and the other chromosome normal, are carriers.
- There is a risk of 50% for sons of a carrier female to be affected.
- The risk is 50% for daughters of a carrier female to be carriers.

Women may express G6PD deficiency under three different conditions:

- If they inherit two variant X chromosomes—one from the mother and the other from an affected father (substantial probability only in populations with an extremely high gene frequency, as in some Mediterranean and Mid-Eastern populations).
- Lyon effect: if significantly more than half of the cells with the active X carry the gene for G6PD deficiency.
- They can be as affected as males if they have a chromosome abnormality including only one X chromosome (e.g., Turner's syndrome—45,X chromosome complement).

CLINICAL NOTES

Men with these deficient G6PD variants are generally asymptomatic unless challenged by one of the environmental stresses.

Drug-induced Hemolysis. The clinical course of primaquine-induced hemolysis has been studied under carefully controlled conditions in men with the A-type of G6PD deficiency. Little or no evidence of hemolysis appears the first two or three days the patient is receiving primaquine, 30 mg/day. However, Heinz bodies appear in the erythrocytes and then the hemoglobin concentration falls and urine begins to darken. Sometimes no other abnormalities are noted by the patient. In more severe cases, weakness and abdominal and back pain occur; the patient becomes jaundiced; and the urine turns nearly black. Sequestration of red cells occurs in both the liver and spleen. Heinz bodies disappear as the hemoglobin concentration of the blood approaches its nadir. Reticulocytosis and polychromatophilia are observed on blood films. Regardless of the severity, the acute hemolytic phase ends spontaneously in about one week.

TABLE 50-2
Causes of Hemolytic Anemia in G6PD Deficiency*

Acetanilid	
Methylene blue	Sulfanilamide
Nalidixic acid (NeGram)	Sulfacetamide
Naphthalene	Sulfapyridine
Niridazole (Ambilhar)	Sulfamethoxazole (Gantanol)
Nitrofurantoin (Furadantin)	Thiazolesulfone
Pamaquine	Toluidine blue
Pentaquine	Trinitrotoluene (TNT)
Phenylhydrazine	
Primaquine	

*Note: Not all sulfa drugs represent a hazard, nor do certain other drugs commonly listed in textbooks or articles as hazardous. Source: Beutler E: *Hemolytic Anemia in Disorders of Red Cell Metabolism*. New York, Plenum Press, 1978, p 77.

A similar sequence is observed with the Mediterranean type of G6PD deficiency (B-). However, the hemolysis is more severe and may not be self-limited. Young cells are susceptible to destruction, as well as older ones.

Beutler has limited the list of drugs claimed to cause hemolysis in G6PD-deficient individuals, eliminating those for which claims could not be confirmed and those which occurred only in the presence of infections that may account for the hemolysis directly. Table 50-2 lists the agents he has identified as clearly causing hemolysis in G6PD patients,* and Table 50-3 provides a list of drugs that are probably safe for them.

Hemolytic Anemia Induced by Infection. Although the prototype for hemolytic anemia in G6PD deficiency is the effect of drugs, hemolysis precipitated by infectious diseases is probably more common. Typically, the anemia is discovered only as the patient is being treated, often with aspirin or phenacetin, as well as antibiotics. The most common associated infection is bacterial pneumonia. Infections with typhoid fever, *Salmonella, Proteus, Escherichia coli,* B-streptococci, staphylococci, tuberculosis, and rickettsiae have been implicated. Viral hepatitis can also cause anemia in G6PD-deficient individuals; it is essential to recognize that high bilirubin levels may be from hemolysis rather than from hepatic necrosis, because the prognosis is quite good.

Hemolytic Anemia in Diabetic Acidosis. Generally mild, the hemolysis disappears when normal metabolic balance is restored.

TABLE 50-3
Drugs Normally Tolerated By G6PD-Deficient Patients*

Acetaminophen (Tylenol)	Phenylbutazone
Acetophenetidin (phenacetin)	Phenytoin
Acetylsalicylic acid (aspirin)	Probenecid (Benemid)
Aminopyrine	Procainamide (Pronestyl)
Antazoline	Pyrimethamine (Daraprim)
Antipyrine	Quinidine
Ascorbic acid (vitamin C)	Quinine
Benzhexol (Artane)	Streptomycin
Chloramphenicol	Sulfacytine
Chlorguanide (Proguanil)	Sulfadiazine
Chloroquine	Sulfaguanidine
Colchicine	Sulfamerazine
Diphenhydramine (Benadryl)	Sulfamethoxypyridazine (Kynex)
Isoniazid	Sulfisoxazole (Gantrisin)
L-Dopa	Trimethoprim
Menadione sodium bisulfite	Tripelennamine (Pyribenzamine)
Menaphthone	Vitamin K
P-Aminobenzoic acid	

*Without nonspherocytic hemolytic anemia. Source: Beutler E: *Hemolytic Anemia in Disorders of Red Cell Metabolism*, New York, Plenum Press, 1978, p 79.

Neonatal Jaundice. A high proportion of Mediterranean and Oriental newborns with jaundice of unknown origin turn out to be G6PD-deficient. They usually have moderate anemia, though severe hemolysis with kernicterus has been reported. Neonatal jaundice is much less likely in black newborns with G6PD A- variant. The effect is greater in premature babies.

Favism. The occurrence of acute hemolysis observed in Mediterranean populations following ingestion of *Vicia fava* broad beans, a staple in these areas, has been recognized since antiquity. Even the inhalation of the pollen can bring on an attack, but most cases result from ingestion of fresh beans. Favism is more common among children than adults and has been observed in nursing infants whose mothers ingested the beans. The quantity required can be extremely small—as little as one seed. Symptoms occur within a few seconds of inhaling pollen and five to 24 hours after ingesting beans. They include malaise, headache, dizziness, nausea, vomiting, chills, pallor, lumbar pain, and fever. Hemoglobinuria and then jaudice appear within 24 to 30 hours. Favism is observed only in some families with G6PD deficiency, indicating that a second factor—inherited as an autosomal recessive trait—may be necessary for susceptibility.

Occupational Exposures. Definite cases of hemolysis in G6PD-deficient individuals have been reported from work exposures to TNT (e.g., Iraqi Jews in Israel). Other occupational and environmental exposures to oxidizing agents may be significant, but have not been studied carefully.

*Lists of dangerous drugs in chapters on pharmacogenetics in many texts are outdated.

PROCEDURES FOR DIAGNOSTIC CONFIRMATION

Fluorescent spot assay
Quantitative assay of G6PD activity
Electrophoresis for G6PD phenotype

CONSIDERATIONS IN MANAGEMENT

Genetic
- Risk counseling
- Consideration of testing for other family members
- Discussions to explain need to be alert to problems for newborns at risk, diabetics, those with infections

Psychosocial, educational, familial
- Education about avoidance of drugs and foods known to be hemolytic
- Assurance that the vast majority of persons with G6PD deficiency never have any problems at all

Medical
- Advice about drugs to avoid (particularly nitrofurantoin, primaquine, sulfapyridine) and those that can be used (see Tables 50-2 and 50-3)
- Treatment of severe anemia or hyperbilirubinemia, as indicated

SUGGESTED READING

Beutler E: *Hemolytic Anemia in Disorders of Red Cell Metabolism,* New York, Plenum Press, 1978.

Chapter 51

Porphyria

Often unrecognized, porphyria should always be suspected in those patients with unexplained abdominal pain and/or acute psychologic crises. The porphyrias are a group of inborn errors of metabolism in the porphyrin-heme biosynthetic pathway. Different enzyme defects result in abnormal production and excretion of intermediary metabolites and in characteristic clinical manifestations, which include varying degrees of acute episodes of visceral pain and psychologic disturbance and/or cutaneous photosensitivity. Accurate prevalence figures are not available, largely because many gene carriers remain asymptomatic. Particular types of porphyria are more prevalent in specific populations (e.g., acute intermittent porphyria in Sweden, variegate porphyria among white South Africans). Worldwide, acute intermittent porphyria is thought to be the most prevalent of all the porphyrias, with estimates ranging from 1.5/100,000 to 1/5,000. However, porphyria cutanea tarda, which has only recently been recognized to have a genetic component, may well be at least as common.

The various porphyrias are classified by the porphyrin-producing tissue involved, into erythropoietic and hepatic forms. Table 51-1 summarizes clinical manifestations and biochemical findings, and lists the modes of inheritance.

Most forms of porphyria can be successfully managed, especially when gene carriers are identified before becoming symptomatic.

ETIOLOGY AND PATHOGENESIS

The basic enzymatic defects of the various porphyrias constitute blocks in the biochemical pathway that begins when glycine and succinate combine to form delta-aminolevulinic acid (ALA), and ends with a heme protein (e.g., hemoglobin, cytochrome, myoglobin). Figure 51-1 is a schematic illustration of this path way, showing the enzymes found to be malfunctioning in the different porphyrias.

The pathogenic process is not as clearly understood as the enzymes involved. Cutaneous manifestations are thought to be related to a fluorescent response of the porphyrins in the skin to long-wave ultraviolet light, and a resulting photochemical damage to adjacent membranes. The neurovisceral attacks in the hepatic forms of porphyria are generally precipitated by endogenous or environmental triggers, such as drugs, hormones, fasting, alcohol, or infections. It is hypothesized that in the presence of these factors an increased production of ALA synthetase, perhaps resulting from an altered negative-feedback control by decreased levels of heme, leads to greater production of intermediate metabolites that can-

FIGURE 51-1 Enzyme Defects in the Porphyrias

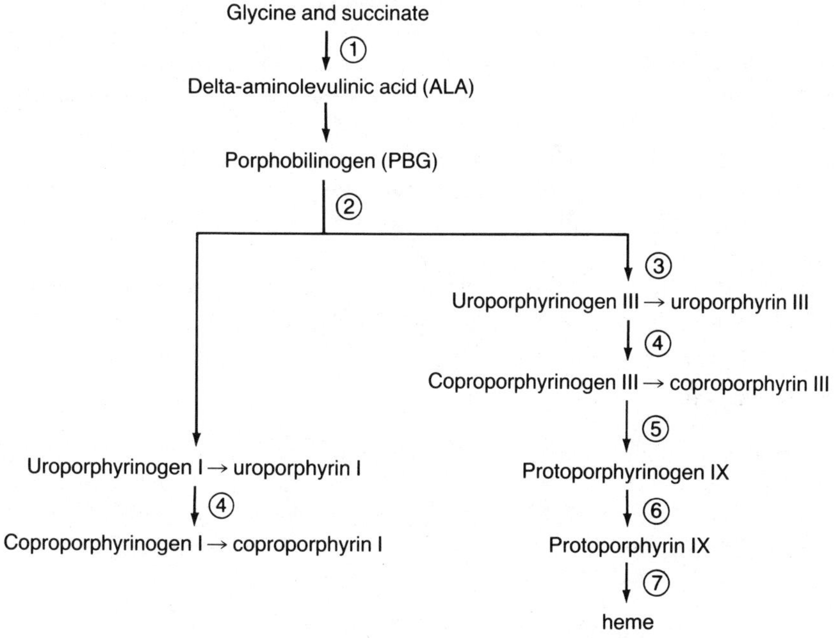

Malfunctioning enzyme		Resulting disorder
①	ALA synthetase	Increased in all acute porphyric attacks
②	PBG deaminase	Acute intermittent porphyria
③	Uroporphyrinogen III cosynthetase	Congenital erythropoietic porphyria
④	Uroporphyrinogen decarboxylase	Porphyria cutanea tarda
⑤	Coproporphyrinogen oxidase	Hereditary coproprorphyria
⑥	Protoporphyrinogen oxidase	Variegate porphyria
⑦	Ferrochelatase	Protoporphyria

not be used properly. These metabolites are thought to lead to neuromuscular toxicity, which in turn produces the clinical symptoms.

GENETIC CHARACTERISTICS AND MODE OF INHERITANCE

Most of the porphyrias are autosomal dominant disorders with a 50% recurrence risk for offspring of individuals carrying one mutant gene.

Since many gene carriers stay asymptomatic for many years, or never become symptomatic, appropriate diagnostic studies on relatives of known porphyria patients are necessary to identify the latent carriers in the family, and to assess their risks for becoming symptomatic and/or to have affected offspring. Identification, in itself, reduces the risk of clinical expression, because it brings asymptomatic gene carriers to medical attention. In most cases clinical symptoms can be prevented by appropriate management for avoidance of precipitating factors.

Not all cases of *porphyria cutanea tarda* have a genetic component. In families where a genetic factor has been shown to exist, it is an autosomal dominant predisposition to the disorder. The gene for the predisposition is passed on to 50% of the offspring, but the overt disorder is expressed only in the presence of strong environmental stimuli, such as liver damage from alcohol abuse.

Congenital erythropoietic porphyria is the only form inherited as an autosomal recessive disorder with a 25% risk of occurrence for each pregnancy, when both parents are carriers.

Carriers of one gene for erythropoietic porphyria do not develop symptoms. They have usually been identified through the birth of an affected child. Carrier testing has been conducted experimentally, as has prenatal diagnosis.

CLINICAL NOTES

As summarized in Table 51-1, the various porphyrias are characterized by somewhat overlapping clinical features and biochemical findings. Diagnostic tests distinguish between them (see Procedures for diagnostic confirmation,). Acute porphyric attacks should also be distinguished from lead poisioning, which may display similar symptoms.

The Hepatic Porphyrias.

Acute intermittent porphyria (AIP) usually presents after puberty, but may remain latent for years or for a lifetime. It is characterized most often by attacks of severe abdominal pain, which may be accompanied by pain in the extremities, weakness, anorexia, nausea, and vomiting. Constipation, tachycardia, and hypertension are common. All these symptoms are thought to have a neurogenic origin, like the clearly neurologic aspects of the attacks, which include confusion, psychosis, seizures, and potentially fatal motor and sensory deficits. Urine may be purple. Between attacks, mood swings and body pains may betray the presence of the gene. This is the only type of porphyria with no cutaneous photosensitivity.

Variegate porphyria (VP) is clinically latent in most individuals who have inherited the gene. If it is expressed, attacks like those in acute intermittent porphyria are likely to begin between the ages of 30 and 40. However, these patients also have photosensitive skin. Slowly healing lesions, which often leave pigmented scars, appear in the second or third decade, after exposure to sunlight.

Hereditary coproporphyria (HC) is clinically similar to variegate porphyria, but is less severe.

AIP, VP and HC may easily go unrecognized, because they are rare and their episodic symptoms are suggestive of other disorders. Since clinical expression is characteristically the result of exposure to a specific precipitating factor, a high index of suspicion for the diagnosis of porphyria is prudent when a patient has acute episodes of abdominal pain and/or psychiatric or neurologic disturbance

TABLE 51-1
Summary of Distinguishing Features of the Porphyrias

	CLINICAL MANIFESTATIONS			
TYPE	CUTANEOUS	VISCERAL/ NEUROLOGIC	SELECTED BIOCHEMICAL FINDINGS	MODE OF INHERITANCE
Hepatic porphyrias				
Acute intermittent porphyria	–	+ + +	↑Urinary porphobilinogen (PBG) ↑ Urinary aminolevulinic acid (ALA) ↓Porphobilinogen deaminase	AD
Variegate porphyria	+ +	+ +	↑Fecal coproporphyrin ↑Fecal protoporphyrin ↑Urinary coproporphyrin ↑Urinary ALA and PBG	AD
Hereditary coproporphyria	+	+	↑Fecal and urinary coproporphyrin ↑Urinary ALA and PBG	AD
Porphyria cutanea tarda	+	–	↑Urinary uroporphyrin ↑Fecal coproporphyrin	AD (in some families)
Erythropoietic porphyrias				
Congenital erythropoietic porphyria	+ + +	–	↑↑Urinary and red cell uroporphyrin I ↑Uroporphyrinogen III cosynthetase ↑Coproporphyrins	AR
Protoporphyria	+	–	↑Red cell and fecal protoporphyrins	AD

Key: AD = autosomal dominant; AR = autosomal recessive

subsequent to such an exposure. Similarly, cutaneous fragility and blistering of light-exposed skin may suggest VP or HC.

Once diagnosed, most patients can remain symptom free through avoidance of risk factors, and/or shielding from sunlight. This holds, as well, for gene carriers identified through screening of relatives. Hematin, a medication that limits the synthesis of porphyrin, is available for patients in whom avoidance of risk factors is not possible (e.g., hormone changes in the menstrual cycle).

The *risk factors* for the visceral and neurologic manifestations include fasting, alcohol consumption, and hormonal changes in puberty and/or the menstrual

TABLE 51-2
Drug Recommendations for Patients With Acute Intermittent Porphyria, Variegate Porphyria, and Hereditary Coproporphyria*

UNSAFE	POTENTIALLY UNSAFE	PROBABLY SAFE	SAFE
Barbiturates	2-Allyloxy-3-methylbenzamide	Diazepam	Narcotic analgesics
Sulfonamide antibiotics	Alfaxolone	Chlordiazepoxide	Chloral hydrate
Sulfonmethane (Sulfonal)	Alfadolone acetate	Dicumarol	Phenothiazines
Sulfonethylmethane (Trional)	Hydralazine	Digoxin	Penicillin and derivatives
Meprobamate	Clonidine	Imipramine	Aspirin
Isopropylmeprobamate	Phenoxybenzamine	Chloramphenicol	Acetaminophen
Glutethimide	o,p[1]-DDD	Nitrofurantoin	Streptomycin
Aminoglutethimide	Tolazamide	Mandelamine	Tetracycline
Methylprylon	Methyldopa	Rauwolfia alkaloids	Glucocorticoids
Ethchlorvynol	Alkylating agents	Diphenhydramine	Propranolol
Carbromal	Tranylcypromine	Vitamin B group	Bromides
Ethinamate	Theophylline	Vitamin C	Insulin
Diphenylhydantoin	Spironolactone	Neostigmine	Atropine
Mephenytoin	Rifampin	Prostigmin	Amitryptyline
Succinimides	Pargyline	Propoxyphene	
Pyrazolone preparations	Bemegride	Propanidid	
Griseofulvin	Nikethamide	Nitrous oxide	
Ergot preparations	Pentylenetetrazole	Methylphenidate	
Danazol	Metyrapone	Guanethidine	
Synthetic estrogens, progestins	Food additives		
Carbamazepine	Pyrazinamide		
Valproic acid	Pentazocine		
Primadone	Chloroquine		
Trimethadione	Heavy metals		
Diclophenac	Fluroxene		
Novobiocin	Etomidate		
N-Butylscopolammonium bromide	Ketamine		
Tolbutamide			
Tolazamide			
Chlorpropamide			

*Drugs are not listed alphabetically; e.g., barbiturates and sulfonamide antibiotics are at the top of the list of unsafe drugs on the basis of extensive reports in the literature.
Source: Adapted from Kappas A, Sassa S, Anderson KE: The porphyrias, in Stanbury JB, Wyngaarden JB, Fredrickson, DS, et al, (eds): *The Metabolic Basis of Inherited Disease*, 5th ed, New York, McGraw-Hill, 1983.

cycle (for the effects of pregnancy, see below). The primary precipitating agents, however, are specific drugs, some of which might otherwise be logical medications for the symptoms of an acute porphyric attack. Most notable among these are the barbiturates. Table 51-2 provides a recently updated list of medications, classified according to their effect on patients with porphyria. Information on medications not included in this list may be obtained from specialized research centers. A warning bracelet or necklace is recommended for patients subject to acute attacks, to provide information in emergencies.

Although the *effects of pregnancy* on the hepatic porphyrias are unpredictable, experience with severe exacerbation in some cases emphasizes that a major danger may exist, and that pregnancies of affected women are prudently considered high-risk.

Porphyria cutanea tarda (PCT), generally seen only in adults, is the least severe of the hepatic porphyrias. It has no neurologic manifestations, being characterized only by skin fragility and photosensitivity under special conditions and by hypertrichosis. Historically, outbreaks of PCT have been linked to large-scale contamination of food supplies or industrial pollution by hepatotoxins. The condition was considered nongenetic. However, recent findings of the enzyme malfunction indicate that, at least in some families, a genetic lesion underlies the expression of the disorder, which appears to be triggered by increased hepatic iron, alcohol abuse, or exposure to one of a few drugs (especially estrogens). Most gene carriers generally do not develop clinical symptoms. When they do, they can usually be relieved by phlebotomy and administration of low doses of chloroquine.

The Erythropoietic Porphyrias.

Congenital erythropoietic porphyria is an extremely rare and devastating disorder. It is marked by severe cutaneous reactions to sunlight, from infancy on. Affected individuals become grossly disfigured by the scars and secondary mutilations (e.g., loss of digits or cartilage from ears or nose) that develop from bullae, vesicles, and infected ulcerations resulting from exposure to even small amounts of sunlight. In addition, hypertrichosis and erythrodontia commonly account for an even more frightening appearance. Splenomegaly may be present, as well as intermittent hemolytic anemia. First clinical suspicion may be raised by an infant's red urine. Survival past middle age is not expected.

Protoporphyria is less rare and less severe than erythropoietic porphyria. Onset is in childhood or adolescence, with cutaneous features limited to temporary burning and stinging or pruritus upon exposure to sunlight, with, perhaps, some edema and/or erythema later. Attacks are restricted to 12 to 14 hours' and generally do not lead to scarring. Administration of beta-carotene increases the tolerance for sunlight. Possible hepatic involvement is a potential complication.

PROCEDURES FOR DIAGNOSTIC CONFIRMATION

Diagnostic evaluation may involve ruling out another form of porphyria, as well as other disorders. The tests below are used to differentiate the porphyrias listed previously.

Acute intermittent porphyria
- Urine that turns red on standing will sometimes make the diagnosis obvious
- Quantitative assay for decreased red cell porphobilinogen deaminase
- Urinalysis for increased porphobilinogen and for increased aminolevulinic acid most useful during an attack
- Urine and stool porphyrin assays

Variegate porphyria and hereditary coproporphyria
- Stool porphyrin determination
- Red cell porphobilinogen deaminase assay (normal)

Porphyria cutanea tarda
- Quantitative urinalysis for uroporphyrin and coproporphyrin
- Quantitative stool analysis for coproporphyrin and protoporphyrin
- Plasma porphyrin levels
- Porphyrin isomer analysis by chromatography

Congenital erythropoietic porphyria
- Assays for marked increase of porphyrins in urine and red cells, including quantitative analysis of types I and II isomers
- Uroporphyrinogen III cosynthase evaluation in cultured fibroblasts for heterozygote detection

Protoporphyria
- Red cell analysis for increased protoporphyrins
- Stool analysis for increased protoporphyrins

CONSIDERATIONS IN MANAGEMENT

Genetic
- Establish diagnosis
- Risk counseling
- Evaluation of first-degree relatives for diagnosis and preventive management
- Discussion of sharing information with extended family to encourage evaluation, counseling, and preventive management
- Consideration of prenatal diagnosis in future pregnancies in families with erythropoietic porphyria

Psychosocial, educational, familial
- Family counseling
- Education for compliance to preventive measures
- Consideration of referral for psychotherapy and/or support, especially in cases involving disfigurement or emotional crises

Medical (usually requires specialty attention)
- Acute porphyric attacks
  - Hospitalization
  - Narcotic analgesics (nonprecipitating; see Table 51-2)
  - Respiratory support
  - Intravenous (IV) fluids, including glucose
  - Hematin, IV (should be used only by a physician experienced in the management of porphyria)
  - General supportive care
- Prevention of acute porphyric attacks
  - Avoidance of drugs known to be harmful
  - High carbohydrate intake
  - Hematin to prevent premenstrual attacks
  - Use of warning bracelet or necklace

Photosensitivity and skin fragility
- Avoidance of, or shielding from, sunlight
- Avoidance of trauma to skin
- Prompt treatment of secondary infections
- In hereditary coproporphyria: cholestyramine to increase tolerance for sunlight
- In protoporphyria: beta-carotene to increase tolerance for sunlight

Porphyria cutanea tarda
- Avoidance of alcohol abuse and exogenous estrogens
- Phlebotomy
- Low doses of chloroquine

SUGGESTED READING

Kappas A, Sassa S, Anderson KE: The Porphyrias, in Stanbury, JB, Wyngaarden JB, Fredrickson DS, et al, (eds): *The Metabolic Basis of Inherited Disease,* 5th ed. New York, McGraw-Hill, 1983.

Muhlbauer JE, Pathak MA, Tishler PV, et al: Variegate porphyria in New England. *JAMA* 1982;247:3095.

Nitowsky HM, Sassa S, Nakagawa A, et al: Prenatal diagnosis of congenital erythropoietic porphyria. *Pediatr Res* 1978;12:455.

Sassa S, Kappas A: Genetic metabolic and biochemical aspects of the porphyrias. *Adv Hum Genet* 11:121, 1981.

Section Three

Concerns in Family Planning

Chapter 52

Reproductive Failure

Reproductive failure can be defined to include all efforts at reproduction that do not result in the birth of a viable infant. About 15% of couples experience infertility, or inability to conceive after a year of unprotected intercourse. This group represents both those cases where conception actually does not occur and those whose pregnancies abort before they are recognized. Among recognized pregnancies, from 15% to 20% are lost in spontaneous abortion; close to 1% of the remainder are stillborn and others are born too prematurely to survive.

The reasons for such reproductive failure are numerous and are not always determined, but both nongenetic and genetic causes are known to play a part.

It is rarely clear at the beginning of an infertility workup whether maternal or paternal factors or a combination of the two are responsible, a fact which argues for a diagnostic workup including both potential parents. Among the etiologies identified, genetic ones may, on occasion, be obvious; more often they come to attention only when reproductive problems are otherwise unexplained.

ETIOLOGY AND PATHOGENESIS

Infertility. Of all the factors contributing to infertility, genetic components are among the most elusive, especially in the absence of an obvious genetic syndrome associated with reduced fertility (see Table 52-1 for a list of such genetic disorders). Of course, reproduction is limited, as well, for patients with genetic disease who are so severely affected that survival to the reproductive age is precluded, or whose disorder is sufficiently disabling or disfiguring that their opportunity for having children is reduced. The genetic conditions discussed in the text are primarily those that seem to confer no phenotypic abnormalities or only such minor ones that they may not come to medical attention before reproductive problems are noted.

Both among males and females, causes of infertility generally considered nongenetic include chronic infections, congenital anatomic anomalies, endocrine disorders, chronic debilitating diseases, certain tumors, nutritional factors, medications, environmental exposure, and substance abuse. Usually, when these have been ruled out, genetic factors are investigated to provide a cause for the inability to reproduce.

Among apparently healthy men presenting for unexplained infertility in genetics clinics, roughly 10% have been reported to have primary hypogonadism due to Klinefelter's syndrome, which occurs in approximately 1/1,000 to 2/1,000

TABLE 52-1
Genetic Disorders With Associated Infertility

DISORDER	MODE OF INHERITANCE
Testicular feminization	XR
Down's syndrome	CH
Cystic fibrosis	AR
Kartagener's syndrome	AR
Hemochromatosis	AR
Klinefelter's syndrome	CH
Laurence-Moon-Biedl syndrome	?AR
Myotonic dystrophy	AD
Prader-Willi syndrome	? (?CH)
Turner's syndrome	CH
Pseudohermaphroditism	AR
True hermaphroditism	Varied
Polycystic ovarian disease	?MF

Key: AD = autosomal dominant; AR = autosomal recessive; XR = X-linked recessive; CH = chromosomal; MF = multifactorial

newborn males. The chromosome constitution in this condition is generally 47,XXY. While phenotypic abnormalities may accompany this chromosome complement (see Chapter 5), they can be so minor that the only one noted is tall stature. Another genetic cause for hypogonadism may be found in the spectrum of male pseudohermaphroditism. Further unrecognized genetic conditions may also underlie the infertility. For instance, immotile sperm may be associated with cystic fibrosis or Kartagener's syndrome, which is characterized by sinusitis, bronchiectasis and situs inversus.

Genetic etiologies for women with *primary amenorrhea* may include an error in gender assignment (see Chapter 27). It has been suggested that 50% of cases of primary amenorrhea are explained in this way. In such a situation the genetic sex is male, with a 46,XY chromosome complement, but inadequate testosterone action has led to an external female phenotype. In testicular feminization, an example of such a disorder, the genetic lesion is a tissue resistance to androgenic stimulation, usually due to absent or abnormal androgen receptors.

Primary amenorrhea may also be the presenting symptom for women with Turner's syndrome (classically a 45,X chromosome complement), who happen not to show most of the Turner's stigmata (see Chapter 5). More common in women with infertility or multiple miscarriage than pure 45,X, are cases of chromosomal mosaicism, such as X/XY, or a partial deletion of one X chromosome. Women in the latter group have been known to conceive, with the pregnancy more likely to end in a spontaneous abortion than with a live birth. They have been identified, as well, through delayed menarche or early menopause.

Since infertility often reflects a very early spontaneous abortion, rather than inability to conceive, the causes also include those that account for loss of unrecognized pregnancies.

Spontaneous Abortion. About 15% to 20% of all recognized pregnancies end in spontaneous abortion, with the reason not always determined. However, although many nongenetic maternal or fetal factors (e.g., infection, endocrine abnormalities, environmental exposure, substance abuse—see Chapter 45—nutritional deficits, medical complications, anatomic anomalies, fibroids) are known to precipitate pregnancy loss, genetic factors play an important role in spontaneous abortion, especially in the first trimester when 85% of the miscarriages occur. Of these early abortions at least 50% are estimated to be chromosomally abnormal, and other genetic mechanisms are thought to be etiologic among the chromosomally normal ones.

A wide range of chromosome abnormalities has been noted in spontaneous abortions. Trisomies of every chromosome except chromosome 1 have been documented and account for 50% of the chromosomally-abnormal abortions. Most of the autosomal trisomies as well as most of the other anomalies are not viable. Trisomies 8, 13, 18, and 21 can go to term and produce a severely abnormal child, but a large proportion of these conceptions also end in spontaneous abortions, as do 98% of monosomy X fetuses. Altogether 90% of chromosomally abnormal conceptions do not survive the pregnancy. The reason for most of the abnormalities is not clearly understood, although it is known that couples who have had one trisomic conception are more likely to have another than couples who have had only chromosomally normal ones.

About 1% to 2% of spontaneous abortions have an unbalanced chromosomal rearrangement. This could have happened in a germ cell of the particular conception. However, it is also possible that such an anomaly is inherited from a phenotypically normal parent with a balanced translocation (no chromosomal material is lost or gained). When such an individual reproduces, unequal distribution of the genetic material can occur, leading to an unbalanced translocation, which can cause apparent inability to conceive, pregnancy loss, or the birth of a child with birth defects and mental retardation. In studies of couples with multiple spontaneous abortions, a balanced chromosomal rearrangement has been found in one of the parents in 6% to 14% of these cases. It has been suggested that different rates reported in diverse studies may, in part, be due to different methods of choosing the study population. Thus, for example, when only early miscarriages are considered, the rate is higher than when all are included. It is especially high when the prior birth of an abnormal child is a study criterion. Other studies suggest that mosaic sex chromosome aneuploidy in a very small proportion of cells may also be etiologic in women with recurrent spontaneous abortions.

Genetic causes of spontaneous abortions with normal chromosomes are thought to include blood-group incompatibilities and single gene disorders that are not viable, or only rarely so. Indeed, specific single-gene defects have been documented in morphologic examination of abortuses (e.g., leprechaunism; limb defects). Multifactorial disorders, as well, account for pregnancy loss. For example, even as mothers of children with neural-tube defects or cleft lip and/or cleft palate are known to have more spontaneous abortions than others, studies have shown a higher rate of these abnormalities in abortuses than in newborns.

Stillbirth and Neonatal Death. Approximately 7% of stillbirths and neonatal deaths have chromosomal abnormalities (not always accompanied by obvious congenital malformations), as compared with 0.5% of all newborns. Similarly, about 20% of stillborns and 21% of infants dying neonatally have congenital abnormalities, while the general newborn incidence is 2% to 3%. These abnormalities may or may not be the reason for nonviability. The etiologies are as diverse as those for congenital anomalies in all newborns, and genetic factors often play a role and impose a risk for recurrence.

GENETIC CHARACTERISTICS AND MODE OF INHERITANCE

Infertility

- Reduced fertility, when associated with another genetic condition, is transmitted in the same way as the underlying disorder, assuming of course, that reproduction is achieved.
- Klinefelter's syndrome (47,XXY or variations) is usually sporadic and there appears to be no increased risk for brothers of men with Klinefelter's syndrome to be equally affected. However, as with trisomies involving other chromosomes, there is an increasing frequency of XXY conceptions with increasing maternal age.
- Testicular feminization is inherited as an X-linked recessive disorder, with a risk of 1 in 2 for XY offspring of carrier females. The unaffected sister of an affected individual has a 50% risk of being a carrier, if the mother is a carrier.
- Turner's syndrome, if it is monosomy X (45,X) or one of a spectrum of mosaic karyotypes, is usually a sporadic occurrence. However, when a deletion of an X chromosome is responsible, women who achieve at least limited reproduction can pass on the deletion as a dominant trait.

Spontaneous Abortion

- In general, when a woman has had one spontaneous abortion, the risk to have another is about 25%—somewhat higher than the population risk of 15% to 20%. After two, it rises to 25% to 30%.
- Habitual aborters, with three or more spontaneous abortions (especially if they occur after the first trimester), are more likely to have chromosomally normal abortions, perhaps caused by maternal factors, than are sporadic aborters, or those who also have term pregnancies.
- The frequency of spontaneous abortions rises with increasing maternal age, as does the likelihood of a trisomic conception (see Chapter 6). When a young woman has had a trisomic conception, whether or not it goes to term, she has risk of about 1% to have another.
- When unbalanced rearrangements in a fetus are not derived from a parental balanced rearrangement, the risk for recurrence is not increased.
- When a balanced rearrangement is found in a parent, the chance for unbalanced offspring rises. Not only does the spontaneous abortion rate increase, but so does the likelihood for a liveborn child with congenital anomalies. Depending on the nature of the rearrangement, the risk for an abnormal full term birth varies. It is usually in the range of a few percent (but can go as high as 100%, for example, in the case of 21/21

translocation.) Most such couples can have phenotypically normal offspring who have inherited the balanced rearrangement or have entirely normal chromosomes. As a corollary, phenotypically normal relatives of rearrangement carriers may also have inherited the balanced arrangement. As they may thus be at risk for abnormal offspring, they may benefit from cytogenetic analysis.

- When spontaneous abortions are due to another genetic condition, morphologic examination or the family history may provide a clue for recurrence risks.

Stillbirths and Neonatal Death.

- The risk for recurrence or for a child with a severe defect may be high, depending, of course, on the etiology. Postmortem examination may be crucial for diagnosis, risk counseling, and management of future pregnancies.

CLINICAL NOTES

In general, some findings raise the suspicion of a chromosomal error in couples with unexplained infertility or with recurrent spontaneous abortions. Disturbed spermatogenesis, amenorrhea, or oligomenorrhea may be present. Anomalies of sexual development or differentiation should be ruled out. The family may have an excess of pregnancy wastage, neonatal death, mental retardation, or inability to conceive. The birth of a previous child with a chromosome abnormality or with unexplained congenital abnormalities may be another indication for possible parental chromosome anomaly.

Cytogenetic studies of the couple and, if possible, of the product of conception, may be highly informative. The latter is not widely available for spontaneous abortions, however, and even when it is, successful analysis cannot yet be assured. Parental chromosome analysis, however, is a standard procedure. Minimal indications for such studies vary, depending on the experience and judgment of different authorities. Generally the analysis is recommended after at least two or three spontaneous abortions; sometimes only if they were early; sometimes only if there was also a term pregnancy, and especially if it was abnormal. Parental chromosomes may also be studied when an extensive infertility workup provides no diagnosis.

When one of a couple is found to have a balanced chromosome rearrangement prenatal diagnosis can be offered if the pregnancy is maintained to the time of the investigation, to rule out an abnormal conception. Even if the parents are normal, prenatal diagnosis may be considered when a prior chromosomally abnormal conception is known or suspected to have occurred.

Since many malformations may otherwise go undetected, postmortem evaluation of any late pregnancy loss or perinatal death is expected to yield important information for current use, and may become invaluable at a later date when reproduction is planned by the parents or by other family members (for a sample protocol see Figure 52-1).

FIGURE 52-1 Protocol for Evaluating Stillborn Infants

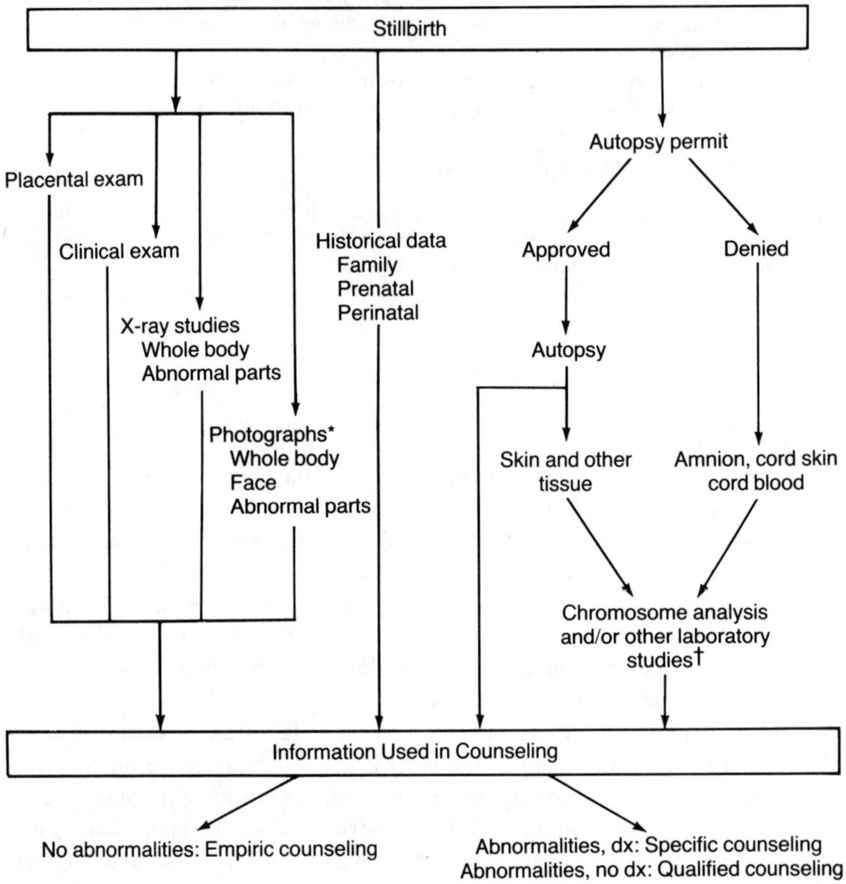

*Parental consent may be required
†Linkage studies may be important for disorders subject to subsequent prenatal diagnosis by recombinant DNA linkage techniques.

PROCEDURES FOR DIAGNOSTIC CONFIRMATION

Investigation of potential nongenetic causes of reproductive failure
Cytogenetic analysis of:
- Abortus, if possible
- Stillborn or infant dying perinatally
- Parental chromosomes

Morphologic studies of abortus
Autopsy of stillborn or infant dying neonatally
Workup for possible environmental causes (see Chapter 45)
Prenatal diagnosis for future pregnancies, as indicated

CONSIDERATIONS IN MANAGEMENT

Genetic
- Pedigree analysis
- Consideration of cytogenetic and other studies, as indicated
- Risk counseling
- Consideration of prenatal diagnosis in future pregnancies, if indicated
- Consideration of donor insemination in cases of male infertility or sterility
- Consideration of adoption, when indicated
- Discussion of sharing information with other family members when indicated, to encourage evaluation and counseling (e.g., balanced rearrangement, testicular feminization)

Psychosocial, educational, familial
- Family counseling:
  - Support for couples undergoing grief reaction after inability to conceive or loss of a pregnancy; consideration of coping mechanisms, e.g., encourage parents to see/touch/hold baby.
  - Prepare parents for reaction of others, discuss significance of fetal evaluation and encourage consent for studies
- After one or two spontaneous abortions, reassurance about the common occurrence of miscarriage and explanation that often this represents a protective mechanism of the body for the rejection of an abnormal conception
- Education for prevention of environmentally caused reproductive failure
- Assessment of community resources and voluntary agencies for appropriate support

Medical
- As indicated by diagnosis

SUGGESTED READING

Byrne JM: *Fetal Pathology; Laboratory Manual,* vol 19. (Birth Defects Original Article Series.) White Plains, New York, March-of-Dimes Birth Defects Foundation, 1983.

Glass RH, Golbus MS: Habitual abortion. *Fertil Steril* 1978;29:257.

MacLeod PM, Dill F, Hardwick DFF: Chromosomes, syndromes, and perinatal deaths: The genetic counseling value of making a diagnosis in a malformed abortus, stillborn and deceased newborn, vol 15 (5A). (Birth Defects Original Article Series.) White Plains, New York, March-of-Dimes Birth Defects Foundation, 1979, pp 105-111.

Mueller RF, Sybert VP, Johnson J, et al: Evaluation of a protocol for post-mortem examination of stillbirths, *N Engl J Med* 1983;309:586.

Porter IH, Hook EB, (eds): *Human Embryonic and Fetal Death,* New York, Academic Press, 1980.

Simpson JL, Jirasek JE, Speroff L, et al: *Disorders of Sexual Differentiation—Etiology and Clinical Delineation,* New York, Academic Press, 1976.

Simpson JL: Genes, chromosomes, and reproductive failure, *Fertil Steril* 1980;33:107.

Tho PT, Byrd JR, McDonough PG: Etiologies and subsequent reproductive performance of 100 couples with recurrent abortion, *Fertil Steril* 1979;32:389.

Chapter 53

Consanguinity

Consanguinity, the mating of two individuals who have one or more recent ancestors in common, occurs with widely varying frequency in different parts of the world and in different communities within individual countries. In the United States, marriages between first- or second-degree relatives are considered incestuous and are illegal; the legality of marriage between third- (or greater) degree relatives varies from state to state, with first cousin marriage often proscribed. Nevertheless, cousin marriages are not rare, and couples concerned about potential implications to offspring are counseled regularly in genetics clinics. Cousin marriages are common and even encouraged in some genetic isolates, which consequently become highly inbred. Children of incestuous parent-child or uncle-niece matings are born more often than might be expected and may come to medical attention through congenital anomalies or mental retardation. Consanguinity is also found in the parents of some probands with known genetic disorders or with unexplained malformation, mental retardation, or other disability.

GENETIC CHARACTERISTICS

Consanguinity in the parents of one of an unrelated couple is not relevant to the health of the couple's children. The potential increase in risk for genetic disease in consanguineous matings applies only to offspring of individuals who are related to each other by common ancestry. The risks vary for different types of disorders, depending on the mode of inheritance.

- There is essentially no increase in risk for autosomal dominant disorders unless both of the partners are actually affected or carry the particular mutant gene without expressing it.
- X-linked recessive disorders pose an increased risk only when the male is affected and the female is a gene carrier. This is, of course, more likely in consanguineous matings, and can produce affected females, but the situation is rarely relevant in counseling for consanguinity, because the affected male would become the primary focus of the counseling
- The main source of potential increased risk lies in autosomal recessive inheritance. It is estimated that, among the thousands of genes in the genome, everyone carries from five to seven deleterious mutant genes in the heterozygous state. Since all of these are likely to be rare, the chance that both members of a couple carry the same mutant gene (which could thus be passed on to a child in a double dose) is very low in the general population. However, if two people are related, the likelihood of having inherited the same gene from a shared ancestor increases with the closeness of the relationship.

TABLE 53-1
Proportion of Genes Shared by Relatives

RELATIONSHIP	GENES (%)
First-degree relatives	
Sibs	50
Parent-child	50
Second-degree relatives	
Half-sibs	25
Grandparents	25
Uncle, aunt-niece, nephew	25
Third-degree relatives	
First cousins	12.5
Half uncle, half aunt-niece, nephew	12.5
Fourth-degree relatives	
First cousins once removed	6.25
Fifth-degree relatives	
Second cousins	3.12

Thus, first cousins share, on the average, 1/8 of their genes, which have been passed on to them from their common set of grandparents (for other relationships see Table 53-1). If one of the shared genes is a deleterious mutant one, there is a risk of 25% in each pregnancy that the offspring will inherit it from both parents and be affected with the disorder in question. Generally, the existence of such a particular risk is not suspected before the birth of an affected child, unless there is a prior history of the disorder in the family or population group.

- Risk for polygenic or multifactorial conditions, which require several genes, each contributing a small effect together with environmental influences, is also slightly increased when part of the genome is shared by the parents.

Genetic Risks in Consanguinity. In the absence of known genetic disease or additional consanguinity in the family history, or membership in a population group with a known increased risk for a specific recessive disorder, the increased risk for cousin matings is lower than most counselees expect. The overall risk for healthy first cousins to have children with genetically determined disorders has been estimated to be no higher than about 3%, as compared with about 2% for children of random matings. For offspring of second cousins, the risk is reported as little, if at all, increased over the population risk. Cousin couples considering marriage usually have been given the prior impression that their risks for abnormal offspring are very high. They may, or may not, feel reassured by the accepted risk figures. There is considerable difference of opinion among professionals, as well, in the perception of the risks for cousin matings. Whether the couple proceed with their plans ultimately depends on how they view the risks.

Of course, if an inherited disorder is already present in the family, consanguinity may influence the risk of recurrence significantly. For example, if first cousins marry and one has a sibling with a rare autosomal recessive disorder, such as homocystinuria, the risk for having an affected child is 1/24.* Had the sibling of the affected individual married an unrelated person, the risk would have been 1/600. The general population risk is 1/40,000. Similarly, if the cousins come from a highly inbred population isolate (e.g., the Hutterite community; Amish population), they may be related in other ways, as well, and may thus have a greatly increased risk for abnormal offspring, especially for those rare disorders that are found only in certain isolates.

In families with extensive consanguinity, inherited recessive disorders may appear to show dominant transmission (parent to child) when one parent is affected and the other has inherited the same gene in the heterozygous state from a common ancestor.

There is a known increased risk, as well, when the couple belongs to one of the ethnic groups with increased incidence of specific disorders, such as sickle cell anemia in blacks, Tay-Sachs disease among Ashkenazi Jews, or thalassemia in people of Mediterranean ancestry. However, with heterozygote screening available for most of these conditions, these specific risks can be ruled out.

The few studies conducted on offspring of incestuous matings have reported a high risk for severely abnormal children. As many as 40% have been estimated to have serious disorders including mental retardation and epilepsy. Consideration of foster care has been recommended for the first year of life, when children of such unions become available for adoption, so that developmental and neurologic status can be evaluated before final placement.

Consanguinity in the Parents of Probands With Genetic Disorders. Consanguinity may be discovered in parents when pedigrees are constructed for probands with autosomal recessive disorders. The rarer the disorder, the likelier it is that the parents are related and inherited the gene from a common source (i.e., the rarer the gene, the lower the chance that two unrelated individuals carrying the same gene should happen to marry). Congenital deafness is one of the conditions where parental consanguinity should be considered, as is severe mental retardation. Both disorders have a number of potential genetic and nongenetic causes. In undiagnosed cases the finding of parental consanguinity suggests an autosomal recessive etiology, but does not constitute proof of such inheritance. However, if it is thought likely in a given case, a 25% recurrence risk for each pregnancy must be considered.

CONSIDERATIONS IN MANAGEMENT

Genetic

- A very thorough pedigree to rule out family history of autosomal recessive disease, multifactorial conditions, or additional consanguinity
- Consideration of carrier testing, if available, for conditions found in the family history or in the population group of the couple
- Consideration of a rare recessive disorder in undiagnosed syndromes with parental consanguinity

---

*A $\frac{2}{3}$ chance that the sibling of an affected individual has the gene × a $\frac{1}{4}$ chance that a first cousin has the gene × a $\frac{1}{4}$ risk for an affected child if both parents are carriers.

- Risk counseling
- Consideration of prenatal diagnosis, if available, for couples identified as being at increased risk for specific disorders

Psychosocial, educational, familial

- Support for whatever decision is reached by related couple considering marriage
- Discussion about dealing with possible reactions of family and friends, if decision is made to proceed with a consanguineous marriage
- Support and services, as indicated, for consanguineous families with affected offspring

Medical

- As indicated by presence of a disorder

SUGGESTED READING

Harper PS: *Practical Genetic Counseling,* Baltimore, University Park Press, 1981.

McCluer JW: Inbreeding and human fetal death, in Porter IH, Hook EB, (eds): *Human Embryonic and Fetal Death,* New York, Academic Press, 1980.

Schull WJ: Empirical risks in consanguineous marriages: Sex ratio, malformation and viability. *Am J Genet* 1958;10:294.

Appendix

General References on Genetic Disorders

Bergsma D (ed): *Birth Defects Compendium,* 2nd ed. New York, Alan R. Liss, 1979.

Over a thousand birth defects and genetic conditions described individually in a standardized format, which includes photographs and information on diagnosis, clinical findings and pathogenesis, as well as genetics.

Emery AEH, Rimoin DL (eds): *Principles and Practice of Medical Genetics,* New York, Churchill Livingstone, 1983.

Compehensive reference provides up-to-date information on all areas of medical genetics in great detail.

Goodman RM, Gorlin RJ: *The Malformed Infant and Child,* New York, Oxford University Press, 1983.

Guidebook for diagnosis, care, and counseling for genetic syndromes, developmental defects and malformations caused by the fetal environment. Standardized format; 200 entries; schematic drawings; appendix on normal parameters. Available in paperback.

Jackson LG, Schimke RN (eds): *Clinical Genetics, A Source Book for Physcians,* New York, John Wiley and Sons, 1979.

Quite academic overview of the entire spectrum of genetic disorders, organized along lines of clinical presentation.

Kessler S (ed): *Genetic Counseling—Psychological Dimensions,* New York, Academic Press, 1979.

The psychodynamics and psychosocial issues that warrant consideration for successful genetic counseling.

McKusick VA: *Mendelian Inheritance in Man,* 7th ed. Baltimore, The John Hopkins University Press, 1986.

Classic catalog of all reported single-gene disorders. Three alphabetic listings cover over 3,300 dominant, recessive, or X-linked conditions. Clinical information is limited, but references are given. An extensive introduction provides further material on genetic fundamentals, such as nomenclature, classification, gene mapping.

Smith DW: *Recognizable Patterns of Human Deformation,* Philadelphia, WB Saunders, 1981.
Well-illustrated review of congenital abnormalities due to intrauterine mechanical forces, which generally have a more favorable prognosis and lower recurrence risks than malformations.

Smith DW: *Recognizable Patterns of Human Malformation,* 3rd ed. Philadelphia, WB Saunders, 1982.
Reference for identification of malformation syndromes. Individual syndromes discussed with liberal use of photographs. Review chapters cover recognition and classification of dysmorphology and normal physical standards.

Stanbury JB, Wyngaarden JB, Fredrickson DS, et al (eds): *The Metabolic Basis of Inherited Disease,* 5th ed. New York, McGraw-Hill, 1983.
Comprehensive review of all manner of genetic metabolic disease gives extensive detail, ranging from biochemical research to therapeutic measures. Chapter abstracts and clear headings facilitate selective reading.

Thompson JS, Thompson MW: *Genetics in Medicine,* 4th ed. Philadelphia, WB Saunders, 1986.
Clear, concise textbook on the principles of medical genetics.

Wright EE, Shaw MV: Legal liability in genetic screening, genetic counseling and prenatal diagnosis. *Clin Obstet Gynecol* 1981;24:1133.

Wyngaarden JB, Smith LH, Jr (eds): *Cecil Textbook of Medicine,* 16th ed. Philadelphia, WB Saunders, 1982.
General textbook of medicine with excellent coverage of many hereditary disorders and of genetics and genetic counseling as a medical discipline.

Glossary of Genetic Terms

Allele. Alternative form of a gene at a given locus on homologous chromosomes. For example, the A, B, and O blood group genes are allelic; they are alternative forms of the gene at the ABO locus.

Amniocentesis. Technique for withdrawing amniotic fluid and fetal cells for prenatal diagnosis.

Aneuploid. Having either more or less than an exact multiple of the haploid number of chromosomes (e.g., trisomy, monosomy).

Anticipation. Apparent earlier occurrence and more severe expression of a disorder in succeeding generations. It is thought not to have a biologic basis and to be the result of earlier detection of additional cases in a family.

A priori. Designates risk calculations unadjusted for modifying factors (e.g., carrier testing, penetrance, age of onset).

Association. Occurrence together of two or more characteristics (in an individual or in family members) more often than would be expected by chance alone.

Assortative mating. See selective mating.

Autosome. Any chromosome other than a sex chromosome. Normal human cells have 22 pairs of autosomes and 1 pair of sex chromosomes.

Banding. Technique of staining chromosomes to bring out characteristic patterns of light and dark horizontal bands, thus allowing more precise identification of each chromosome.

Barr body (sex chromatin). Darkly stained mass of inactive X-chromosome material found in the nucleus of interphase cells. The number of Barr bodies in a cell is equal to the number of X chromosomes minus one., A normal female cell contains one Barr body, and a normal male cell contains none.

Bayesian analysis. A mathematical method for adjusting theoretical recurrence risks of genetic disorders by taking into account available information on modifying factors (e.g., pattern of affected and unaffected relatives in pedigree; inconclusive, but suggestive laboratory results; average age of onset).

Carrier (heterozygote). Individual who has one normal gene and one abnormal (mutant) gene at a given locus. The abnormal gene is usually not expressed, although it may be detectable by appropriate laboratory tests. The term is used in the description of both autosomal and X-linked recessive inheritance, and also in autosomal dominant inheritance for clinically asymptomatic carriers of a mutant gene.

Chorionic Villus Sampling (CVS). Procedure for obtaining cells from chorionic villi for early prenatal diagnosis (eight to 10 weeks' gestation); also known as transcervical chorionic villus biopsy.

Chromosome. Structure in the cell nucleus containing a linear thread of DNA, which transmits genetic information.

Codominant. Measurable expression in a heterozygote of both alleles of a pair, for example, alleles for blood groups A and B, yielding blood type AB.

Compound heterozygote. Genotype in which two different mutant alleles are present at corresponding loci on homologous chromosomes, for example, an individual with alleles for both hemoglobin S and hemoglobin C: producing hemoglobin SC. Also called genetic compound.

Congenital. Condition present at birth; may be of genetic or environmental etiology, or a combination of both.

Consanguineous mating. Mating between blood relatives.

Cytogenetics. Microscopic study of chromosome structure and behavior during cell division.

Deformation. Abnormality caused primarily by mechanical prenatal forces, although maternal uterine environment or underlying fetal pathology may be under genetic control.

Degrees of relatedness. Closeness of the genetic connection between two people, based on the proportion of shared genes. First-degree relatives share one-half of their genes: parents, siblings, offspring. Second-degree relatives share one-fourth of their genes: grandparents, aunts, uncles, nieces, nephews, grandchildren. Third-degree relatives share one-eighth of their genes: great-grandparents, great-aunts and great-uncles, first cousins.

Deletion. Absence of a segment of a chromosome (can be as small as one DNA base or may involve many genes), leading to a deficit of genetic material.

Dermatoglyphics. Study of skin ridge patterns on the fingers, palms, toes and soles.

Diploid. Number of chromosomes in most somatic cells; in man, 46. This is double the number in gametes (2n or twice the haploid number).

Dominant. The expressed gene, when the two genes of any pair are different alleles and only one is expressed. The risk of transmitting a disorder determined by a dominant gene is 50%.

DNA probe. Laboratory-synthesized radioactively labeled single-stranded DNA, which has the ability to bond with a complementary segment of natural DNA. Also called gene probe.

Duplication. Presence of an extra piece of a chromosome resulting in a redundant copy of a gene or series of genes.

Empiric risk. Estimate of risk derived from experience (population studies) rather than from calculations based on the mechanism by which the trait is produced.

Expressivity. Form and degree to which a heritable trait is manifest in an individual carrying the gene(s) for that trait; the variation in a phenotype associated with a particular genotype.

Fetoscopy. Technique of in utero visualization of the fetus, using a fiberoptic endoscope, for prenatal diagnosis or fetal treatment.

Fibroblast. Connective-tissue cell. Can be obtained from skin biopsy or amniotic fluid and grown in culture for genetic analyses.

First-degree relatives. See degrees of relatedness.

Gamete. Mature germ cell (ovum or sperm). A human gamete normally contains 23 chromosomes, the haploid number.

Gene. Unit of heredity. At present a gene is equated with a chromosomal unit of function, for example, the sequence of DNA required to code for one polypeptide.

Gene frequency. Proportion in which a particular allele is present in a specified population.

Genetic disorder. Disorder influenced by genes or chromosomes; usually, but not necessarily, hereditary. Examples of nonheritable genetic conditions include trisomies due to advanced maternal age and monosomy X Turner's syndrome.

Genome. Total genetic material in a cell. Can refer more specifically to the haploid complement.

Genotype. The genetic make up of an individual (or cell) for the trait(s) of particular interest.

Haploid. Containing only one chromosome of each type; the chromosome number of a normal gamete. In man the haploid number (n) is 23.

Hemizygous. Individual or cell having only one allele at a given locus. Applicable to the genes on the X chromosome in males, since males have only one X chromosome, and to individuals in whom there is a deletion of a chromosome (XO Turner's syndrome) or chromosomal segment (e.g., 18q-).

Hereditary. Genetically transmitted to offspring from parent.

Heritable. Capable of being transmitted genetically.

Heterogeneity. Phenomenon of multiple etiologies for the same (or very similar) condition(s). The heterogeneity may involve different genetic mechanisms, as well as nongenetic causes.

Heterozygous. Having different alleles at a given locus on homologous chromosomes.

Heterozygote manifestation. Clinically observable, though usually subtle, manifestation of a recessive trait in a carrier; more common for X-linked traits.

HLA system (human leukocyte antigens). Group of cell surface antigens coded by a series of closely linked genes on chromosome 6; involved in histocompatibility reactions; also thought to be implicated in predisposition to a host of disorders.

HLA haplotype. Set of closely linked HLA alleles found on one of the pair of chromosome 6; usually inherited as a unit.

Homologous chromosome. One chromosome of a matched pair, similar with respect to gene loci, different with respect to parental origin.

Homozygous. Having two identical alleles at a given locus on homologous chromosomes.

Incidence. Frequency of a trait at birth.

Isolated (abnormality). Not associated with any other abnormality(ies).

Karyotype. Usually refers to the photomicrographs of chromosomes arranged by pairs, according to size, shape and banding patterns to allow laboratory assessment for chromosomal abnormalities.

Linkage. Assocation of two or more loci on a single chromosome such that alleles at these loci have a tendency to be inherited together; observed when loci are close to one another. See linkage analysis.

Linkage analysis. Procedure to aid in diagnosis or prenatal diagnosis, using family studies of genotypes and marker traits (often restriction fragment length polymorphisms [RFLPs] or blood groups which have been linked to the trait being studied); includes pedigree analysis to study transmission of markers and trait in family.

Locus. Site of a gene on a chromosome (pl: loci).

Lyon hypothesis. In females, the random inactivation of one X chromosome in early embryonic cells, with fixed inactivation in descendant cells, leading to a mixed cell population of maternally and paternally derived inactive chromosomes.

Malformation. Birth defect resulting from abnormal or incomplete organ development, often familial, but may be of teratogenic etiology.

Meiosis. Type of cell division occurring in the germ cells by which gametes containing the haploid chromosome number are produced from diploid cells.

Mendelian. Transmission of genes and resultant traits or disorders according to specific patterns and proportions as put forth by Gregor Mendel. Disorders transmitted in this manner are also referred to as "single gene" disorders.

Mitosis. Process of division in somatic cells, which results in the formation of two daughter cells, each with the same (diploid) chromosomal complement as the parent cell.

Monosomy. Presence of only one member of a pair of chromosomes (e.g., XO Turner's syndrome).

Mosaic. An individual or tissue derived from a single zygote, but with two or more cell lines, differing in genotype or karyotype.

Multifactorial. A trait determined by the interaction of one or more genes and environmental factors.

Mutagen. An agent capable of inducing a genetic change or mutation.

Mutation. A permanent heritable change (loss, gain, or exchange) in the genetic material.

Mutation rate. Frequency of mutations at a given locus per generation.

Nondisjunction. Failure of paired chromosomes to separate normally (i.e., one member of the pair going to each new cell) during cell division. Nondisjunction during ovum or sperm production may result in an aneuploid embryo; during cell division in the developing embryo, it may produce mosaicism.

Obligate carrier. An asymptomatic individual carrying one gene for a genetic condition, which may or may not be identifiable through a laboratory test. Carrier state is inferred when family-history analysis indicates that the individual must be carrying the gene (e.g., the parent of a child with a recessive disorder; the daughter of a man with an X-linked disorder).

Pedigree. A diagram of the genetic relationships and medical history in a family.

Penetrance. Proportion of all of the individuals carrying a particular gene, that manifest the trait determined by that gene. When penetrance is less than 100% in a dominant disorder, some individuals who inherit and transmit the gene do not manifest the disorder themselves.

Phenocopy. A nongenetic trait or disorder that mimics a genetic one.

Phenotype. Observable and/or measurable expression of a gene or genes.

Pleiotropy. Production of apparently different phenotypic effects by a single allele or pair of alleles.

Polygenic. A trait determined by multiple genes at different loci.

Prevalence. Relative frequency of a trait in a population at any one point in time.

Proband. Individual who first brings a family to attention for genetic evaluation and counseling; also called the index case or propositus.

Recessive. A trait or gene that is expressed only if the individual is homozygous for a given allele.

Restriction fragment-length polymorphism (RFLP). Normal variation in length of fragments created via restriction enzyme site specific clevage of DNA; can be used for diagnosis when linked to a mutant gene in a family.

Second-degree relatives. See degrees of relatedness.

Selective mating (assortative mating). Nonrandom mating; preference for a mate with a particular trait.

Sex chromatin. See Barr body.

Sex-limited. A (genetic) trait, not coded on the X chromosome, which affects individuals of one sex only.

Sex-linked. See X-linked.

Single gene disorders. Disorders determined by one gene or gene pair. See Mendelian.

Somatic cell. Any cell not part of the germ line.

Sporadic. Single occurrence of a disorder in a family, either of nongenetic etiology or the first appearance of a genetic disorder in that family.

Teratogen. An agent that is capable of causing a defect in the embryo or fetus.

Third-degree relatives. See Degrees of relatedness.

Translocation. A chromosome abnormality in which a chromosome, or a segment thereof becomes attached to another chromosome. A translocation is balanced when no chomosome material is lost or gained in the rearrangement.

Trisomy. Presence of three copies of a particular chromosome instead of the usual pair (e.g., Down's Syndrome - trisomy 21).

X-chromosome inactivation. See Lyon's hypothesis.

X-linked. Genes on the X chromosome or traits determined by these genes. Also called sex-linked.

Zygote. Diploid cell formed from fusion of ovum and sperm; the first cell of a new organism.

Resources in Clinical Genetics

When genetic disease strikes, genetic counseling and medical management only begin to answer the many needs that suddenly descend on the whole family. The deleterious effects that accompany a serious genetic disorder have many long-term implications. They may break up a marriage, devastate siblings, exhaust caretakers, and deplete financial resources. This chapter is devoted to helping the physician locate appropriate referral agencies to address these problems, complementing the care provided in primary practice.

REFERRAL FOR GENETIC SERVICES

To begin with, a workup in a genetics clinic may be indicated for a definitive diagnosis and/or for initial genetic counseling. Genetics clinics are well established in most major medical centers, especially in those that are university affiliated. In addition to offering workups for a wide range of genetic disorders, most have specific research interests in which they can provide highly sophisticated expertise. For the physician, who does not have proximity to a genetics clinic, or who wishes to refer a patient to a center that specializes in a particular type of disorder or service, referral information is available from:

The National Genetics Foundation
555 West 57th Street
New York, N.Y. 10019
(212) 586-5800

The National Center for Education
in Maternal and Child Health (NCEMCH)
38th and R Streets, N.W.
Washington, D.C. 20057
(202) 625-8400

SUPPORT SERVICES

Once a family has had genetic counseling, a diagnosis has been explained, options have been discussed, and the implications of a genetic disorder have begun to sink in, psychologic reactions take over. Referral for psychotherapy may help ease acceptance, but a pragmatic approach to problems of day-to-day coping often is also needed for families that may face years of dealing with a handicapped child, or who may have to live with onerous risks for serious illness.

In addition to hospital-associated social service departments, a profusion of agencies exist to fill this need. Among them are public and private services, as well as national and regional organizations devoted to giving support. The help they offer includes professional services (e.g., infant stimulation groups, educational facilities, and institutional placement); financial support for research; referral and funding for specialized medical care; provision of information for physicians and patients; legal advocacy; and mutual support for coping.

These organizations can be roughly divided into state- and community-operated public facilities, categoric disease foundations, university-affiliated programs, protection and advocacy agencies, and self-help mutual-support groups.

Information about nearby government agencies and academic programs devoted to assisting patients and their families is generally available from hospital social service departments, local health departments, and public or medical libraries.

Lists of voluntary groups and foundations, though often appended to articles and books, are seldom comprehensive enough for all readers and tend to go out of date, since many of these organizations are headquartered at the home of the current president. To facilitate up-to-date referrals, we are providing below sources that specialize in keeping lists current and publishing them on a regular basis. Most organizations listed are national ones. They, in turn, will often be able to provide referral to local or regional chapters.

One caveat: While referral to categoric disease groups may be very helpful, it should be made only after a specific diagnosis has been established and primary counseling has been provided. When offered prematurely or inappropriately, or when the diagnosis is uncertain, their publications providing specific medical and genetic information may be frightening or even misleading.

A list of easily obtained and useful directories follows:

The National Center for Education in Maternal and Child Health (38th and R Sts., N.W., Washington, D.C. 20057; Phone: (202) 625-8400) has a number of publications that list voluntary organizations and resources for dealing with genetic disease and birth defects. These are available free or for a minimal charge. Among them:

Reaching Out, A Directory of Voluntary Organizations in Maternal and Child Health

This is an alphabetic listing of name, address, telephone number, and contact person for organizations in the United States, Canada, Europe, Australia and South Africa. Many of the groups listed are specifically concerned with rare genetic disorders. The publication also lists self-help clearinghouses by state, and identifies societies and professional associations active in dealing with maternal and child health.

A Guide to Selected National Genetic Voluntary Organizations
Although this volume lists many of the same organizations as *Reaching Out,* with emphasis on the genetic ones, it does includes additional information. A "statement of purpose" explains at a glance the type of assistance provided. Some groups may emphasize fund raising and research above all, and others are more concerned with treatment, direct support for families, dissemination of information, or advocacy. This paragraph is a useful guide for appropriate referrals. Equally helpful is a listing of educational materials available from each organization.

State Treatment Centers for Metabolic Disorders
As the title suggests, this publication includes a geographic listing of clinical treatment centers for metabolic disorders, along with the names of their directors. It also lists the same information for newborn-screening centers.

The National Self-Help Clearinghouse (Graduate School and University Center of the City University of New York, 33 West 42nd Street, Room 1227, New York, N.Y. 10036; Phone: (212) 840-1259), along with regional subsidiaries, keeps track of existing self-help groups for people who share any kind of common experience and wish to assist each other. It also provides guidance for starting new groups, which may be a welcome service to families with very rare genetic disorders.

Index